METHODS IN MOLECULAR BIOLOGY

For further volumes:
http://www.springer.com/series/7651

Muscle Stem Cells

Methods and Protocols

Edited by

Eusebio Perdiguero

Department of Experimental & Health Sciences, University Pompeu Fabra (UPF), Barcelona, Spain

DDW Cornelison

Division of Biological Sciences and Christopher S. Bond Life Sciences Center, University of Missouri, Columbia, MO, USA

Humana Press

Editors
Eusebio Perdiguero
Cell Biology Group
Department of Experimental and Health Sciences
University Pompeu Fabra (UPF)
Barcelona, Spain

DDW Cornelison
Division of Biological Sciences
and Christopher S. Bond Life Sciences Center
University of Missouri
Columbia, MO, USA

ISSN 1064-3745 ISSN 1940-6029 (electronic)
Methods in Molecular Biology
ISBN 978-1-4939-6769-8 ISBN 978-1-4939-6771-1 (eBook)
DOI 10.1007/978-1-4939-6771-1

Library of Congress Control Number: 2016962638

Printed on acid-free paper

This Humana Press imprint is published by Springer Nature
The registered company is Springer Science+Business Media LLC
The registered company address is: 233 Spring Street, New York, NY 10013, U.S.A.

Preface

In the last decade, skeletal muscle stem cells have joined hematopoietic, intestinal, and skin stem cells as a standard model to study stem cell function in healthy and aging tissue and the role of stem cells in tissue-specific diseases. *Muscle Stem Cells: Methods and Protocols* provides leading-edge protocols in the study of the molecular and cellular biology of muscle stem cells. Protocols representing current and updated methods for muscle stem cell isolation, culture, molecular analysis, cellular analysis, and reintroduction in vivo as well as protocols for studying myogenic stem cells in non-mammalian model systems. Due to the application of techniques such as these, the potential for exploring muscle stem cell biology continues to expand, as does the state of our understanding of this critical stem cell population.

Barcelona, Spain *Eusebio Perdiguero*
Columbia, MO, USA *DDW Cornelison*

Contents

KRZYSZTOF JAGLA • *GReD (Genetics, Reproduction and Development laboratory), INSERM 1103, CNRS 6293, University of Clermont Auvergne, Clermont-Ferrand, France*
PAUL KEIRE • *Department of Biological Structure, School of Medicine, University of Washington, Seattle, WA, USA*
JOHNNY KIM • *Department of Cardiac Development and Remodelling, Max Planck Institute for Heart and Lung Research, Bad Nauheim, Germany*
GUILLAUME LAVERGNE • *GReD (Genetics, Reproduction and Development laboratory), INSERM 1103, CNRS 6293, University of Clermont Auvergne, Clermont-Ferrand, France*
MARCELA LOW • *Biomedical Research Centre, University of British Columbia, Vancouver, BC, Canada; Faculty of Medicine, University of British Columbia, Vancouver, BC, Canada*
DANE K. LUND • *Division of Biological Sciences and Christopher S. Bond Life Sciences Center, University of Missouri, Columbia, MO, USA; Developmental Biology Program, Sloan Kettering Institute, New York, NY, USA*
BARBORA MALECOVA • *Development, Aging and Regeneration Program, Sanford Burnham Prebys Medical Discovery Institute, La Jolla, CA, USA; Regulus Therapeutics, Science Center Dr., San Diego, CA, USA*
MARTA MARTÍNEZ-VICENTE • *Neurodegenerative Diseases Research Group, Vall d'Hebron Research Institute-CIBERNED, Barcelona, Spain*
PATRICK MCANULTY • *Division of Biological Sciences and Christopher S. Bond Life Sciences Center, University of Missouri, Columbia, MO, USA; The Kidney Institute, University of Kansas Medical Center, Kansas City, KS, USA*
GRAZIELLA MESSINA • *Department of Biosciences, University of Milan, Milan, Italy*
RÉMI MOUNIER • *Institut NeuroMyoGène, INMG, Université Claude Bernard Lyon 1, Villeurbanne, France*
LOUISE A. MOYLE • *Department of Cell and Developmental Biology, University College London, London, UK*
PURA MUÑOZ-CÁNOVES • *Cell Biology Group, Department of Experimental and Health Sciences, Pompeu Fabra University (UPF), CIBER on Neurodegenerative diseases (CIBERNED), Barcelona, Spain; Institució Catalana de Recerca i Estudis Avançats (ICREA), Barcelona, Spain; Tissue Regeneration Laboratory, Centro Nacional de Investigaciones Cardiovasculares (CNIC), Madrid, Spain*
PHONG D. NGUYEN • *Australian Regenerative Medicine Institute, Monash University, Clayton, VIC, Australia*
BRADLEY B. OLWIN • *Department of Molecular, Cellular and Developmental Biology, University of Colorado, Boulder, CO, USA*
FRANCESCA PALA • *Department of Developmental and Stem Cell Biology, Stem Cells and Development, Institut Pasteur, CNRS URA 2578, Paris, France*
BRADLEY T. PAWLIKOWSKI • *Department of Molecular, Cellular and Developmental Biology, University of Colorado, Boulder, CO, USA*
XIANLU PENG • *Department of Chemical Pathology, The Chinese University of Hong Kong, Hong Kong, China; Li Ka Shing Institute of Health Sciences, The Chinese University of Hong Kong, Hong Kong, China*
EUSEBIO PERDIGUERO • *Cell Biology Group, Department of Experimental and Health Sciences (DCEXS), Pompeu Fabra University (UPF) and CIBER on Neurodegenerative Diseases (CIBERNED), Barcelona, Spain*

PIER LORENZO PURI • *Development, Aging and Regeneration Program, Sanford Burnham Prebys Medical Discovery Institute, La Jolla, CA, USA; IRCCS Fondazione Santa Lucia, Rome, Italy*
FABIO ROSSI • *Biomedical Research Centre and Faculty of Medicine, University of British Columbia, Vancouver, BC, Canada; Faculty of Medicine, University of British Columbia, Vancouver, BC, Canada*
GIULIANA ROSSI • *Department of Biosciences, University of Milan, Milan, Italy*
MICHAEL A. RUDNICKI • *Regenerative Medicine Program, Sprott Center For Stem Cell Research, Ottawa Hospital Research Institute, Ottawa, ON, Canada; Faculty of Medicine, Department of Cellular and Molecular Medicine, University of Ottawa, Ottawa, ON, Canada*
MOHAMMED RUMMAN • *Institute for Stem Cell Biology and Regenerative Medicine, Bangalore, India; Manipal University, Manipal, India*
JAMES G. RYALL • *Stem Cell Metabolism and Regenerative Medicine Group, Basic and Clinical Myology Laboratory, Department of Physiology, The University of Melbourne, Parkville, VIC, Australia*
ALESSANDRA SACCO • *Development, Aging and Regeneration Program, Sanford Children's Health Research Center, Sanford Burnham Prebys Medical Discovery Institute, La Jolla, CA, USA*
MARIELLE SACLIER • *Department of Biosciences, University of Milan, Milan, Italy*
HIROSHI SAKAI • *Department of Developmental and Stem Cell Biology, Stem Cells and Development, Institut Pasteur, CNRS URA 2578, Paris, France*
ASHLEY L. SIEGEL • *Division of Biological Sciences and Christopher S. Bond Life Sciences Center, University of Missouri, Columbia, MO, USA; Elemental Enzymes, St. Louis, MO, USA*
MARIE CLAUDE SINCENNES • *Regenerative Medicine Program, Sprott Center For Stem Cell Research, Ottawa Hospital Research Institute, Ottawa, ON, Canada; Faculty of Medicine, Department of Cellular and Molecular Medicine, University of Ottawa, Ottawa, ON, Canada*
CEDRIC SOLER • *GReD (Genetics, Reproduction and Development laboratory), INSERM 1103, CNRS 6293, University of Clermont-Ferrand Auvergne, Clermont-Ferrand, France*
KRISHNAMOORTHY SREENIVASAN • *Department of Cardiac Development and Remodelling, Max Planck Institute for Heart and Lung Research, Bad Nauheim, Germany*
PASCAL STUELSATZ • *Department of Biological Structure, School of Medicine, University of Washington, Seattle, WA, USA*
HAO SUN • *Department of Chemical Pathology, The Chinese University of Hong Kong, Hong Kong, China; Li Ka Shing Institute of Health Sciences, The Chinese University of Hong Kong, Hong Kong, China*
KUN SUN • *Department of Chemical Pathology, The Chinese University of Hong Kong, Hong Kong, China; Li Ka Shing Institute of Health Sciences, The Chinese University of Hong Kong, Hong Kong, China*
SHAHRAGIM TAJBAKHSH • *Department of Developmental and Stem Cell Biology, Stem Cells and Development, Institut Pasteur, CNRS URA 2578, Paris, France*
FRANCESCO SAVERIO TEDESCO • *Department of Cell and Developmental Biology, University College London, London, UK*
MARINE THERET • *Institut NeuroMyoGène, INMG, Université Claude Bernard Lyon 1, Villeurbanne, France; Université Paris Descartes, Sorbonne Paris Cité, Paris, France*

Matthew Tierney • *Graduate School of Biomedical Sciences, Sanford Burnham Prebys Medical Discovery Institute, La Jolla, CA, USA; Development, Aging and Regeneration Program, Sanford Children's Health Research Center, Sanford Burnham Prebys Medical Discovery Institute, La Jolla, CA, USA*
Rossana Tonlorenzi • *INSPE - Institute of Experimental Neurology, San Raffaele Scientific Institute, Milano, Italy*
Nisha Venugopal • *CSIR-Centre for Cellular and Molecular Biology, Hyderabad, India*
Huating Wang • *Department of Orthopaedics and Traumatology, The Chinese University of Hong Kong, Hong Kong, China; Li Ka Shing Institute of Health Sciences, Prince of Wales Hospital, The Chinese University of Hong Kong, Hong Kong, China*
Yu Xin Wang • *Sprott Center For Stem Cell Research, Regenerative Medicine Program, Ottawa Hospital Research Institute, Ottawa, ON, Canada; Faculty of Medicine, Department of Cellular and Molecular Medicine, University of Ottawa, Ottawa, ON, Canada*
Zipora Yablonka-Reuveni • *Department of Biological Structure, School of Medicine, University of Washington, Seattle, WA, USA*
Jiajian Zhou • *Department of Chemical Pathology, The Chinese University of Hong Kong, Hong Kong, China; Li Ka Shing Institute of Health Sciences, The Chinese University of Hong Kong, Hong Kong, China*
Monika Zmojdzian • *GReD (Genetics, Reproduction and Development laboratory), INSERM 1103, CNRS 6293, University of Clermont Auvergne, Clermont-Ferrand, France*

Part I

Overview

Chapter 1

Muscle Stem Cells: A Model System for Adult Stem Cell Biology

DDW Cornelison and Eusebio Perdiguero

Abstract

Skeletal muscle stem cells, originally termed satellite cells for their position adjacent to differentiated muscle fibers, are absolutely required for the process of skeletal muscle repair and regeneration. In the last decade, satellite cells have become one of the most studied adult stem cell systems and have emerged as a standard model not only in the field of stem cell-driven tissue regeneration but also in stem cell dysfunction and aging. Here, we provide background in the field and discuss recent advances in our understanding of muscle stem cell function and dysfunction, particularly in the case of aging, and the potential involvement of muscle stem cells in genetic diseases such as the muscular dystrophies.

Key words Skeletal muscle, Stem cells, Muscle regeneration, Heterogeneity, Aging, Muscular dystrophies

1 Introduction

Stem cells are most simply defined as precursor cells which guarantee sustained maintenance and regeneration of a tissue or organ by providing a population of replacement cells in case of loss or damage. If this regenerative property is to be maintained over the lifetime of the organism, stem cells of each particular tissue should be capable of proliferating to both give rise to a large number of differentiated progeny and to self-renew their population. In addition to germline stem cells, the first compelling experiments demonstrating the existence of tissue-specific adult stem cells were done in studies investigating the origin of the hematopoietic system [1–5], establishing hematopoietic stem cells as the prototypical model for stem cell studies.

Adult stem cells may be classified into two major groups: those from tissues with high turnover and those from tissues with low turnover [6]. In tissues with high turnover rates, such as the skin, the hematopoietic system, and the intestine, stem cells are also responsible for repair after damage or wounding, but are most

Eusebio Perdiguero and DDW Cornelison (eds.), *Muscle Stem Cells: Methods and Protocols*, Methods in Molecular Biology, vol. 1556,
DOI 10.1007/978-1-4939-6771-1_1, © Springer Science+Business Media LLC 2017

important in maintaining tissue homeostasis on an ongoing basis. In tissues with low turnover, such as brain and skeletal muscle, the activity of stem cells is more restricted to specific stimuli such as establishing new neural circuits during learning [7] or responses to episodes of injury or disease in the skeletal muscle [8–11]. In the last decade, skeletal muscle stem cells have joined hematopoietic, intestinal, and skin stem cells as a standard model to study stem cell function in healthy and aging tissue and the role of stem cells in tissue-specific diseases.

2 Muscle Stem Cells in the Adult Skeletal Muscle

Adult skeletal muscle is a postmitotic tissue whose primary function is contraction, which allows organisms to generate force and facilitates voluntary movement. This tissue is also essential for the regulation of whole-body metabolism. Skeletal muscles are composed of muscle fibers (myofibers) which are multinucleated syncytial cells formed during the developmental process of myogenesis. Each myofiber is composed of numerous myofibrils, structures that extend along the complete length of each fiber cell and hold many sequentially ordered subunits of the functional unit of contraction, the sarcomere, where the interaction of actin and myosin protein filaments translate ATP and Ca^{2+} into force and thus motion [12]. Since all myofiber nuclei are terminally postmitotic, they are unable to contribute effectively to skeletal muscle growth and repair, which is instead provided by a unique population of muscle stem cells, the satellite cells [8–11].

By the late 1800s and early 1900s, several researchers had noticed that skeletal muscle development and regeneration appeared to be fueled by proliferative cells [10], which were termed myoblasts [13]. In 1961, an electron microscopy study [14] identified mononucleated muscle cells residing beneath the basal lamina of muscle fibers, therefore, termed satellite cells, which were proposed to be a muscle stem/progenitor cell. Since this first description, satellite cells in healthy, uninjured muscle tissue have been characterized as quiescent cells with high nuclear-to-cytoplasmic ratio, reduced organelle content, increased amounts of heterochromatin, and reduced transcriptional activity compared to myofiber nuclei and are functionally defined as muscle precursor cells which provide myoblasts for postnatal muscle growth and regeneration [10].

Satellite cells normally reside in a quiescent state until activated by damage or growth signals; in this quiescent state, they do not express significant levels of the myogenic regulatory transcription factors (MRFs: MYOD, MYF5, Myogenin, and MYF6) [15], but express high levels of the paired box protein Pax-7 (PAX7), considered to be the definitive marker for satellite cells [16–19]. Apart from PAX7, a growing number of proteins are known to be expressed in quiescent satellite cells and have been used to isolate them by fluorescence-activated cell sorting (FACS) or to label them for microscopic observation, including attachment and adhesion molecules such as integrin alpha-7, M-cadherin, and CD34; cell-surface receptors such as HGF receptor/c-Met, CXCR4, and calcitonin receptor; the heparan sulfate proteoglycans syndecan-3 and syndecan-4; caveolae-forming protein caveolin-1; and to a lesser extent transcription factors such as PAX3 [8]. Transplantation studies have thoroughly demonstrated that satellite cells are capable of self-renewal to replenish the stem cell pool as well as generating committed myoblasts that will proliferate and differentiate into new myofibers to orchestrate tissue repair [20–26]. Finally, several groups simultaneously demonstrated that skeletal muscle regeneration fails after genetic ablation of satellite cells [27–29], proving that satellite cells are the bona fide muscle stem cells and indispensable for skeletal muscle repair.

Molecular factors regulating satellite cell quiescence and activation are a focus of extensive study in the field. The notch signaling pathway in particular has been demonstrated to play a key role in the maintenance of quiescence [30–33]. Notch expression in satellite cells is induced by the transcription factor forkhead box protein O3 (FOXO3), which is necessary to maintain quiescence [34]. Upon binding of a membrane-bound or soluble ligand, notch is proteolytically cleaved and the intracellular domain (NICD) translocates to the nucleus, where it interacts with the recombining binding protein suppressor of hairless (RBP-J kappa). The NICD–RBPJ kappa complex represses target genes, including those encoding the HES and HEY families of transcription factors, which inhibit MYOD and induce PAX7 expression [35–38]. Quiescence is also actively maintained by the expression of cell cycle inhibitors such as p21 and p27Kip1 [39, 40].

Upon muscle injury, quiescent satellite cells are exposed to many pro-proliferative signals produced by the surrounding milieu of infiltrating inflammatory cells and other resident cells including fibroblast growth factor 2 (FGF-2), hepatocyte growth factor (HGF), insulin-like growth factor 1 (IGF-1), and tumor necrosis factor (TNF-a), among others [40–47]. These growth factors and cytokines have been shown to activate intracellular signaling cascades, being the p38 MAPK cascade among the more prominent ones [48]. MAPK 14/p38α has been shown to be essential for distinct myogenic stages, including activation from quiescence, cessation of proliferation, and differentiation of satellite cells [49–53]. At early stages of satellite cell activation, p38 MAPKs are implicated in the asymmetric division machinery [54] and lead to stabilization of *MyoD* mRNA through the inhibition of the RNA-binding destabilizing protein tristetraprolin [55, 56]. Similarly, this signaling may be linked to transcriptional and posttranscriptional regulation of *Myf5*, a crucial step for satellite cell activation [57, 58]. MYF5 and MYOD transcription factors control the gene expression program of activated satellite cells [59–61], with upregulation of genes implicated in cell cycle progression, response to immune system, and chemotaxis [39, 62, 63]. These transcriptome differences between quiescent and activated satellite cells are associated with profound changes in the epigenetic landscape [63] linked with fluctuations in the amount of facultative heterochromatin [64], probably mediated by differential activity of the polycomb repressive complex 2 (PRC2) [63, 65] and repression of the histone methyltransferases of the Suv4-20h family [64]. PRC2 activity is also controlled by p38 MAPKs and impinges in essential genes like *Pax7* [66].

Although satellite cells were considered until the early 2000s to be a homogeneous population of stem cells, mounting evidence has demonstrated that satellite cells are in fact heterogeneous, containing subpopulations with distinct gene expression profiles and different propensities for self-renewal or differentiation (reviewed in [67]). Many of these satellite cell subpopulations were identified based on differential expression of markers such as PAX3, CD34, M-cadherin, or MYF5 [18, 23, 25, 26, 68], which w then shown to correlate with marked functional differences. Heterogeneity in gene expression is also likely to underlie the ability of a subset of satellite cells to undergo asymmetric division [69, 70]. This has led to the hypothesis that there are at least two subpopulations of satellite cells, one less committed and thus with increased stemness and one more committed for differentiation. Different methods have been used to quantify the percentage of satellite "stem" cells, with different results. Satellite stem cell representation in the general population is about 40 % based on the expression of the cell-fate determinant numb in vitro [70]. It is about 7 % based on co-segregation of template DNA strands in vivo, a phenomenon of asymmetric segregation of chromatids ensuring that one daughter cell contains only the old intact DNA (the template) and the other carries chromatids composed exclusively of the newly synthesized DNA, remaining the template or so-called "immortal" DNA strand in the truly stem cell [69, 70]. It is about 10 % based on the number of cells which lack of a history of MYF5 expression: these cells are proposed to be the subpopulation able to undergo asymmetric division, while MYF5 lineage-positive cells divide symmetrically, expand, and follow the differentiation path [23], being therefore a more committed progenitor. This percentage was shown to vary in different lineage tracing models [71], and interestingly, satellite cells derived from *Myf5* heterozygous mice have a higher self-renewal capacity [72]. Around 5 % of satellite cells labeled with PKH26 dye have slow divisions after activation and retained long-term self-renewal ability [73], correlating with previous observations during development [74]. Finally, around 10 % of satellite cells with high PAX7 expression (based on a transgenic mouse model in which GFP is controlled by *Pax7* promoter) were shown to manifest more stemlike traits such as lower levels of muscle-specific gene expression, increased dormancy, and lower metabolism [75]. Future studies in which these various criteria for the more stemlike cells in the satellite cell population are correlated with each other as well as with self-renewal, and myoblast generating capacity should better define the nature of this potential subpopulation; in vivo work in particular will be informative.

While satellite cells are the obligate stem cells of skeletal muscle, other cells present in muscle tissue have been implicated in skeletal muscle homeostasis and regeneration. These skeletal muscle interstitial cell populations do not meet the classical criteria for satellite cells (e.g.,, they are outside the basal lamina of the myofiber and do not generally express MYOD or PAX family transcription factors), although some of them have been described to be able to undergo myogenesis upon tissue injury (*see* Tedesco et al. in this volume) [76]. In particular, PW1-expressing interstitial cells (also known as PICs) have been shown to contribute to regenerated muscle upon engraftment [77, 78], and pericytes or pericyte-derived mesoangioblasts have also been shown to have myogenic capacity [79–81]. The relative contribution of these cell types in normal muscle regeneration and their relationship and/or hierarchy with satellite cells are still being studied. While *Pax7*-driven diphtheria toxin-mediated satellite cell elimination has demonstrated that *Pax7*-expressing satellite cells are absolutely required for skeletal muscle regeneration [27–29], one study also noted a requirement for *Tcf4*-expressing muscle connective tissue fibroblasts [28], suggesting that satellite cell interactions with these and possibly other interstitial cells are required for the regeneration process.

3 Satellite Cells as a Model for Stem Cell Aging

Aging is a process that affects all organs and tissues leading to a decline in their homeostatic maintenance and function as well as reduced regenerative response after injury or disease [82]. Reduced or blunted stem cell function is generally considered to be responsible for this decline. Aging is an enormous challenge for biomedical sciences and much work has been concentrated on discovering the complex role of different stem cell types in this process and identifying the mechanisms involved, in order to prevent or reverse them. In the last decade, muscle stem cells have arisen as an excellent model for the study of the role of stem cell aging in tissue dysfunction in that they provide a tractable model to investigate the mechanisms of age-related dysfunction of stem cells, including whether these mechanisms are cell extrinsic, cell intrinsic, or both.

Skeletal muscle aging is associated with a reduction in muscle mass and performance that is clinically described as sarcopenia [83–85]. Sarcopenia is known to increase the predisposition to muscle injury [86]; concomitantly, aging is associated with a progressive decline in satellite cell number and function resulting in defective tissue regeneration in rodents and humans [87–97]. Several extrinsic factors have been proposed that may influence satellite cell function during aging: often they are increased in aged skeletal muscle fibers (the cellular niche of satellite cells) and/or systemically in the circulation. FGF2 released by aged muscle fibers has been shown to downregulate Sprouty1, a self-renewal promoting factor [98], in a subset of aged satellite cells, and to compromise their quiescence even in undamaged tissue [88]. Transforming growth factor beta-1 (TGFβ1) was demonstrated to enhance SMAD transcription factor activation and cyclin-dependent kinase inhibitor expression in aged satellite cells, a phenomenon that is countered by notch signaling [99]. Increased activity of Wnt signaling cascade has been implicated in reduced differentiation of aged satellite cells and increased transdifferentiation to a fibrogenic lineage [100]. In agreement with this, a sequential switch from notch to Wnt signaling was shown to be required for effective satellite cell activation and differentiation [101], and inhibition of notch signaling in young satellite cells induced age-associated regenerative defects, while activation of notch restored muscle regeneration in aged mice [93, 94, 102].

Suggestions that extrinsic or systemic influences may have a dominant role over cell-autonomous factors have come from experiments that improved muscle regeneration in aged muscle through rejuvenation of the aged environment. To ask whether a young environment would promote robust regeneration of aged muscle and vice versa, several groups (most notably Bruce Carlson's) performed heterochronic whole-muscle and muscle cell transplants [103–107], although without conclusive results [108, 109]. More recently, testing the effects of the environment on aged cells has been done by heterochronic parabiosis, wherein two mice of different ages are surgically joined such that they share the same circulatory system [94, 100, 110]. In validation studies, regeneration has been rescued in aged muscle through modulation of some of the known extrinsic pathways (notch, Wnt, TGFβ1) [99, 111] or with newly identified circulating factors revealed in the parabiosis model such as oxytocin [112] and growth/differentiation factor 11 (GDF11) [113]. However, the relevance of GDF11 as a general rejuvenating factor has recently been disputed [114–118]; future studies will hopefully clarify this interesting scientific kerfuffle.

Impairment of satellite cell function during aging has also been proposed to involve cell-intrinsic or cell-autonomous mechanisms that are consequences of irreversible damage to old or very old (geriatric) cells. Indeed, many of the defined primary hallmarks of aging [119] as well as deregulation of many key transcriptional circuits necessary for stem cell identity and activity have already been demonstrated in satellite cells. For example, telomere attrition and increased DNA damage has been demonstrated in aged and geriatric satellite cells [89, 113, 120–125]. Similarly, myoblasts deficient in Ku80, a component of the nonhomologous end joining (NHEJ) pathway, showed accelerated aging due to telomere shortening and accumulated DNA damage [126]. Global epigenetic alterations [63] and depression of key loci such as INK4a [89, 95, 97, 99, 112], which shows alterations chromatin marks such as K3K4me3 and H3K27me3 in old satellite cells [112] and H2AK119ub in geriatric satellite cells [89, 95], are also a hallmark of aging. These are known to lead to elevated p16-INK4a expression and a pre-senescent state turning into full cellular senescence upon proliferative signals. Satellite cell aging has also been linked to activation of the p38 MAPK pathway, probably as a response to increased FGF signaling [97, 127]. Deregulation of several key transcriptional circuits such as Rb/E2F [89] (leading to full senescence) and JAK/STAT [128, 129] (leading to defective asymmetric division) have also been described. Finally, loss of proteostasis has been also observed in aged satellite cells: old and geriatric satellite cells have a reduced autophagic flux resulting in accumulation of damaged proteins, mitochondrial dysfunction, and increased oxidative stress which leads to entry into a pre-senescent state [95]. Consistent with all these findings, genetic or pharmacological inhibition of p16-INK4a [89], STAT3 [128, 129], p38 MAPK [97], or refueling of the autophagy flux [95] rejuvenates the satellite cell population in aged mice. These results hold a growing promise for antiaging treatments, although the translation of specific gene inhibition in mice to human therapies may be difficult if not impossible to implement with current technologies. An interesting potential exception may be caloric restriction, a known inducer of autophagy [130], which has been shown to improve the engraftment capacity of adult satellite cells and the expansion of adult and aged satellite cells in vitro [131], suggesting that metabolic reprogramming of aged satellite cells could promote their survival and enhance their regenerative functions.

4 Old and New Implications of Satellite Cells in Muscular Dystrophies

Muscular dystrophies are a heterogeneous group of diseases characterized by skeletal muscle degeneration and wasting accompanied by inflammatory response [132]. In the most severe dystrophies, such as Duchenne muscular dystrophy (DMD, caused by a deficiency in the dystrophin protein), the failure of muscle maintenance eventually leads to death by respiratory insufficiency and cardiomyopathy [133]. In DMD, the muscle tissue is in a chronic state of damage due to the weakness of the dystrophin-deficient myofibers, leading to chronic activation of the regeneration response and, potentially, exhaustion of the satellite cell population [134, 135], although conflicting data exist between mice and humans [136]. The disease pathology also leads to incomplete muscle regeneration and progressive replacement of muscle by fibrotic and fat tissue [137].

Were satellite cell dysfunction a causative factor in the later stages of disease progression, one would expect to note reduced differentiation capacity of satellite cells and myoblasts from DMD patients. Indeed, several reports indicated that myoblasts derived from DMD patients and from the mouse model of DMD, the mdx mouse, showed aberrant differentiation [138–140]. However, recent evidence from engraftment of satellite cells from mdx mice into wild-type hosts demonstrated that their regenerative capacity was similar to that of the healthy muscle host [141], implying that (like many factors described above in the case of aging) the defect may be non-cell-autonomous. To add to the debate, a recent study demonstrated that (contrary to popular belief) dystrophin is expressed in satellite cells, where it is necessary for regulation of their polarity and asymmetric division [142]: in the absence of dystrophin, cells attempting to undergo asymmetric division instead experience mitotic catastrophe due to nucleation of multiple spindle poles [142].

Finally, satellite cells from DMD patients [143] and mdx mice [144] have shorter telomeres, which could possibly lead to a senescence-like phenotype, similar to that observed in geriatric satellite cells [89] or to cells from young mice deficient in polycomb repressive complex proteins such as Bmi1 [145] or Ezh2 [65]. In agreement with this hypothesis, mdx mice lacking telomerase activity develop a very aggressive phenotype that worsens with age, similar to human children [146]. Interestingly, premature senescence is also an underlying pathogenic feature of satellite cells from a mouse model of LGMD2H (limb-girdle muscular dystrophy type 2H) [147]. These questions will also ideally be resolved by further experimentation into the interplay between intrinsic and extrinsic factors affecting satellite cell function in the dystrophic environment.

5 Concluding Remarks

Skeletal muscle regeneration is fueled by satellite cells, the bona fide somatic stem cells of the muscle, although in the last decade other interstitial stem or progenitor cells from diverse origins have been shown to collaborate with satellite cells during tissue repair (*see* Tedesco et al. in this volume) [76]. Data generated in the last 10 years using sophisticated genetic models and new technologies have indicated that, as in many somatic tissues like blood, skin, and intestine [148], the skeletal muscle stem cell system is also heterogeneous, comprising different subsets of satellite cells which differ in their capacity for proliferation, differentiation, and self-renewal. However, as noted above, significant unsolved questions still remain in the areas of satellite cell heterogeneity, contribution of non-satellite cells to muscle regeneration, and the role of satellite cells themselves vs. their local and systemic environments in the case of aging or disease.

Although satellite cells have thus far not fulfilled their initial promise in regenerative medicine approaches to treat muscular diseases, the development and implementation of new protocols for their isolation, the selection of subpopulations with extended stemness ex vivo, the use of single-cell or clonal population analysis, and the implementation of innovative bioengineering techniques may initiate a new era for satellite cells as therapeutic tools for sarcopenia and muscular dystrophies.

The protocols included in this volume "Muscle Stem Cells: Methods and Protocols" of the *Methods in Molecular Biology* series represent current and updated methods for muscle stem cell isolation (Gayraud-Morel et al., Tonlorenzi et al., Stuelsatz et al., Sincennes et al., and Low et al.) [149–153], culture (Saclier et al., Arora et al., Davoudi and Gilbert) [154–156], molecular analysis (Sreenivasan et al., Gatto et al., and Peng et al.) [157–159], cellular analysis (Ryall, Garcia-Prat et al. and Lund et al.) [160–162], and reintroduction in vivo (Hall et al., Tierney and Sacco) [163, 164] as well as protocols for studying myogenic stem cells in nonmammalian model systems (Lavergne et al., Nguyen and Currie) [165, 166]. Due to the application of techniques such as these, the potential for exploring satellite cell biology continues to expand, as does the state of our understanding of this critical stem cell population.

Acknowledgments

We thank Drs. Pura Muñoz-Cánoves and Antonio Serrano (UPF, Barcelona) and Elena Rebollo (MIP, IBMB, Barcelona) for their insightful discussions. E.P. acknowledges the funding from MINECO, Spain (SAF2015-67369-R and "María de Maeztu" Programme for Units of Excellence in R&D MDM-2014-0370), AFM, CIBERNED (IntraCIBER 2015-2/06, InterCIBER PIE14/00061). DDWC was funded by National Institutes of Health grant #AR067450.

References

1. Till JE, McCulloch CE (1961) A direct measurement of the radiation sensitivity of normal mouse bone marrow cells. Radiat Res 14:213–222
2. Till JE, Mak TW, Price GB, Senn JS, McCulloch EA (1976) Cellular subclasses in human leukemic hemopoiesis. Hamatol Bluttransfus 19:33–45
3. Sabin FR, Doan CA, Forkner CE (1932) The production of osteogenic sarcomata and the effects on lymph nodes and bone marrow of intravenous injections of radium chloride and mesothorium in rabbits. J Exp Med 56(2):267–289
4. Ramalho-Santos M, Willenbring H (2007) On the origin of the term "stem cell". Cell Stem Cell 1(1):35–38. doi:10.1016/j.stem.2007.05.013
5. Becker AJ, McCulloch CE, Till JE (1963) Cytological demonstration of the clonal nature of spleen colonies derived from transplanted mouse marrow cells. Nature 197:452–454
6. Hsu YC, Fuchs E (2012) A family business: stem cell progeny join the niche to regulate homeostasis. Nat Rev Mol Cell Biol 13(2):103–114. doi:10.1038/nrm3272
7. Bond AM, Ming GL, Song H (2015) Adult mammalian neural stem cells and neurogenesis: five decades later. Cell Stem Cell 17(4):385–395. doi:10.1016/j.stem.2015.09.003
8. Yin H, Price F, Rudnicki MA (2013) Satellite cells and the muscle stem cell niche. Physiol Rev 93(1):23–67. doi:10.1152/physrev.00043.2011
9. Chang NC, Rudnicki MA (2014) Satellite cells: the architects of skeletal muscle. Curr Top Dev Biol 107:161–181. doi:10.1016/B978-0-12-416022-4.00006-8
10. Scharner J, Zammit PS (2011) The muscle satellite cell at 50: the formative years. Skelet Muscle 1(1):28. doi:10.1186/2044-5040-1-28
11. Sambasivan R, Tajbakhsh S (2015) Adult skeletal muscle stem cells. Results Probl Cell Differ 56:191–213. doi:10.1007/978-3-662-44608-9_9
12. Frontera WR, Ochala J (2015) Skeletal muscle: a brief review of structure and function. Calcif Tissue Int 96(3):183–195. doi:10.1007/s00223-014-9915-y
13. Tello JF (1917) Genesis de las terminaciones nerviosas motrices y sensitivas. Trab Lab Invest biol Univ Madr 15:99
14. Mauro A (1961) Satellite cells of skeletal fibers. J Biophys Biochem Cytol 9:493–495
15. Sartorelli V, Caretti G (2005) Mechanisms underlying the transcriptional regulation of skeletal myogenesis. Curr Opin Genet Dev 15(5):528–535. doi:10.1016/j.gde.2005.04.015
16. Gros J, Manceau M, Thome V, Marcelle C (2005) A common somitic origin for embryonic muscle progenitors and satellite cells. Nature 435(7044):954–958. doi:10.1038/nature03572
17. Kassar-Duchossoy L, Giacone E, Gayraud-Morel B, Jory A, Gomes D, Tajbakhsh S (2005) Pax3/Pax7 mark a novel population of primitive myogenic cells during development. Genes Dev 19(12):1426–1431. doi:10.1101/gad.345505
18. Relaix F, Montarras D, Zaffran S, Gayraud-Morel B, Rocancourt D, Tajbakhsh S, Mansouri A, Cumano A, Buckingham M (2006) Pax3 and Pax7 have distinct and overlapping functions in adult muscle progenitor cells. J Cell Biol 172(1):91–102. doi:10.1083/jcb.200508044
19. Seale P, Sabourin LA, Girgis-Gabardo A, Mansouri A, Gruss P, Rudnicki MA (2000) Pax7 is required for the specification of myogenic satellite cells. Cell 102(6):777–786
20. Cerletti M, Jurga S, Witczak CA, Hirshman MF, Shadrach JL, Goodyear LJ, Wagers AJ (2008) Highly efficient, functional engraftment of skeletal muscle stem cells in dystrophic muscles. Cell 134(1):37–47. doi:10.1016/j.cell.2008.05.049
21. Collins CA, Olsen I, Zammit PS, Heslop L, Petrie A, Partridge TA, Morgan JE (2005) Stem cell function, self-renewal, and behavioral heterogeneity of cells from the adult muscle satellite cell niche. Cell 122(2):289–301. doi:10.1016/j.cell.2005.05.010
22. Gunther S, Kim J, Kostin S, Lepper C, Fan CM, Braun T (2013) Myf5-positive satellite cells contribute to Pax7-dependent long-term maintenance of adult muscle stem cells. Cell Stem Cell 13(5):590–601. doi:10.1016/j.stem.2013.07.016
23. Kuang S, Kuroda K, Le Grand F, Rudnicki MA (2007) Asymmetric self-renewal and commitment of satellite stem cells in muscle. Cell 129(5):999–1010. doi:10.1016/j.cell.2007.03.044
24. Lepper C, Conway SJ, Fan CM (2009) Adult satellite cells and embryonic muscle progenitors have distinct genetic requirements. Nature 460(7255):627–631. doi:10.1038/nature08209
25. Montarras D, Morgan J, Collins C, Relaix F, Zaffran S, Cumano A, Partridge T, Buckingham M (2005) Direct isolation of satellite cells for skeletal muscle regeneration. Science

309(5743):2064–2067. doi:10.1126/science.1114758
26. Sacco A, Doyonnas R, Kraft P, Vitorovic S, Blau HM (2008) Self-renewal and expansion of single transplanted muscle stem cells. Nature 456(7221):502–506. doi:10.1038/nature07384
27. Lepper C, Partridge TA, Fan CM (2011) An absolute requirement for Pax7-positive satellite cells in acute injury-induced skeletal muscle regeneration. Development 138(17):3639–3646. doi:10.1242/dev.067595
28. Murphy MM, Lawson JA, Mathew SJ, Hutcheson DA, Kardon G (2011) Satellite cells, connective tissue fibroblasts and their interactions are crucial for muscle regeneration. Development 138(17):3625–3637. doi:10.1242/dev.064162
29. Sambasivan R, Yao R, Kissenpfennig A, Van Wittenberghe L, Paldi A, Gayraud-Morel B, Guenou H, Malissen B, Tajbakhsh S, Galy A (2011) Pax7-expressing satellite cells are indispensable for adult skeletal muscle regeneration. Development 138(17):3647–3656. doi:10.1242/dev.067587
30. Kopan R, Nye JS, Weintraub H (1994) The intracellular domain of mouse Notch: A constitutively activated repressor of myogenesis directed at the basic helix-loop-helix region of MyoD. Development 120:2385–2396
31. Mourikis P, Sambasivan R, Castel D, Rocheteau P, Bizzarro V, Tajbakhsh S (2012) A critical requirement for notch signaling in maintenance of the quiescent skeletal muscle stem cell state. Stem Cells 30(2):243–252. doi:10.1002/stem.775
32. Bjornson CR, Cheung TH, Liu L, Tripathi PV, Steeper KM, Rando TA (2012) Notch signaling is necessary to maintain quiescence in adult muscle stem cells. Stem Cells 30(2):232–242. doi:10.1002/stem.773
33. Vasyutina E, Lenhard DC, Wende H, Erdmann B, Epstein JA, Birchmeier C (2007) RBP-J (Rbpsuh) is essential to maintain muscle progenitor cells and to generate satellite cells. Proc Natl Acad Sci U S A 104(11):4443–4448. doi:10.1073/pnas.0610647104
34. Gopinath SD, Webb AE, Brunet A, Rando TA (2014) FOXO3 promotes quiescence in adult muscle stem cells during the process of self-renewal. Stem Cell Reports 2(4):414–426. doi:10.1016/j.stemcr.2014.02.002
35. Buas MF, Kabak S, Kadesch T (2010) The Notch effector Hey1 associates with myogenic target genes to repress myogenesis. J Biol Chem 285(2):1249–1258. doi:10.1074/jbc.M109.046441
36. Wen Y, Bi P, Liu W, Asakura A, Keller C, Kuang S (2012) Constitutive Notch activation upregulates Pax7 and promotes the self-renewal of skeletal muscle satellite cells. Mol Cell Biol 32(12):2300–2311. doi:10.1128/MCB.06753-11
37. Kuroda K, Tani S, Tamura K, Minoguchi S, Kurooka H, Honjo T (1999) Delta-induced Notch signaling mediated by RBP-J inhibits MyoD expression and myogenesis. J Biol Chem 274(11):7238–7244
38. Pisconti A, Cornelison DD, Olguin HC, Antwine TL, Olwin BB (2010) Syndecan-3 and Notch cooperate in regulating adult myogenesis. J Cell Biol 190(3):427–441. doi:10.1083/jcb.201003081
39. Fukada S, Uezumi A, Ikemoto M, Masuda S, Segawa M, Tanimura N, Yamamoto H, Miyagoe-Suzuki Y, Takeda S (2007) Molecular signature of quiescent satellite cells in adult skeletal muscle. Stem Cells 25(10):2448–2459. doi:10.1634/stemcells.2007-0019
40. Chakkalakal JV, Christensen J, Xiang W, Tierney MT, Boscolo FS, Sacco A, Brack AS (2014) Early forming label-retaining muscle stem cells require p27kip1 for maintenance of the primitive state. Development 141(8):1649–1659. doi:10.1242/dev.100842
41. Allen RE, Boxhorn LK (1989) Regulation of skeletal muscle satellite cell proliferation and differentiation by transforming growth factor-beta, insulin-like growth factor I, and fibroblast growth factor. J Cell Physiol 138(2):311–315. doi:10.1002/jcp.1041380213
42. Chen SE, Gerken E, Zhang Y, Zhan M, Mohan RK, Li AS, Reid MB, Li YP (2005) Role of TNF-{alpha} signaling in regeneration of cardiotoxin-injured muscle. Am J Physiol Cell Physiol 289(5):C1179–C1187. doi:10.1152/ajpcell.00062.2005
43. Chen SE, Jin B, Li YP (2007) TNF-alpha regulates myogenesis and muscle regeneration by activating p38 MAPK. Am J Physiol Cell Physiol 292(5):C1660–C1671. doi:10.1152/ajpcell.00486.2006
44. Mourkioti F, Rosenthal N (2005) IGF-1, inflammation and stem cells: interactions during muscle regeneration. Trends Immunol 26(10):535–542. doi:10.1016/j.it.2005.08.002
45. Schiaffino S, Mammucari C (2011) Regulation of skeletal muscle growth by the IGF1-Akt/PKB pathway: insights from genetic models. Skelet Muscle 1(1):4. doi:10.1186/2044-5040-1-4
46. Sheehan SM, Allen RE (1999) Skeletal muscle satellite cell proliferation in response to members of the fibroblast growth factor family and hepatocyte growth factor. J Cell Physiol 181(3):499–506. doi:10.1002/(SICI)1097-4652(199912)181:3<499::AID-JCP14>3.0.CO;2-1

47. Tatsumi R, Anderson JE, Nevoret CJ, Halevy O, Allen RE (1998) HGF/SF is present in normal adult skeletal muscle and is capable of activating satellite cells. Dev Biol 194:114–128
48. Lluis F, Perdiguero E, Nebreda AR, Munoz-Canoves P (2006) Regulation of skeletal muscle gene expression by p38 MAP kinases. Trends Cell Biol 16(1):36–44. doi:10.1016/j.tcb.2005.11.002
49. Wu Z, Woodring PJ, Bhakta KS, Tamura K, Wen F, Feramisco JR, Karin M, Wang JY, Puri PL (2000) p38 and extracellular signal-regulated kinases regulate the myogenic program at multiple steps. Mol Cell Biol 20(11):3951–3964
50. Takaesu G, Kang JS, Bae GU, Yi MJ, Lee CM, Reddy EP, Krauss RS (2006) Activation of p38alpha/beta MAPK in myogenesis via binding of the scaffold protein JLP to the cell surface protein Cdo. J Cell Biol 175(3):383–388. doi:10.1083/jcb.200608031
51. Perdiguero E, Ruiz-Bonilla V, Gresh L, Hui L, Ballestar E, Sousa-Victor P, Baeza-Raja B, Jardi M, Bosch-Comas A, Esteller M, Caelles C, Serrano AL, Wagner EF, Munoz-Canoves P (2007) Genetic analysis of p38 MAP kinases in myogenesis: fundamental role of p38alpha in abrogating myoblast proliferation. EMBO J 26(5):1245–1256. doi:10.1038/sj.emboj.7601587
52. Ruiz-Bonilla V, Perdiguero E, Gresh L, Serrano AL, Zamora M, Sousa-Victor P, Jardi M, Wagner EF, Munoz-Canoves P (2008) Efficient adult skeletal muscle regeneration in mice deficient in p38beta, p38gamma and p38delta MAP kinases. Cell Cycle 7(14):2208–2214
53. Brien P, Pugazhendhi D, Woodhouse S, Oxley D, Pell JM (2013) p38alpha MAPK regulates adult muscle stem cell fate by restricting progenitor proliferation during postnatal growth and repair. Stem Cells 31(8):1597–1610. doi:10.1002/stem.1399
54. Troy A, Cadwallader AB, Fedorov Y, Tyner K, Tanaka KK, Olwin BB (2012) Coordination of satellite cell activation and self-renewal by par-complex-dependent asymmetric activation of p38alpha/beta MAPK. Cell Stem Cell 11(4):541–553. doi:10.1016/j.stem.2012.05.025
55. Hausburg MA, Doles JD, Clement SL, Cadwallader AB, Hall MN, Blackshear PJ, Lykke-Andersen J, Olwin BB (2015) Post-transcriptional regulation of satellite cell quiescence by TTP-mediated mRNA decay. Elife 4:e03390. doi:10.7554/eLife.03390
56. Sachidanandan C, Sambasivan R, Dhawan J (2002) Tristetraprolin and LPS-inducible CXC chemokine are rapidly induced in presumptive satellite cells in response to skeletal muscle injury. J Cell Sci 115(Pt 13):2701–2712
57. Crist CG, Montarras D, Buckingham M (2012) Muscle satellite cells are primed for myogenesis but maintain quiescence with sequestration of Myf5 mRNA targeted by microRNA-31 in mRNP granules. Cell Stem Cell 11(1):118–126. doi:10.1016/j.stem.2012.03.011
58. Diao Y, Guo X, Li Y, Sun K, Lu L, Jiang L, Fu X, Zhu H, Sun H, Wang H, Wu Z (2012) Pax3/7BP is a Pax7- and Pax3-binding protein that regulates the proliferation of muscle precursor cells by an epigenetic mechanism. Cell Stem Cell 11(2):231–241. doi:10.1016/j.stem.2012.05.022
59. Cooper RN, Tajbakhsh S, Mouly V, Cossu G, Buckingham M, Butler-Browne GS (1999) In vivo satellite cell activation via Myf5 and MyoD in regenerating mouse skeletal muscle. J Cell Sci 112(Pt 17):2895–2901
60. Blum R, Vethantham V, Bowman C, Rudnicki M, Dynlacht BD (2012) Genome-wide identification of enhancers in skeletal muscle: the role of MyoD1. Genes Dev 26(24):2763–2779. doi:10.1101/gad.200113.112
61. Cao Y, Yao Z, Sarkar D, Lawrence M, Sanchez GJ, Parker MH, MacQuarrie KL, Davison J, Morgan MT, Ruzzo WL, Gentleman RC, Tapscott SJ (2010) Genome-wide MyoD binding in skeletal muscle cells: a potential for broad cellular reprogramming. Dev Cell 18(4):662–674. doi:10.1016/j.devcel.2010.02.014
62. Pallafacchina G, Francois S, Regnault B, Czarny B, Dive V, Cumano A, Montarras D, Buckingham M (2010) An adult tissue-specific stem cell in its niche: a gene profiling analysis of in vivo quiescent and activated muscle satellite cells. Stem Cell Res 4(2):77–91. doi:10.1016/j.scr.2009.10.003
63. Liu L, Cheung TH, Charville GW, Hurgo BM, Leavitt T, Shih J, Brunet A, Rando TA (2013) Chromatin modifications as determinants of muscle stem cell quiescence and chronological aging. Cell Rep 4(1):189–204. doi:10.1016/j.celrep.2013.05.043
64. Boonsanay V, Zhang T, Georgieva A, Kostin S, Qi H, Yuan X, Zhou Y, Braun T (2016) Regulation of skeletal muscle stem cell quiescence by Suv4-20h1-dependent facultative heterochromatin formation. Cell Stem Cell 18(2):229–242. doi:10.1016/j.stem.2015.11.002
65. Juan AH, Derfoul A, Feng X, Ryall JG, Dell'Orso S, Pasut A, Zare H, Simone JM, Rudnicki MA, Sartorelli V (2011) Polycomb

EZH2 controls self-renewal and safeguards the transcriptional identity of skeletal muscle stem cells. Genes Dev 25(8):789–794. doi:10.1101/gad.2027911
66. Palacios D, Mozzetta C, Consalvi S, Caretti G, Saccone V, Proserpio V, Marquez VE, Valente S, Mai A, Forcales SV, Sartorelli V, Puri PL (2010) TNF/p38alpha/polycomb signaling to Pax7 locus in satellite cells links inflammation to the epigenetic control of muscle regeneration. Cell Stem Cell 7(4):455–469. doi:10.1016/j.stem.2010.08.013
67. Tierney MT, Sacco A (2016) Satellite cell heterogeneity in skeletal muscle homeostasis. Trends Cell Biol 26(6):434–444. doi:10.1016/j.tcb.2016.02.004
68. Beauchamp JR, Heslop L, Yu DS, Tajbakhsh S, Kelly RG, Wernig A, Buckingham ME, Partridge TA, Zammit PS (2000) Expression of CD34 and Myf5 defines the majority of quiescent adult skeletal muscle satellite cells. J Cell Biol 151(6):1221–1234
69. Conboy MJ, Karasov AO, Rando TA (2007) High incidence of non-random template strand segregation and asymmetric fate determination in dividing stem cells and their progeny. PLoS Biol 5(5):e102. doi:10.1371/journal.pbio.0050102
70. Shinin V, Gayraud-Morel B, Gomes D, Tajbakhsh S (2006) Asymmetric division and cosegregation of template DNA strands in adult muscle satellite cells. Nat Cell Biol 8(7):677–687. doi:10.1038/ncb1425
71. Comai G, Sambasivan R, Gopalakrishnan S, Tajbakhsh S (2014) Variations in the efficiency of lineage marking and ablation confound distinctions between myogenic cell populations. Dev Cell 31(5):654–667. doi:10.1016/j.devcel.2014.11.005
72. Gayraud-Morel B, Chretien F, Jory A, Sambasivan R, Negroni E, Flamant P, Soubigou G, Coppee JY, Di Santo J, Cumano A, Mouly V, Tajbakhsh S (2012) Myf5 haploinsufficiency reveals distinct cell fate potentials for adult skeletal muscle stem cells. J Cell Sci 125(Pt 7):1738–1749. doi:10.1242/jcs.097006
73. Ono Y, Masuda S, Nam HS, Benezra R, Miyagoe-Suzuki Y, Takeda S (2012) Slow-dividing satellite cells retain long-term self-renewal ability in adult muscle. J Cell Sci 125(Pt 5):1309–1317. doi:10.1242/jcs.096198
74. Schultz E (1996) Satellite cell proliferative compartments in growing skeletal muscles. Dev Biol 175(1):84–94. doi:10.1006/dbio.1996.0097
75. Rocheteau P, Gayraud-Morel B, Siegl-Cachedenier I, Blasco MA, Tajbakhsh S (2012) A subpopulation of adult skeletal muscle stem cells retains all template DNA strands after cell division. Cell 148(1–2):112–125. doi:10.1016/j.cell.2011.11.049
76. Tedesco FS, Moyle L, Perdiguero E (2017) Muscle interstitial cells: a brief field guide to non-satellite cell populations in skeletal muscle. Methods Mol Biol 1556:129–148
77. Mitchell KJ, Pannerec A, Cadot B, Parlakian A, Besson V, Gomes ER, Marazzi G, Sassoon DA (2010) Identification and characterization of a non-satellite cell muscle resident progenitor during postnatal development. Nat Cell Biol 12(3):257–266. doi:10.1038/ncb2025
78. Pannerec A, Formicola L, Besson V, Marazzi G, Sassoon DA (2013) Defining skeletal muscle resident progenitors and their cell fate potentials. Development 140(14):2879–2891. doi:10.1242/dev.089326
79. Dellavalle A, Maroli G, Covarello D, Azzoni E, Innocenzi A, Perani L, Antonini S, Sambasivan R, Brunelli S, Tajbakhsh S, Cossu G (2011) Pericytes resident in postnatal skeletal muscle differentiate into muscle fibres and generate satellite cells. Nat Commun 2:499. doi:10.1038/ncomms1508
80. Dellavalle A, Sampaolesi M, Tonlorenzi R, Tagliafico E, Sacchetti B, Perani L, Innocenzi A, Galvez BG, Messina G, Morosetti R, Li S, Belicchi M, Peretti G, Chamberlain JS, Wright WE, Torrente Y, Ferrari S, Bianco P, Cossu G (2007) Pericytes of human skeletal muscle are myogenic precursors distinct from satellite cells. Nat Cell Biol 9(3):255–267. doi:10.1038/ncb1542
81. Sampaolesi M, Blot S, D'Antona G, Granger N, Tonlorenzi R, Innocenzi A, Mognol P, Thibaud JL, Galvez BG, Barthelemy I, Perani L, Mantero S, Guttinger M, Pansarasa O, Rinaldi C, Cusella De Angelis MG, Torrente Y, Bordignon C, Bottinelli R, Cossu G (2006) Mesoangioblast stem cells ameliorate muscle function in dystrophic dogs. Nature 444(7119):574–579. doi:10.1038/nature05282
82. Kirkwood TB (2005) Understanding the odd science of aging. Cell 120(4):437–447. doi:10.1016/j.cell.2005.01.027
83. Alway SE, Myers MJ, Mohamed JS (2014) Regulation of satellite cell function in sarcopenia. Front Aging Neurosci 6:246. doi:10.3389/fnagi.2014.00246
84. Glass D, Roubenoff R (2010) Recent advances in the biology and therapy of muscle wasting. Ann N Y Acad Sci 1211:25–36. doi:10.1111/j.1749-6632.2010.05809.x
85. Mitchell WK, Williams J, Atherton P, Larvin M, Lund J, Narici M (2012) Sarcopenia, dynapenia, and the impact of advancing age on human skeletal muscle size and strength; a

quantitative review. Front Physiol 3:260. doi:10.3389/fphys.2012.00260
86. Faulkner JA, Brooks SV, Zerba E (1995) Muscle atrophy and weakness with aging: contraction-induced injury as an underlying mechanism. J Gerontol A Biol Sci Med Sci 50 Spec No:124–129
87. Brack AS, Bildsoe H, Hughes SM (2005) Evidence that satellite cell decrement contributes to preferential decline in nuclear number from large fibres during murine age-related muscle atrophy. J Cell Sci 118(Pt 20):4813–4821. doi:10.1242/jcs.02602
88. Chakkalakal JV, Jones KM, Basson MA, Brack AS (2012) The aged niche disrupts muscle stem cell quiescence. Nature 490(7420):355–360. doi:10.1038/nature11438
89. Sousa-Victor P, Gutarra S, Garcia-Prat L, Rodriguez-Ubreva J, Ortet L, Ruiz-Bonilla V, Jardi M, Ballestar E, Gonzalez S, Serrano AL, Perdiguero E, Munoz-Canoves P (2014) Geriatric muscle stem cells switch reversible quiescence into senescence. Nature 506(7488):316–321. doi:10.1038/nature13013
90. Verdijk LB, Dirks ML, Snijders T, Prompers JJ, Beelen M, Jonkers RA, Thijssen DH, Hopman MT, Van Loon LJ (2012) Reduced satellite cell numbers with spinal cord injury and aging in humans. Med Sci Sports Exerc 44(12):2322–2330. doi:10.1249/MSS.0b013e3182667c2e
91. Verdijk LB, Snijders T, Drost M, Delhaas T, Kadi F, van Loon LJ (2014) Satellite cells in human skeletal muscle; from birth to old age. Age 36(2):545–547. doi:10.1007/s11357-013-9583-2
92. Zwetsloot KA, Childs TE, Gilpin LT, Booth FW (2013) Non-passaged muscle precursor cells from 32-month old rat skeletal muscle have delayed proliferation and differentiation. Cell Prolif 46(1):45–57. doi:10.1111/cpr.12007
93. Conboy IM, Conboy MJ, Smythe GM, Rando TA (2003) Notch-mediated restoration of regenerative potential to aged muscle. Science 302(5650):1575–1577. doi:10.1126/science.1087573
94. Conboy IM, Conboy MJ, Wagers AJ, Girma ER, Weissman IL, Rando TA (2005) Rejuvenation of aged progenitor cells by exposure to a young systemic environment. Nature 433(7027):760–764. doi:10.1038/nature03260
95. Garcia-Prat L, Martinez-Vicente M, Perdiguero E, Ortet L, Rodriguez-Ubreva J, Rebollo E, Ruiz-Bonilla V, Gutarra S, Ballestar E, Serrano AL, Sandri M, Munoz-Canoves P (2016) Autophagy maintains stemness by preventing senescence. Nature 529(7584):37–42. doi:10.1038/nature16187
96. Collins CA, Zammit PS, Ruiz AP, Morgan JE, Partridge TA (2007) A population of myogenic stem cells that survives skeletal muscle aging. Stem Cells 25(4):885–894. doi:10.1634/stemcells.2006-0372
97. Cosgrove BD, Gilbert PM, Porpiglia E, Mourkioti F, Lee SP, Corbel SY, Llewellyn ME, Delp SL, Blau HM (2014) Rejuvenation of the muscle stem cell population restores strength to injured aged muscles. Nat Med 20(3):255–264. doi:10.1038/nm.3464
98. Shea KL, Xiang W, LaPorta VS, Licht JD, Keller C, Basson MA, Brack AS (2010) Sprouty1 regulates reversible quiescence of a self-renewing adult muscle stem cell pool during regeneration. Cell Stem Cell 6(2):117–129. doi:10.1016/j.stem.2009.12.015
99. Carlson ME, Hsu M, Conboy IM (2008) Imbalance between pSmad3 and Notch induces CDK inhibitors in old muscle stem cells. Nature 454(7203):528–532. doi:10.1038/nature07034
100. Brack AS, Conboy MJ, Roy S, Lee M, Kuo CJ, Keller C, Rando TA (2007) Increased Wnt signaling during aging alters muscle stem cell fate and increases fibrosis. Science 317(5839):807–810. doi:10.1126/science.1144090
101. Brack AS, Conboy IM, Conboy MJ, Shen J, Rando TA (2008) A temporal switch from notch to Wnt signaling in muscle stem cells is necessary for normal adult myogenesis. Cell Stem Cell 2(1):50–59. doi:10.1016/j.stem.2007.10.006
102. Wagers AJ, Conboy IM (2005) Cellular and molecular signatures of muscle regeneration: current concepts and controversies in adult myogenesis. Cell 122(5):659–667. doi:10.1016/j.cell.2005.08.021
103. Gutmann E, Carlson BM (1976) Regeneration and transplantation of muscles in old rats and between young and old rats. Life Sci 18(1):109–114
104. Carlson BM, Faulkner JA (1989) Muscle transplantation between young and old rats: age of host determines recovery. Am J Physiol 256(6 Pt 1):C1262–C1266
105. Roberts P, McGeachie JK, Grounds MD (1997) The host environment determines strain-specific differences in the timing of skeletal muscle regeneration: cross-transplantation studies between SJL/J and BALB/c mice. J Anat 191(Pt 4):585–594
106. Harrison DE (1983) Long-term erythropoietic repopulating ability of old, young, and fetal stem cells. J Exp Med 157(5):1496–1504

107. Carlson BM, Faulkner JA (1983) The regeneration of skeletal muscle fibers following injury: a review. Med Sci Sports Exerc 15(3):187–198
108. Shavlakadze T, McGeachie J, Grounds MD (2010) Delayed but excellent myogenic stem cell response of regenerating geriatric skeletal muscles in mice. Biogerontology 11(3):363–376. doi:10.1007/s10522-009-9260-0
109. Smythe GM, Shavlakadze T, Roberts P, Davies MJ, McGeachie JK, Grounds MD (2008) Age influences the early events of skeletal muscle regeneration: studies of whole muscle grafts transplanted between young (8 weeks) and old (13-21 months) mice. Exp Gerontol 43(6):550–562. doi:10.1016/j.exger.2008.02.005
110. Villeda SA, Luo J, Mosher KI, Zou B, Britschgi M, Bieri G, Stan TM, Fainberg N, Ding Z, Eggel A, Lucin KM, Czirr E, Park JS, Couillard-Despres S, Aigner L, Li G, Peskind ER, Kaye JA, Quinn JF, Galasko DR, Xie XS, Rando TA, Wyss-Coray T (2011) The ageing systemic milieu negatively regulates neurogenesis and cognitive function. Nature 477(7362):90–94. doi:10.1038/nature10357
111. Naito AT, Sumida T, Nomura S, Liu ML, Higo T, Nakagawa A, Okada K, Sakai T, Hashimoto A, Hara Y, Shimizu I, Zhu W, Toko H, Katada A, Akazawa H, Oka T, Lee JK, Minamino T, Nagai T, Walsh K, Kikuchi A, Matsumoto M, Botto M, Shiojima I, Komuro I (2012) Complement C1q activates canonical Wnt signaling and promotes aging-related phenotypes. Cell 149(6):1298–1313. doi:10.1016/j.cell.2012.03.047
112. Elabd C, Cousin W, Upadhyayula P, Chen RY, Chooljian MS, Li J, Kung S, Jiang KP, Conboy IM (2014) Oxytocin is an age-specific circulating hormone that is necessary for muscle maintenance and regeneration. Nat Commun 5:4082. doi:10.1038/ncomms5082
113. Sinha M, Jang YC, Oh J, Khong D, Wu EY, Manohar R, Miller C, Regalado SG, Loffredo FS, Pancoast JR, Hirshman MF, Lebowitz J, Shadrach JL, Cerletti M, Kim MJ, Serwold T, Goodyear LJ, Rosner B, Lee RT, Wagers AJ (2014) Restoring systemic GDF11 levels reverses age-related dysfunction in mouse skeletal muscle. Science 344(6184):649–652. doi:10.1126/science.1251152
114. Egerman MA, Cadena SM, Gilbert JA, Meyer A, Nelson HN, Swalley SE, Mallozzi C, Jacobi C, Jennings LL, Clay I, Laurent G, Ma S, Brachat S, Lach-Trifilieff E, Shavlakadze T, Trendelenburg AU, Brack AS, Glass DJ (2015) GDF11 increases with age and inhibits skeletal muscle regeneration. Cell Metab 22(1):164–174. doi:10.1016/j.cmet.2015.05.010
115. Brun CE, Rudnicki MA (2015) GDF11 and the mythical fountain of youth. Cell Metab 22(1):54–56. doi:10.1016/j.cmet.2015.05.009
116. Rodgers BD, Eldridge JA (2015) Reduced circulating GDF11 is unlikely responsible for age-dependent changes in mouse heart, muscle, and brain. Endocrinology 156(11):3885–3888. doi:10.1210/en.2015-1628
117. Smith SC, Zhang X, Zhang X, Gross P, Starosta T, Mohsin S, Franti M, Gupta P, Hayes D, Myzithras M, Kahn J, Tanner J, Weldon SM, Khalil A, Guo X, Sabri A, Chen X, MacDonnell S, Houser SR (2015) GDF11 does not rescue aging-related pathological hypertrophy. Circ Res 117(11):926–932. doi:10.1161/CIRCRESAHA.115.307527
118. Poggioli T, Vujic A, Yang P, Macias-Trevino C, Uygur A, Loffredo FS, Pancoast JR, Cho M, Goldstein J, Tandias RM, Gonzalez E, Walker RG, Thompson TB, Wagers AJ, Fong YW, Lee RT (2016) Circulating growth differentiation factor 11/8 levels decline with age. Circ Res 118(1):29–37. doi:10.1161/CIRCRESAHA.115.307521
119. Lopez-Otin C, Blasco MA, Partridge L, Serrano M, Kroemer G (2013) The hallmarks of aging. Cell 153(6):1194–1217. doi:10.1016/j.cell.2013.05.039
120. Bortoli S, Renault V, Eveno E, Auffray C, Butler-Browne G, Piétu G. Gene expression profiling of human satellite cells during muscular aging using cDNA arrays. Gene. 2003 Dec 4;321:145–54. http://dx.doi.org/10.1016/j.gene.2003.08.025
121. Zhu CH, Mouly V, Cooper RN, Mamchaoui K, Bigot A, Shay JW, Di Santo JP, Butler-Browne GS, Wright WE (2007) Cellular senescence in human myoblasts is overcome by human telomerase reverse transcriptase and cyclin-dependent kinase 4: consequences in aging muscle and therapeutic strategies for muscular dystrophies. Aging Cell 6(4):515–523. doi:10.1111/j.1474-9726.2007.00306.x
122. Kadi F, Ponsot E (2010) The biology of satellite cells and telomeres in human skeletal muscle: effects of aging and physical activity. Scand J Med Sci Sports 20(1):39–48. doi:10.1111/j.1600-0838.2009.00966.x
123. Shadrach JL, Wagers AJ (2011) Stem cells for skeletal muscle repair. Philos Trans R Soc Lond B Biol Sci 366(1575):2297–2306. doi:10.1098/rstb.2011.0027
124. Kondo H, Kim HW, Wang L, Okada M, Paul C, Millard RW, Wang Y (2016) Blockade of senescence-associated microRNA-195 in aged skeletal muscle cells facilitates repro-

gramming to produce induced pluripotent stem cells. Aging Cell 15(1):56–66. doi:10.1111/acel.12411
125. Renault V, Thornell LE, Eriksson PO, Butler-Browne G, Mouly V (2002) Regenerative potential of human skeletal muscle during aging. Aging Cell 1(2):132–139
126. Didier N, Hourde C, Amthor H, Marazzi G, Sassoon D (2012) Loss of a single allele for Ku80 leads to progenitor dysfunction and accelerated aging in skeletal muscle. EMBO Mol Med 4(9):910–923. doi:10.1002/emmm.201101075
127. Bernet JD, Doles JD, Hall JK, Kelly Tanaka K, Carter TA, Olwin BB (2014) p38 MAPK signaling underlies a cell-autonomous loss of stem cell self-renewal in skeletal muscle of aged mice. Nat Med 20(3):265–271. doi:10.1038/nm.3465
128. Price FD, von Maltzahn J, Bentzinger CF, Dumont NA, Yin H, Chang NC, Wilson DH, Frenette J, Rudnicki MA (2014) Inhibition of JAK-STAT signaling stimulates adult satellite cell function. Nat Med 20(10):1174–1181. doi:10.1038/nm.3655
129. Tierney MT, Aydogdu T, Sala D, Malecova B, Gatto S, Puri PL, Latella L, Sacco A (2014) STAT3 signaling controls satellite cell expansion and skeletal muscle repair. Nat Med 20(10):1182–1186. doi:10.1038/nm.3656
130. Lopez-Lluch G, Navas P (2015) Calorie restriction as an intervention in ageing. J Physiol 594(8):2043–2060. doi:10.1113/JP270543
131. Cerletti M, Jang YC, Finley LW, Haigis MC, Wagers AJ (2012) Short-term calorie restriction enhances skeletal muscle stem cell function. Cell Stem Cell 10(5):515–519. doi:10.1016/j.stem.2012.04.002
132. Wallace GQ, McNally EM (2009) Mechanisms of muscle degeneration, regeneration, and repair in the muscular dystrophies. Annu Rev Physiol 71:37–57. doi:10.1146/annurev.physiol.010908.163216
133. Emery AE (2002) The muscular dystrophies. Lancet 359(9307):687–695
134. Heslop L, Morgan JE, Partridge TA (2000) Evidence for a myogenic stem cell that is exhausted in dystrophic muscle. J Cell Sci 113(Pt 12):2299–2308
135. Webster C, Blau HM (1990) Accelerated age-related decline in replicative life-span of Duchenne muscular dystrophy myoblasts: implications for cell and gene therapy. Somat Cell Mol Genet 16(6):557–565
136. Kottlors M, Kirschner J (2010) Elevated satellite cell number in Duchenne muscular dystrophy. Cell Tissue Res 340(3):541–548. doi:10.1007/s00441-010-0976-6
137. Mann CJ, Perdiguero E, Kharraz Y, Aguilar S, Pessina P, Serrano AL, Munoz-Canoves P (2011) Aberrant repair and fibrosis development in skeletal muscle. Skelet Muscle 1(1):21. doi:10.1186/2044-5040-1-21
138. Delaporte C, Dehaupas M, Fardeau M (1984) Comparison between the growth pattern of cell cultures from normal and Duchenne dystrophy muscle. J Neurol Sci 64(2):149–160
139. Melone MA, Peluso G, Petillo O, Galderisi U, Cotrufo R (1999) Defective growth in vitro of Duchenne Muscular Dystrophy myoblasts: the molecular and biochemical basis. J Cell Biochem 76(1):118–132
140. Blau HM, Webster C, Pavlath GK (1983) Defective myoblasts identified in Duchenne muscular dystrophy. Proc Natl Acad Sci U S A 80(15):4856–4860
141. Boldrin L, Zammit PS, Morgan JE (2015) Satellite cells from dystrophic muscle retain regenerative capacity. Stem Cell Res 14(1):20–29. doi:10.1016/j.scr.2014.10.007
142. Dumont NA, Wang YX, von Maltzahn J, Pasut A, Bentzinger CF, Brun CE, Rudnicki MA (2015) Dystrophin expression in muscle stem cells regulates their polarity and asymmetric division. Nat Med 21(12):1455–1463. doi:10.1038/nm.3990
143. Decary S, Hamida CB, Mouly V, Barbet JP, Hentati F, Butler-Browne GS (2000) Shorter telomeres in dystrophic muscle consistent with extensive regeneration in young children. Neuromuscul Disord 10(2):113–120
144. Lund TC, Grange RW, Lowe DA (2007) Telomere shortening in diaphragm and tibialis anterior muscles of aged mdx mice. Muscle Nerve 36(3):387–390. doi:10.1002/mus.20824
145. Sousa-Victor P, Perdiguero E, Munoz-Canoves P (2014) Geroconversion of aged muscle stem cells under regenerative pressure. Cell Cycle 13(20):3183–3190. doi:10.4161/15384101.2014.965072
146. Sacco A, Mourkioti F, Tran R, Choi J, Llewellyn M, Kraft P, Shkreli M, Delp S, Pomerantz JH, Artandi SE, Blau HM (2010) Short telomeres and stem cell exhaustion model Duchenne muscular dystrophy in mdx/mTR mice. Cell 143(7):1059–1071. doi:10.1016/j.cell.2010.11.039
147. Kudryashova E, Kramerova I, Spencer MJ (2012) Satellite cell senescence underlies myopathy in a mouse model of limb-girdle muscular dystrophy 2H. J Clin Invest 122(5):1764–1776. doi:10.1172/JCI59581
148. Goodell MA, Nguyen H, Shroyer N (2015) Somatic stem cell heterogeneity: diversity in the blood, skin and intestinal stem cell com-

partments. Nat Rev Mol Cell Biol 16(5):299–309. doi:10.1038/nrm3980
149. Gayraud-Morel B, Pala, F., Sakai, H., Tajbakhsh S. (2017) Isolation of muscle stem cells from mouse skeletal muscle. Methods Mol Biol 1556:23–40
150. Tonlorenzi R, Rossi G, Messina G (2017) Isolation and characterization of vessel-associated stem/progenitor cells from skeletal muscle. Methods Mol Biol 1556:149–170
151. Stuelsatz P, Keire P, Yablonka-Reuveni Z (2017) Isolation, culture, and immunostaining of skeletal muscle myofibers from wild-type and Nestin-GFP mice as a means to analyze satellite cells. Methods Mol Biol 1556:51–102
152. Sincennes MC, Wang YX, Rudnicki MA (2017) Primary mouse myoblast purification using magnetic cell separation. Methods Mol Biol 1556:41–50
153. Low M, Eisner C, Rossi FM (2017) Fibro/adipogenic progenitors (FAPs): isolation by FACS and culture. Methods Mol Biol 1556:179–190
154. Saclier M, Theret M, Mounier R, Chazaud B (2017) Effects of macrophage conditioned-medium on murine and human muscle cells: analysis of proliferation, differentiation and fusion. Methods Mol Biol 1556:317–328
155. Arora R, Rumman M, Venugopal N, Gala H, Dhawan J (2017) Mimicking muscle stem cell quiescence in culture: methods for synchronization in reversible arrest. Methods Mol Biol 1556:283–302
156. Davoudi S, Gilbert PM (2017) Optimization of satellite cell culture through biomaterials. Methods Mol Biol 1556:329–342
157. Gatto S, Puri PL, Malecova B (2017) Single cell gene expression profiling of skeletal muscle-derived cells. Methods Mol Biol 1556:191–221
158. Sreenivasan K, Braun T, Kim J (2017) Systematic identification of genes regulating muscle stem cell self-renewal and differentiation. Methods Mol Biol 1556:343–354
159. Peng X, Sun K, Zhou J, Sun H, Wang H (2017) Bioinformatics for novel large intergenic non-coding RNA (lincRNA) identification in skeletal muscle cells. Methods Mol Biol 1556:355–362
160. Ryall JG (2017) Simultaneous measurement of mitochondrial and glycolytic activity in quiescent muscle stem cells. Methods Mol Biol 1556:245–254
161. Garcia-Prat L, Munoz-Canoves P, Martinez-Vicente M (2017) Monitoring autophagy in muscle stem cells. Methods Mol Biol 1556:255–281
162. Lund DK, McAnulty P, Siegel AL, Cornelison DD (2016) Methods for observing and quantifying muscle satellite cell motility and invasion in vitro. Methods Mol Biol 1556: 303–316
163. Tierney D, Sacco A (2017) Engraftment of FACS isolated muscle stem cells into injured skeletal muscle. Methods Mol Biol 1556:223–236
164. Hall MN, Hall JK, Cadwallader AB, Pawlikowski BT, Doles JD, Elston TC, Olwin BB (2017) Transplantation of Skeletal Muscle Stem cells. Methods Mol Biol 1556:237–244
165. Nguyen PD, Currie PD (2017) Using transgenic zebrafish to study muscle stem/progenitor cells. Methods Mol Biol 1556:117–127
166. Lavergne G, Soler C, Zmojdzian M, Jagla K (2016) Characterization of Drosophila muscle stem cell-like adult muscle precursors. Methods Mol Biol 1556:103–116

Part II

Muscle Stem Cells (Satellite cells)

Chapter 2

Isolation of Muscle Stem Cells from Mouse Skeletal Muscle

Barbara Gayraud-Morel, Francesca Pala, Hiroshi Sakai, and Shahragim Tajbakhsh

Abstract

Isolation of muscle stem cells from skeletal muscle is a critical step for the study of skeletal myogenesis and regeneration. Although stem cell isolation has been performed for decades, the emergence of flow cytometry with defined cell surface markers, or transgenic mouse models, has allowed the efficient isolation of highly enriched stem cell populations. Here, we describe the isolation of mouse muscle stem cells using two different combinations of enzyme treatments allowing the release of mononucleated muscle stem cells from their niche. Mouse muscle stem cells can be further isolated as a highly enriched population by flow cytometry using fluorescent reporters or cell surface markers. We will present advantages and drawbacks of these different approaches.

Key words Satellite cells, Muscle stem cell isolation, Enzymatic dissociation, FACS

Abbreviations

FACS	Fluorescent-activated cell sorting
TA	Tibialis anterior
GFP	Green fluorescent protein
FSC	Forward scatter
SSC	Side scatter
C/T	Collagenase D/Trypsin
C/D	Collagenase A/Dispase II
FBS	Fetal bovine serum
CD	Cluster of differentiation

1 Introduction

Biochemistry and molecular or cell biology studies on specific organs require the isolation of highly purified cell types to assess stem cell properties and their role in growth and regeneration.

Eusebio Perdiguero and DDW Cornelison (eds.), *Muscle Stem Cells: Methods and Protocols*, Methods in Molecular Biology, vol. 1556, DOI 10.1007/978-1-4939-6771-1_2, © Springer Science+Business Media LLC 2017

However, such isolation of mononucleated cells from solid tissues and organs requires enzymatic treatments that ultimately result in the destruction of the stem cell niche and the differential stripping of cell surface molecules. One example is skeletal muscle satellite (stem) cells that lie on myofibers, located between the sarcolemma and its surrounding basement membrane [1]. Most satellite cells are quiescent during homeostasis. Following intense exercise or muscle injuries, satellite cells activate, proliferate, and differentiate to renew damaged myofibers [2, 3]. Molecular markers are used to distinguish quiescent satellite cells from their progeny [4, 5]. Furthermore, several studies have suggested that quiescent satellite cells constitute a heterogeneous cell population [6–8]. To date, the transcription factor Pax7 is the most reliable marker known that identifies all quiescent satellite cells [9]. *Pax7* expression marks the upstream myogenic population from mid-embryogenesis to adulthood. Its expression persists in activated and cycling satellite cells, but it is downregulated during myogenic commitment and differentiation, as the differentiation transcription factor myogenin is upregulated. Therefore, the *Pax7* locus has been a target of choice to generate knock-in and transgenic animals, to introduce GFP reporter or Cre recombinase genes that permit the prospective isolation and characterization of satellite cells.

Several genetically modified mice with a GFP reporter, which recapitulate *Pax7* expression, have been generated, for example, *Tg:Pax7-nGFP* [10, 11] and *Pax7-ZsGreen* [12]. More recently, the generation of four tamoxifen-inducible *Pax7-CreERT2* mouse lines [13–16] has also opened the possibility to label most satellite cells with mouse reporter lines such as $R26^{mT/mG}$ or $R26^{eYFP}$. In these situations, the specificity and efficiency of the *Cre* mouse lines and the efficiency of the reporter line need to be closely examined. These four lines have different characteristics; the *Tg:Pax7-CreERT2* is a transgenic that does not affect the endogenous *Pax7* locus [15]. The knock-in $Pax7^{CreERT2}$ has the *Cre* sequence inserted in the 3′UTR of the *Pax7* locus [16], similar in strategy but distinct from another knock-in [13]. These knock-ins result in *Pax7* expression due to the IRES sequence. Finally, the $Pax7^{CE}$ is a knock-in/knockout, where the *Cre* gene is inserted in the first exon of the *Pax7* locus resulting in a null allele [14].

Other mouse models have been generated, but they do not mark the entire satellite cell population. While *Pax3* is expressed in all body (except the head) myogenic cells during development, in the adult, its expression, reported with the $Pax3^{GFP}$ mouse, is restricted essentially to some trunk muscles [17]. In the $Myf5^{nGFP/+}$ mouse, the nGFP (nuclear GFP) reporter is expressed in a subpopulation of satellite cells [18], and the two $Myf5^{Cre}$ alleles allow the detection of about 90 %, but not all satellite cells in young adult mice [19–21, 10, 22]. Another GFP reporter line, *Nestin-GFP*, has been used to isolate satellite cells [23]. New genetic tools will continue to emerge allowing isolation of subpopulations of muscle stem cells in quiescence and their purification on further step down the myogenic differentiation program.

In addition to isolation of muscle stem cells by flow cytometry, preparation of isolated single myofibers remains an efficient method to isolate and follow individual satellite cells and their fate [24, 25]. This method provides the advantage of isolating satellite cells within their niche, with minimal artificial stress induced by other methods such as FACS. However, the amount of cells that are collected at quiescence is limited, and isolation of activated or proliferating satellite cells needs to be performed ex vivo.

In this chapter, we describe the isolation of satellite cells by FACS where these cells are marked by a fluorescent reporter such as the previously described *Tg:Pax7-nGFP* reporter mice [11]. The benefit of fluorescent reporter mice is the high yield of satellite cells collected in a reasonable amount of time, without the use of antibody staining for surface markers that can be compromised following enzymatic treatment. Muscle stem cells isolated by enzymatic treatments remain functionally competent and are able to proliferate and generate robust myogenic fibers in vitro and in vivo. Transplantation of a single or a population of satellite cells is able to contribute to regenerating myofibers and self-renew efficiently [18, 26, 27].

However, in many cases, the crossing of a fluorescent reporter with other genetically modified mice is time-consuming and cumbersome; therefore, the use of surface markers becomes mandatory. Currently, there is no single cell surface marker that can be universally used to identify quiescent or activated satellite cells. Most of the strategies rely on the combination of at least two positive markers for satellite cells and several negative markers for exclusion of non-myogenic cells. For example, the protocol proposed in this chapter is based on two cell surface markers: α7-integrin and CD34 [21, 28]. Other examples reported previously use different cell surface marker combinations including CXCR4, VCAM, SM/C2.6 [29–32], or Syndecan4 [33].

It should be noted that a major drawback of stem cell isolation and FACS is the stress that is imposed on cells, the consequences of which remain unknown. These have yet to be quantified but should be taken into account particularly in assays such as the measure of metabolic activity. Furthermore, the quiescent status of stem cells is immediately compromised as soon as muscles are isolated. Therefore, satellite cells isolated on single fibers or by FACS have already initiated the G0/G1 transition [34]. To overcome the problem of stress induced by FACS to the cells, an alternative method relies on magnetic activated cell sorting (MACS) [35].

For human muscles, a different strategy is required as some of the cell surface receptors of quiescent satellite cells are different. For example, the cell surface receptor CD34 is not expressed in quiescent human satellite cells [36]. Instead, CD56 is expressed in some quiescent satellite cells and more robustly in activated cells derived from cultured satellite cells. Therefore, CD56 is commonly used to enrich the human myogenic cells following expansion in culture. Different approaches have been developed recently to isolate human muscle stem cells, from manually dissected muscle fiber fragments [37] to FACS using α7-integrin. CXCR4 or CD29 cell surface receptors are also used as positive markers [38–40].

Here, we describe two approaches for the isolation of satellite cells:

1. *Isolation of mouse skeletal muscle stem cells with collagenase D and trypsin*: Collagenase D cleaves native collagen. It has a high collagenase activity and a low contaminating trypic activity. Trypsin is a serine endopeptidase; it cleaves peptide bonds at the carboxylique side of the basic amino acids Arg and Lys. Due to the rapid and broad range of action of the trypsin, this procedure should not be employed extensively if the isolation of muscle stem cell requires the recognition of cell surface antigens. However, this protocol has proven to be efficient for skeletal muscles when fluorescent reporter mice are used to mark satellite cells.
2. *Isolation of mouse muscle satellite cells with collagenase A and dispase II*: Collagenase A degrades native collagen and has a balanced ratio of contaminating enzyme activities. Dispase II is a neutral protease that hydrolyzes the N-terminal peptide bonds of nonpolar amino acid residues. This enzyme combination allows an efficient release of satellite cells from the tissue and it minimizes the cleavage of cell surface receptors necessary for immunodetection and cell sorting [41].

2 Materials

2.1 Isolation of Mouse Muscle Stem Cells with Collagenase D and Trypsin (C/T)

1. Dulbecco's modified Eagle's medium (DMEM, Life Technologies, ref. 31966021) with penicillin/streptomycin (Life Technologies, ref. 15140122).
2. Trypsin stock solution (Life Technologies, ref. 15090-046). Make 15 ml tube aliquots of 2.5 % trypsin stock solution and store at −20 °C.
3. Collagenase D stock solution (Roche, ref. 1108882001). Collagenase D powder is resuspended with cold DMEM to make a 1 % stock solution (250 ml DMEM for 2.5 g collagenase D). Make 15 ml tube aliquots and store at −20 °C.
4. DNase I stock solution (Roche, ref. 11284932001). DNase I is resuspended in cold DMEM to make a 10 mg/ml stock solution. Make aliquots and store at −20 °C. Use at a final concentration of 0.1 mg/ml (*see* **Note 1**).

5. Fetal bovine serum (FBS): serum is used to block trypsin activity once digestion is complete. We do not heat inactivate FBS.
6. Collagenase D/Trypsin working solution (C/T): add 4 ml of collagenase D stock solution and 2 ml of trypsin stock solution to 44 ml of DMEM to obtain final concentrations of 0.08 % collagenase D and 0.1 % trypsin. Add 50 μl of stock DNase I solution to the final 50 ml of C/T solution. This solution is prepared extemporaneously (*see* **Note 2**).
7. Cell strainers; 100, 70 (Miltenyi Biotec, ref. 130-098-463, 130-098-462), and 40 μm (BD Falcon, ref. 352340).
8. Dissecting tools are cleaned and sterilized by autoclaving or 70 % alcohol.

2.2 Isolation of Mouse Muscle Stem Cells with Collagenase A and Dispase II (C/D)

1. Dispase II (Roche, ref. 04942078001): the number of U/mg is provided on each commercial bottle. Dispase II is weighed extemporaneously and used at a final concentration of 2.4 U/ml (*see* Subheading 2.2, **item 4**).
2. Collagenase A (Roche, ref. 11088793001): collagenase A is weighted extemporaneously to be used at a final concentration of 0.2 % (*see* Subheading 2.2, **item 4**).
3. Hank's Balanced Salt Solution (HBSS) (Life Technologies, ref. 24020091) with penicillin/streptomycin (Life Technologies, ref. 15140122).
4. For 10 ml of working collagenase A/dispase II solution (C/D): weigh 20 mg of collagenase A and the appropriate quantity of dispase II (e.g., 24 mg for a dispase II at 1 U/mg), and resuspend in 10 ml of HBSS. DNase I (stock 10 mg/ml; *see* Subheading 2.1) is added to the solution to a final concentration of 0.1 mg/ml. The solution is filtered through a 22 μm filter and kept at room temperature.
5. Washing solution: HBSS with penicillin/streptomycin and 2 % fetal bovine serum.

3 Methods

3.1 Isolation of Mouse Muscle Stem Cells with Collagenase D and Trypsin (C/T)

1. Before dissection, fill 50 ml Falcon tubes with 5 ml of FBS, and place in an ice bucket. Place 100 and 70 μm cell strainers on top.
2. Skeletal muscles are dissected with small scissors from body parts of interest (*see* **Note 3**). Exclude as much as possible adipose tissue (white fat), nerves (like the sciatic nerve running through the hind limb), and tendons. Collect the dissected muscles in a small volume (1 ml) of DMEM in a petri dish (*see* **Note 4**).

3. Before mincing the tissue, remove the excess DMEM with a pipette or by placing the petri dish on an angle. Removing the excess of liquid will facilitate mincing the tissue.
4. Use fine dissecting scissors (Moria, ref. 4878) to mince the tissue until a slurry forms with no more large muscle pieces (*see* **Note 5**).
5. Transfer minced tissue to a 50 ml tube filled with DMEM-Pen/Strep. Mix by inverting the 50 ml tube several times to resuspend the tissue. Leave the tube on ice for 10 min to allow the muscle to sediment, while the fat tissue will stay in suspension. When all the muscle tissue is sedimented, remove the excess of DMEM (*see* **Note 6**).
6. Resuspend the sedimented tissue with 10 ml of collagenase D/trypsin working solution (*see* **Note 7**).
7. Incubate the tube at 37 °C for 25 min in an agitating water bath. To increase the surface between the tissue and enzymes, tubes are incubated in a horizontal position with a gentle agitation (120 rpm). This is a critical step that can impact on the yield of satellite cells.
8. After 25 min of incubation, stop the agitation, and place the tubes in a tube holder in a vertical position to allow tissue sedimentation for 3–5 min.
9. Collect the supernatant by decanting or with a pipette.
10. Filter the supernatant through cell strainers 100 and 70 μm consecutively. The filtered supernatant is collected in 50 ml tubes prepared earlier with serum on ice (*see* **Note 8**).
11. Resuspend the tissue with 10 ml of collagenase D/trypsin working solution. Repeat **steps 6–10** until all tissue is digested (*see* **Note 9**).
12. Centrifuge all collected supernatant tubes for 10 min at 50 × *g* at 4 °C. Transfer the supernatant in a new 50 ml tube, and discard the pellet (*see* **Note 10**).
13. Centrifuge the collected supernatant for 15 min at 550 × *g* at 4 °C.
14. Discard the supernatant by decanting or with a pipette. Keep the pellet.
15. Wash the pellet by gently resuspending with 40 ml of cold DMEM.
16. Centrifuge for 15 min at 550 × *g* at 4 °C (*see* **Note 11**).
17. Discard the supernatant by decanting or with a pipette. Keep the pellet.
18. Resuspend the pellet in 40 ml of cold DMEM, and filter it through a 40 μm cell strainer.
19. Centrifuge for 15 min at 550 × *g* at 4 °C.

20. Resuspend the pellet in cold DMEM with 2 % FBS. The cell suspension is ready to use for sorting. Keep cells on ice until used.
21. Cells can be collected out of the FACS in DMEM/2 % FBS for further molecular analysis and cell culture or directly in lysis buffer for RNA preparation. If a precise number of cells are required for the following experiments, we recommend to verify the actual number of cells collected by FACS with a hemocytometer (e.g., Malassez counting chamber).

3.2 Isolation of Mouse Muscle Stem Cells with Collagenase A and Dispase II (C/D)

1. Repeat **steps 1–5**, from Subheading 3.1, replacing DMEM by HBSS.
2. Resuspend the sedimented tissue with 10 ml of collagenase A/dispase II working solution (*see* **Note 12**).
3. Incubate the tube at 37 °C for 90–120 min in an agitating water bath. To increase the surface between the tissue and enzymes, tubes are incubated in a horizontal position with a gentle agitation (*see* **Note 13**). This is a critical step that can impact on the yield of satellite cells.
4. When muscles are fully dissociated, stop the digestion.
5. Add 30 ml of HBSS to the cell suspension, and filter consecutively through 100, 70, and 40 μm filters.
6. Centrifuge for 15 min at 550 × *g* at 4 °C.
7. Remove the supernatant. At this point, the pellet can be processed for antibody staining (*see* Subheading 3.3) or resuspended in washing solution if the sample is directly used for FACS (*see* **Note 14**).

3.3 Cell Surface Receptor Staining for Isolation of Satellite Cells by Flow Cytometry

We recommend performing cell surface receptor staining following the collagenase A/dispase II protocol (*see* Subheading 3.2) which preserves to a greater extent the surface antigens:

1. A set of controls is required to establish a correct gating of cell populations during FACS acquisition.

   Recommended controls for this protocol are the following; 1/30 of the cell preparation is necessary to carry out each of the controls (*see* **Note 15**):

   (a) Negative control: keep a small aliquot of the sample unstained. It is used to set up the voltage of the lasers and the threshold between positivity and negativity of different fluorochrome-labeled cell populations.

   (b) Single conjugated antibody staining: perform an individual staining for each fluorescent dye-coupled antibody on a small fraction of the sample. These staining are necessary to perform the compensation. Compensation is a technique used to eliminate false signal that results from spectral overlap between fluorescent dyes when used in multicolor staining panels.

(c) Fluorescence minus one (FMO) staining: FMO control is used to properly interpret flow cytometry data. In this control, all conjugated antibodies are included except one which is necessary to discriminate your final population (in the proposed protocol, we excluded CD34).

(d) Isotype control: fluorochrome-coupled antibodies with the same isotype as the primary antibodies used for the surface staining are important to confirm the specificity of primary antibody binding and help to assess the level of background staining (*see* **Note 16**).

2. Prepare the mix of conjugated antibodies at the indicated concentrations (*see* Table 1) in 500 μl of washing solution (HBSS/2 % FBS).
3. Resuspend the cell pellet in the conjugated antibody mix, and incubate on ice for 30 min (*see* **Note 17**).
4. Add three volumes of washing solution and centrifuge for 15 min at 550 × *g* at 4 °C.
5. Remove the supernatant and resuspend the cells in 2 ml of washing solution.
6. Centrifuge for 15 min at 550 × *g* at 4 °C.
7. Resuspend the pellet in an adequate volume of washing solution (100–500 μl), and keep the cells on ice until sorting.

3.4 Profiles of Satellite Cells Isolated by FACS

FACS profiles of quiescent and activated satellite cells isolated from *TgPax7-nGFP* mice are shown in Fig. 1. Quiescent satellite cells are clustered as a homogeneous population, characterized by a small size (FSC) and a low granulosity (SSC). However, activated satellite cells analyzed 40 h after cardiotoxin injury of the muscle present a larger and heterogeneous size with an increased granulosity.

Table 1
Dilution of primary antibodies for cell surface staining of quiescent mouse muscle stem cells

Antibody	Clone	Conjugated dye	Dilution	Isotype	Source
CD45	30-F11	PE-Cy7	1/400	Rat IgG2b, κ	eBioscience
CD31	MEC 13.3	PE	1/50	Rat IgG2a, κ	BD Pharmingen
Sca-1	D7	PE-Cy7	1/100	Rat IgG2a, κ	eBioscience
CD34	RAM34	eFluor 450	1/20	Rat IgG2a, κ	eBioscience
α7-Integrin	α7-Integrin	Alexa 647	1/1000	Rat IgG2b	AbLab

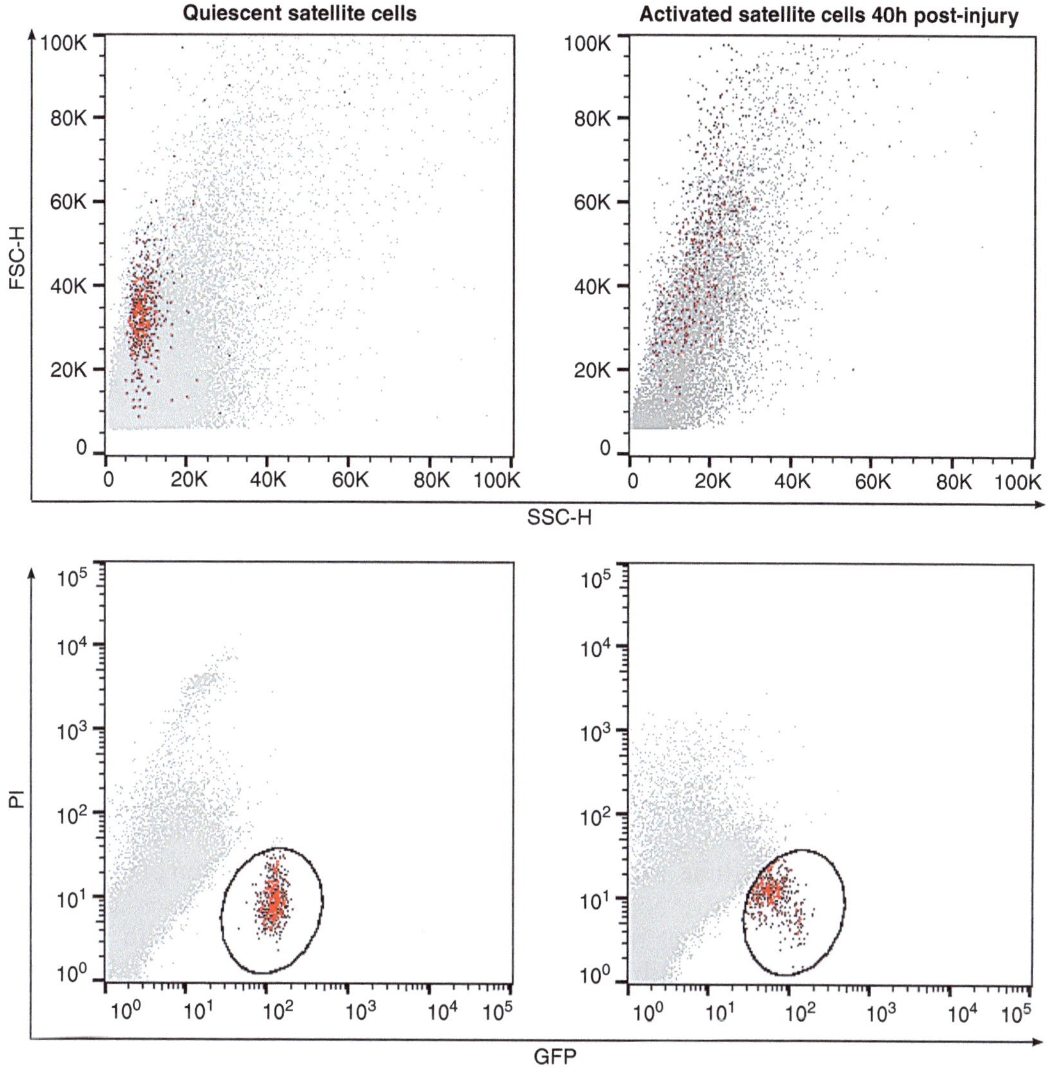

Fig. 1 FACS profiles for isolation of quiescent and activated muscle stem cells from *TgPax7-nGFP* muscles. The GFP+ cell population (*red*) was backgated on scatter gates to highlight small size and low granulosity of quiescent muscle stem cells, while activated muscle stem cells are heterogeneous in size and in granulosity. Fluorescence-activated cell sorting was performed on an Aria III (BD Biosciences) or MoFlow (BD Biosciences), and data were analyzed using FlowJo (BD Biosciences)

Cells are displayed on a plot with GFP on the *X* axis (FITC-A, fluorescein isothiocyanate area) and propidium iodide on the *Y* axis (PE-Texas Red, phycoerythrin area) to detect dead cells. From our experience, the proportion of dead cells within the Pax7-nGFP+ cell population after the digestion protocol is very low, generally not exceeding 0.05 %. Representative yields of Pax7-nGFP+ cells collected from different muscles are presented in Table 2.

As described in this chapter, muscle stem cells can be isolated by flow cytometry using cell surface markers. Figure 2 shows the

Table 2
Representative number of Pax7-nGFP+ cells extracted with collagenase D/trypsin from *TgPax7-nGFP* mouse muscles from different anatomical locations. These values are indicative of the relative cell numbers; actual numbers can be higher

	Pax7-nGFP+ sorted cells/muscle
EDL	2000–3000
TA	4000–7000
Soleus	8000–9000
Quadriceps	30,000–40,000
Leg (hind limb)	120,000–180,000
Abdomen	100,000–120,000
Diaphragm	20,000–30,000
EOM	6000–8000
Tongue	30,000–50,000
Masseter	20,000–30,000
Back	22,000–30,000

gating strategy, starting by FSC and SSC, followed by selection of the negative population for CD45, Sca-1, and CD31 used to exclude non-myogenic cells, and finally the selection of the double-positive population α7-integrin+ and CD34+ chosen as positive myogenic markers. As indicated, a sample treated with all conjugated antibodies except one (fluorescence minus one (FMO)) is necessary to set and select the appropriate population of interest.

Both sets of enzymes presented in this protocol, collagenase D/trypsin or collagenase A/dispase II, can be efficiently used to collect muscle stem cells from transgenic models such as *Tg:Pax7-nGFP* (Fig. 3a, Table 3). However, the cell surface receptor staining strategy, due to additional steps for the staining, the required controls, and the gating strategy, reduces greatly the amount of cells collected compared to the genetically marked muscle stem cells (Table 3). Table 3 shows the susceptibility of cell surface receptors to enzymatic treatment by the collagenase D/trypsin digestion (mainly due to the trypsin activity) compared to the cells extracted with collagenase A/dispase II. The endothelial cell receptor, recognized by CD31, is particularly susceptible to proteolytic cleavage.

Fig. 2 (continued) stained with α7-integrin, which recognizes all satellite cells but also some non-myogenic cells. To enrich for satellite cells, the suspension is double stained with CD34. These double positive α7-integrin+/CD34+ cells correspond largely to satellite cells

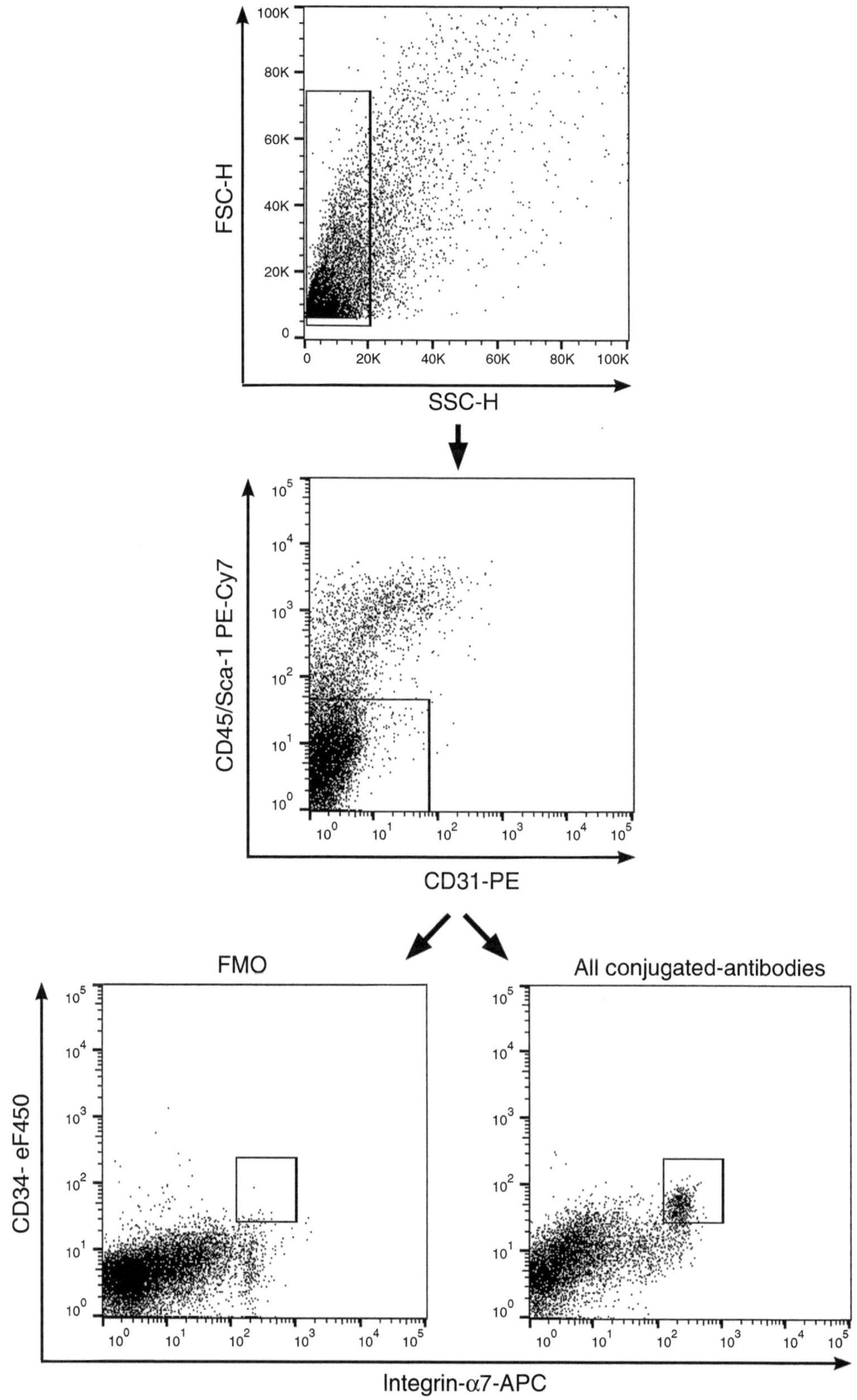

Fig. 2 FACS profile of α7-integrin^{+}CD34^{+} satellite cells. The strategy described here for isolating satellite cells with cell surface markers relies on the elimination of non-myogenic cells: CD45^{+} hematopoietic cells, CD31^{+} endothelial cells, and Sca-1^{+} non-myogenic cells including endothelial cells and fibroblasts. Satellite cells are

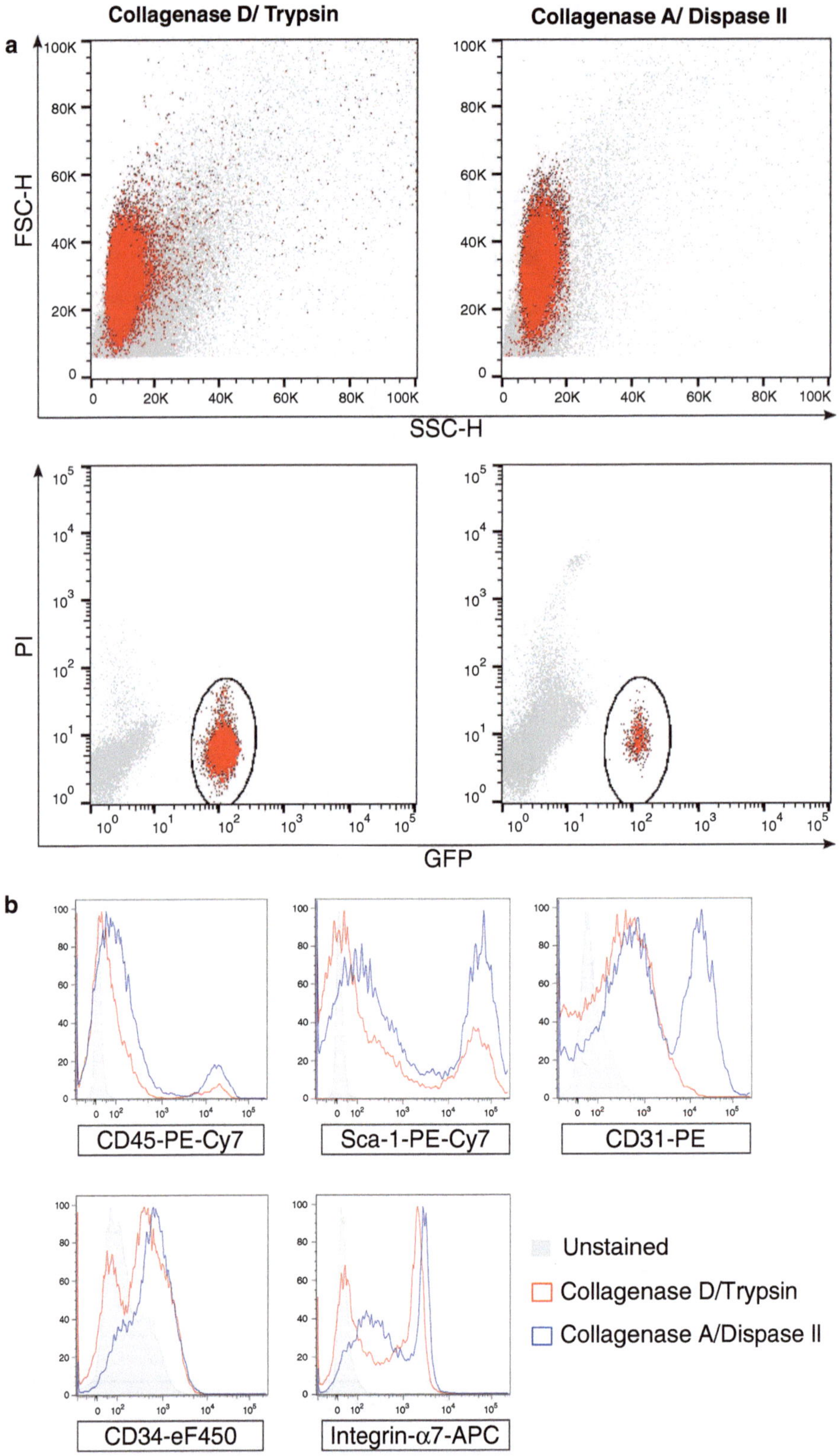

Fig. 3 FACS profiles following enzymatic digestions of skeletal muscles. (**a**) Profile of isolated muscle stem cells from collagenase D/trypsin or collagenase A/dispase II enzymatic treatments does not differ when isolating

Table 3
Comparison of the number of cells collected following collagenase D/trypsin or collagenase A/dispase II digestion. For a given muscle (here, quadriceps), the number of Pax7-nGFP+ cells collected with the two sets of enzymes is similar. However, supplementary steps required for antibody staining of cell surface receptors and FACS diminish greatly the number of cells collected

	Number of Pax7-nGFP+ cells/quadriceps	Number of Pax7-nGFP+ cells/1 leg (hind limb)	Number of CD34+/Itg7+cells/1 leg (hind limb)
Collagenase D/trypsin FACS on GFP expression	32,000	170,000	n.d.
Collagenase A/dispase II FACS on GFP expression	35,000	n.d.	n.d.
Collagenase A/dispase II + FACS on cell surface markers	n.d.	n.d.	60,000

4 Notes

1. DNase I is required especially when preparing muscle stem cells from injured muscle, where a high level of inflammatory cells and dead cells release their DNA during the extraction protocol. This released DNA will interfere later with cell sorting.
2. Prepare working collagenase D/trypsin solution with room temperature DMEM. This minimizes the time to reach 37 °C during digestion. Keep stock solutions on ice, especially trypsin which can self-degrade at 37 °C.
3. The choice of which skeletal muscles are to be used for muscle stem cell isolation has to be considered given the reported heterogeneity among skeletal muscle groups. Head and trunk muscles have different development origins and are governed by distinct genetic networks. Limb and trunk muscles originate from different somitic regions. Moreover, skeletal muscles support different physiological functions (slow or fast twitch) and different metabolisms (oxidative or glycolytic) [42, 43].

Fig. 3 (continued) Pax7-nGFP+ cells. (**b**) Histogram of single antibody staining of muscle stem cells isolated with the two different sets of enzymes shows the susceptibility of some cell surface receptors to collagenase D/trypsin treatment compared to collagenase A/dispase II treatment. The CD31 receptor, and to a lesser extent Sca-1, displays a high susceptibility to trypsin digestion

4. If several mice need to be dissected, perform dissections in DMEM, and keep on ice before proceeding to the following steps.
5. Fine curved scissors (Moria, ref. 8142A) are more appropriate than straight fine scissors.
6. This step can be omitted if the starting material is inferior to 1 g of tissue.
7. 10 ml of collagenase D/trypsin working solution is appropriate for 2 g of dissected muscles. Increase proportionally the amount of collagenase D/trypsin working solution with the amount of dissected muscles (20 ml of C/T for 4 g of muscle).
8. Collection of supernatant can be performed directly through 40 μm cell strainer if starting material is inferior to about 0.3 g, for example, for one or two *tibialis anterior* muscles.
9. Four rounds of digestion are usually enough to digest 2–4 limb muscles. If large pieces of tissue are still visible after the third round of digestion, the pellet can be transferred into a petri dish and minced with scissors before being resuspended in fresh collagenase D/trypsin working solution.
10. This first gentle centrifugation pellet contains essentially debris, whereas satellite cells remain in the supernatant. This step should not be performed if the starting material is inferior to about 0.5 g of tissue.
11. If the pellet is still large, a second wash with DMEM can be performed as mentioned in **steps 14** and **15**.
12. A maximum of 2.5 g of tissue should be digested per tube otherwise the suspension becomes too viscous.
13. If digesting a large amount of material, the digestion can be split in two rounds of 45 min, with addition of fresh enzyme solution to the pellet as described in Subheading 3.1.
14. If antibody staining is not considered, repeat washing by adding 30 ml of HBSS, followed by centrifugation of 15 min.
15. If a small sample, for example a TA, needs to be processed for antibody staining to isolate muscle stem cells, controls can be performed on a muscle from other locations (any other limb muscle).
16. For convenience, controls performed in small volumes of sample can be easily performed in V-bottom 96 well plates (Corning, ref. 3897).
17. Alternative cell surface markers for quiescent satellite cell staining and cell surface markers for non-myogenic cells (Table 4).

Table 4
Alternative cell surface markers for quiescent satellite cell staining and cell surface markers for non-myogenic cells

Positive cell surface satellite cell markers	
SM/C2.6[a]	[31]
VCAM[a]	[29, 30]
CXCR4	[32]
Exclusion markers	
CD11b	Macrophage
Ter119	Red blood cell

[a]SM/C2.6 antigen corresponds to VCAM

Acknowledgments

We acknowledge the funding support from the Institut Pasteur, Centre National pour la Recherche Scientifique, Association Française contre les Myopathies, Agence Nationale de la Recherche (Laboratoire d'Excellence Revive, Investissement d'Avenir; ANR-10-LABX-73), Association pour la Recherche sur le Cancer, EU Advanced ERC grant, and Fondation pour la Recherche Médicale. H. Sakai is funded by the ERC and F. Pala by the LabEx Revive/Pasteur PPU program.

References

1. Mauro A (1961) Satellite cell of skeletal muscle fibers. J Biophys Biochem Cytol 9:493–495
2. Relaix F, Zammit PS (2012) Satellite cells are essential for skeletal muscle regeneration: the cell on the edge returns centre stage. Development 139(16):2845–2856. doi:10.1242/dev.069088
3. Yablonka-Reuveni Z (2011) The skeletal muscle satellite cell: still young and fascinating at 50. J Histochem Cytochem 59(12):1041–1059. doi:10.1369/0022155411426780
4. Fukada S, Uezumi A, Ikemoto M, Masuda S, Segawa M, Tanimura N, Yamamoto H, Miyagoe-Suzuki Y, Takeda S (2007) Molecular signature of quiescent satellite cells in adult skeletal muscle. Stem Cells 25(10):2448–2459. doi:2007-0019 [pii]5.1634/stemcells.2007-0019
5. Kuang S, Rudnicki MA (2008) The emerging biology of satellite cells and their therapeutic potential. Trends Mol Med 14(2):82–91
6. Motohashi N, Asakura A (2014) Muscle satellite cell heterogeneity and self-renewal. Front Cell Dev Biol 2:1. doi:10.3389/fcell.2014.00001
7. Rocheteau P, Vinet M, Chretien F (2015) Dormancy and quiescence of skeletal muscle stem cells. Results Probl Cell Differ 56:215–235. doi:10.1007/978-3-662-44608-9_10
8. Sambasivan R, Tajbakhsh S (2015) Adult skeletal muscle stem cells. Results Probl Cell Differ 56:191–213. doi:10.1007/978-3-662-44608-9_9
9. Seale P, Sabourin LA, Girgis-Gabardo A, Mansouri A, Gruss P, Rudnicki MA (2000) Pax7 is required for the specification of myogenic satellite cells. Cell 102(6):777–786
10. Sambasivan R, Comai G, Le Roux I, Gomes D, Konge J, Dumas G, Cimper C, Tajbakhsh S (2013) Embryonic of adult muscle stem cells are primed by the determination gene Mrf4. Dev Biol 381(1):241–255. doi:10.1016/j.ydbio.2013.04.018

11. Sambasivan R, Gayraud-Morel B, Dumas G, Cimper C, Paisant S, Kelly RG, Tajbakhsh S (2009) Distinct regulatory cascades govern extraocular and pharyngeal arch muscle progenitor cell fates. Dev Cell 16(6):810–821
12. Bosnakovski D, Xu Z, Li W, Thet S, Cleaver O, Perlingeiro RC, Kyba M (2008) Prospective isolation of skeletal muscle stem cells with a Pax7 reporter. Stem Cells 26(12):3194–3204
13. Gunther S, Kim J, Kostin S, Lepper C, Fan CM, Braun T (2013) Myf5-positive satellite cells contribute to Pax7-dependent long-term maintenance of adult muscle stem cells. Cell Stem Cell 13(5):590–601. doi:10.1016/j.stem.2013.07.016
14. Lepper C, Fan CM (2010) Inducible lineage tracing of Pax7-descendant cells reveals embryonic origin of adult satellite cells. Genesis 48(7):424–436. doi:10.1002/dvg.20630
15. Mourikis P, Sambasivan R, Castel D, Rocheteau P, Bizzarro V, Tajbakhsh S (2012) A critical requirement for notch signaling in maintenance of the quiescent skeletal muscle stem cell state. Stem Cells 30(2):243–252. doi:10.1002/stem.775
16. Murphy MM, Lawson JA, Mathew SJ, Hutcheson DA, Kardon G (2011) Satellite cells, connective tissue fibroblasts and their interactions are crucial for muscle regeneration. Development 138(17):3625–3637. doi:10.1242/dev.064162
17. Relaix F, Rocancourt D, Mansouri A, Buckingham M (2005) A Pax3/Pax7-dependent population of skeletal muscle progenitor cells. Nature 435(7044):948–953
18. Gayraud-Morel B, Chretien F, Jory A, Sambasivan R, Negroni E, Flamant P, Soubigou G, Coppee JY, Di Santo J, Cumano A, Mouly V, Tajbakhsh S (2012) Myf5 haploinsufficiency reveals distinct cell fate potentials for adult skeletal muscle stem cells. J Cell Sci 125(Pt 7):1738–1749. doi:10.1242/jcs.097006
19. Comai G, Sambasivan R, Gopalakrishnan S, Tajbakhsh S (2014) Variations in the efficiency of lineage marking and ablation confound distinctions between myogenic cell populations. Dev Cell 31(5):654–667. doi:10.1016/j.devcel.2014.11.005
20. Haldar M, Hancock JD, Coffin CM, Lessnick SL, Capecchi MR (2007) A conditional mouse model of synovial sarcoma: insights into a myogenic origin. Cancer Cell 11(4):375–388. doi:10.1016/j.ccr.2007.01.016
21. Kuang S, Kuroda K, Le Grand F, Rudnicki MA (2007) Asymmetric self-renewal and commitment of satellite stem cells in muscle. Cell 129(5):999–1010
22. Tallquist MD, Weismann KE, Hellstrom M, Soriano P (2000) Early myotome specification regulates PDGFA expression and axial skeleton development. Development 127(23):5059–5070
23. Day K, Shefer G, Richardson JB, Enikolopov G, Yablonka-Reuveni Z (2007) Nestin-GFP reporter expression defines the quiescent state of skeletal muscle satellite cells. Dev Biol 304(1):246–259. doi:10.1016/j.ydbio.2006.12.026
24. Keire P, Shearer A, Shefer G, Yablonka-Reuveni Z (2013) Isolation and culture of skeletal muscle myofibers as a means to analyze satellite cells. Methods Mol Biol 946:431–468. doi:10.1007/978-1-62703-128-8_28
25. Shinin V, Gayraud-Morel B, Tajbakhsh S (2009) Template DNA-strand co-segregation and asymmetric cell division in skeletal muscle stem cells. Methods Mol Biol 482:295–317. doi:10.1007/978-1-59745-060-7_19
26. Rocheteau P, Gayraud-Morel B, Siegl-Cachedenier I, Blasco MA, Tajbakhsh S (2012) A subpopulation of adult skeletal muscle stem cells retains all template DNA strands after cell division. Cell 148(1–2):112–125. doi:10.1016/j.cell.2011.11.049
27. Sacco A, Doyonnas R, Kraft P, Vitorovic S, Blau HM (2008) Self-renewal and expansion of single transplanted muscle stem cells. Nature 456(7221):502–506
28. Montarras D, Morgan J, Collins C, Relaix F, Zaffran S, Cumano A, Partridge T, Buckingham M (2005) Direct isolation of satellite cells for skeletal muscle regeneration. Science 309(5743):2064–2067
29. Chakkalakal JV, Jones KM, Basson MA, Brack AS (2012) The aged niche disrupts muscle stem cell quiescence. Nature 490(7420):355–360. doi:10.1038/nature11438
30. Cheung TH, Quach NL, Charville GW, Liu L, Park L, Edalati A, Yoo B, Hoang P, Rando TA (2012) Maintenance of muscle stem-cell quiescence by microRNA-489. Nature 482(7386):524–528. doi:10.1038/nature10834
31. Fukada S, Higuchi S, Segawa M, Koda K, Yamamoto Y, Tsujikawa K, Kohama Y, Uezumi A, Imamura M, Miyagoe-Suzuki Y, Takeda S, Yamamoto H (2004) Purification and cell-surface marker characterization of quiescent satellite cells from murine skeletal muscle by a novel monoclonal antibody. Exp Cell Res 296(2):245–255. doi:10.1016/j.yexcr.2004.02.018
32. Sherwood RI, Christensen JL, Conboy IM, Conboy MJ, Rando TA, Weissman IL, Wagers AJ (2004) Isolation of adult mouse myogenic progenitors: functional heterogeneity of cells

within and engrafting skeletal muscle. Cell 119(4):543–554
33. Tanaka KK, Hall JK, Troy AA, Cornelison DD, Majka SM, Olwin BB (2009) Syndecan-4-expressing muscle progenitor cells in the SP engraft as satellite cells during muscle regeneration. Cell Stem Cell 4(3):217–225
34. Rodgers JT, King KY, Brett JO, Cromie MJ, Charville GW, Maguire KK, Brunson C, Mastey N, Liu L, Tsai CR, Goodell MA, Rando TA (2014) mTORC1 controls the adaptive transition of quiescent stem cells from G0 to G(Alert). Nature 510(7505):393–396. doi:10.1038/nature13255
35. Motohashi N, Asakura Y, Asakura A (2014) Isolation, culture, and transplantation of muscle satellite cells. J Vis Exp (86). doi:10.3791/50846
36. Pisani DF, Dechesne CA, Sacconi S, Delplace S, Belmonte N, Cochet O, Clement N, Wdziekonski B, Villageois AP, Butori C, Bagnis C, Di Santo JP, Kurzenne JY, Desnuelle C, Dani C (2010) Isolation of a highly myogenic CD34-negative subset of human skeletal muscle cells free of adipogenic potential. Stem Cells 28(4):753–764. doi:10.1002/stem.317
37. Marg A, Escobar H, Gloy S, Kufeld M, Zacher J, Spuler A, Birchmeier C, Izsvak Z, Spuler S (2014) Human satellite cells have regenerative capacity and are genetically manipulable. J Clin Invest 124(10):4257–4265. doi:10.1172/JCI63992
38. Bareja A, Holt JA, Luo G, Chang C, Lin J, Hinken AC, Freudenberg JM, Kraus WE, Evans WJ, Billin AN (2014) Human and mouse skeletal muscle stem cells: convergent and divergent mechanisms of myogenesis. PLoS One 9(2):e90398. doi:10.1371/journal.pone.0090398
39. Castiglioni A, Hettmer S, Lynes MD, Rao TN, Tchessalova D, Sinha I, Lee BT, Tseng YH, Wagers AJ (2014) Isolation of progenitors that exhibit myogenic/osteogenic bipotency in vitro by fluorescence-activated cell sorting from human fetal muscle. Stem Cell Reports 2(1):92–106. doi:10.1016/j.stemcr.2013.12.006
40. Tamaki T, Uchiyama Y, Hirata M, Hashimoto H, Nakajima N, Saito K, Terachi T, Mochida J (2015) Therapeutic isolation and expansion of human skeletal muscle-derived stem cells for the use of muscle-nerve-blood vessel reconstitution. Front Physiol 6:165. doi:10.3389/fphys.2015.00165
41. Mitchell KJ, Pannerec A, Cadot B, Parlakian A, Besson V, Gomes ER, Marazzi G, Sassoon DA (2010) Identification and characterization of a non-satellite cell muscle resident progenitor during postnatal development. Nat Cell Biol 12(3):257–266
42. Sambasivan R, Yao R, Kissenpfennig A, Van Wittenberghe L, Paldi A, Gayraud-Morel B, Guenou H, Malissen B, Tajbakhsh S, Galy A (2011) Pax7-expressing satellite cells are indispensable for adult skeletal muscle regeneration. Development 138(17):3647–3656
43. Schiaffino S, Reggiani C (2011) Fiber types in mammalian skeletal muscles. Physiol Rev 91(4):1447–1531. doi:10.1152/physrev.00031.2010

Chapter 3

Primary Mouse Myoblast Purification using Magnetic Cell Separation

Marie Claude Sincennes, Yu Xin Wang, and Michael A. Rudnicki

Abstract

Primary myoblasts can be isolated from mouse muscle cell extracts and cultured in vitro. Muscle cells are usually dissociated manually by mincing with razor blades or scissors in a collagenase/dispase solution. Primary myoblasts are then gradually enriched by pre-plating on collagen-coated plates, based on the observation that mouse fibroblasts attach quickly to collagen-coated plates, and are less adherent. Here, we describe an automated muscle dissociation protocol. We also propose an alternative to pre-plating using magnetic bead separation of primary myoblasts, which improve myoblast purity by minimizing fibroblast contamination.

Key words Primary myoblasts, Satellite cells, Magnetic separation, Fibroblast contamination, Magnetic-activated cell sorting, Muscle cells, Pre-plating

1 Introduction

Satellite cells are quiescent, mononucleated, myogenic cells located between the basal lamina and sarcolemma of muscle fibers [1]. Following injury or other stimuli, satellite cells can temporarily leave their quiescent state and become activated. They differentiate into myogenic progenitor cells, or myoblasts, a more committed muscle precursor that is highly proliferative and can further differentiate into myotubes to sustain muscle regeneration [2]. Myoblasts can be expanded in vitro and subjected to classical molecular and cellular biology techniques that necessitate large numbers of cells. Primary myoblasts in culture are considered as one of the most representative in vitro models of satellite cell biology. The first step toward the prospective isolation of a primary myoblast culture is to dissociate muscle tissue into a single-cell suspension. Terminally differentiated cells cannot proliferate and be maintained in culture, whereas satellite cells can differentiate into myoblasts and undergo several rounds of cell division [2]. However, fibroblasts from the muscle tissue can also proliferate in culture. Different strategies

Eusebio Perdiguero and DDW Cornelison (eds.), *Muscle Stem Cells: Methods and Protocols*, Methods in Molecular Biology, vol. 1556, DOI 10.1007/978-1-4939-6771-1_3, © Springer Science+Business Media LLC 2017

have been undertaken to favor myoblast growth at the expense of fibroblasts, among others the choice of growth media [3]. One of the most common procedures to eliminate fibroblast contamination is based on the different adhesion properties of myoblasts vs. fibroblasts. Fibroblasts adhere quickly and strongly to tissue culture dishes, whereas primary myoblasts are more loosely attached. It is thus possible to enrich for myoblasts by pre-plating cells on tissue culture dishes for a short period of time and collecting the cells that did not stick to the plate [4]. However, mice from specific genotypes could generate fewer myoblasts, could contain more fibroblasts, or could lead to myoblasts with impaired proliferation. In all these cases, it can be more laborious to eliminate fibroblast contamination from the culture mix. One well-characterized example in the field is the widely used dystrophin-deficient *mdx* mouse, modeling Duchenne muscular dystrophy. Skeletal muscles from dystrophic mice present fibrosis [5–7], which often compromises, or at least challenges, the isolation of pure primary *mdx* myoblasts using pre-plating procedure. Here, we describe an alternative protocol for primary myoblast purification. First, we propose a faster, automated protocol for muscle dissociation, leading to a better homogenized single-cell suspension with increased cell viability. Second, we optimized a magnetic cell separation protocol based on satellite cell surface markers. Satellite cells specifically express different cell surface markers such as α7-integrin [8], M-cadherin [9, 10], CD34 [10, 11], and Vcaml [12]. They do not express Sca-1 neither the lineage (lin) markers Cd11b, Ter-119, Cd45, and Cd31 [13–15]. We determined that the use of a subset of these markers (α7-integrin$^+$, Cd11b$^-$, Sca1$^-$, Cd45$^-$, Cd31$^-$) was sufficient to allow isolation of a pure myoblast population, deprived of fibroblasts. Using biotin-conjugated antibodies and streptavidin-conjugated magnetic microbeads, we can eliminate cells expressing Sca1 and lineage markers (negative selection). As a second step, we stain lin$^-$ cells with a biotin-conjugated α7-integrin antibody to purify satellite cells/myoblasts (positive selection). This protocol was particularly efficient for the purification of myoblasts from WT and *mdx* mice.

2 Materials

1. Dissection tools and scissors.
2. Sterile tissue culture dishes (60 × 15 mm preferably).
3. Sterile phosphate buffered saline (PBS).
4. Container with ice.
5. Sterile tissue culture hood.
6. Automated cell dissociator (gentleMACS Octo Dissociator with Heaters).

7. gentleMACS CTubes.
8. Collagenase/dispase solution: 10 g/L collagenase B and 4 g/L dispase I in HAM'S F-10 media, filter-sterilized.
9. Delicate task wipers (Kimwipes).
10. Fetal bovine serum (FBS).
11. 70 μm nylon mesh that fits to 50 ml conical tubes.
12. Conical 15 ml and 50 ml tubes.
13. Tabletop centrifuge.
14. 1.5 ml tubes (Eppendorf).
15. MACS buffer: 2 mM EDTA and 0.5 % BSA in PBS.
16. Biotin-conjugated anti-Sca1 antibody.
17. Biotin-conjugated anti-Cd45 antibody.
18. Biotin-conjugated anti-Cd11b antibody.
19. Biotin-conjugated anti-Cd31 antibody.
20. Biotin-conjugated anti-α7-integrin antibody.
21. MACS streptavidin microbeads.
22. VarioMACS™ separator with LS and MS column adaptors.
23. MACS LD and MS columns.
24. Myoblast growth media: HAM'S F-10 medium supplemented with 20 % FBS, 2.5 ng/ml β-FGF, and antibiotics (penicillin/streptomycin).
25. Humidified 37 °C, 5 % CO_2 incubator.
26. Sterile collagen-coated tissue culture dishes (60 mm in size and higher): Solution containing 0.01 % collagen and 0.2 % acetic acid, sterilized. Put enough solution to cover all the culture dish surface for 1 h, remove excess liquid, and let dry.

3 Methods

3.1 Muscle Cell Dissociation

1. Sacrifice one to three mice per genotype, and remove the posterior limbs using scissors (*see* **Note 1**). Dissect the muscles away from the skin and bone. Remove as much as non-muscle tissue (i.e., fat, tendons, blood vessels) as possible. Place dissected muscle tissue in a 60 × 15 mm culture dish filled with sterile, cold PBS and keep on ice (*see* **Note 2**). It is preferable to perform the subsequent steps into a sterile tissue culture hood.
2. In a gentleMACS CTube, add 5 ml of the collagenase/dispase solution (*see* **Note 3**).
3. Transfer muscle tissue in a new, empty petri dish. Remove excess PBS using a delicate task wiper. Transfer muscle tissue

into the CTube containing the collagenase/dispase. Using sterile scissors, mince muscles directly into the tube, to generate pieces of ~2 mm in size.

4. Close the CTube securely and turn it upside down. Ensure that all muscle pieces came down in the collagenase/dispase solution.
5. Insert the CTube into the gentleMACS Octo Dissociator with Heaters, and insert the heater according to the manufacturer's instructions.
6. Use the customized programs displayed in Table 1 for (a) mechanical and (b) enzymatic dissociation of the muscle tissue.
7. Remove the heater and the CTube according to the manufacturer's instruction. A homogenous cell suspension without any aggregates or clumps should be obtained (*see* **Note 4**).
8. Into a sterile tissue culture hood, open the CTube and add 5 ml of FBS. Mix with a pipette.
9. Prepare a 70 μm nylon mesh over a conical 50 ml tube. Pre-wet with 2 ml of FBS.
10. Add cell suspension gradually to filter.
11. Rinse the empty CTube twice with 8–9 ml of PBS, and add to filter. Collect total effluent.

Table 1
Suggested automated muscle dissociation program

	Mechanical dissociation program		Enzymatic dissociation program
1	Temp ON	1	Temp ON
2	Loop 10×	2	Spin 60 rpm 3 min
3	Spin 360 rpm 5 s	3	Spin −30 rpm 9 min
4	Spin −360 rpm 5 s	4	Loop 10×
5	End loop	5	Spin 360 rpm 5 s
6	End	6	Spin −360 rpm 5 s
		7	End loop
		8	Spin −30 rpm 12 min
		9	End

"temp ON" indicates heating at 37 °C
"+" or "−" signs indicate the direction of the circular motion
"loop 10×" indicates that the steps between "loop" and "end loop" will be repeated ten times

12. Evenly transfer filtered cell suspension to two conical 15 ml tubes (*see* **Note 5**).
13. Centrifuge at 350×*g* for 10 min (*see* **Note 6**). Aspirate off the supernatant.
14. Resuspend each pellet with 150 μl MACS buffer and transfer in one single 1.5 ml tube (final volume of 300 μl).

3.2 Negative Selection of lin+ Cells

1. Add 5 μl of each of the following antibodies in the 1.5 ml tube: Cd31-biotin, Cd45-biotin, Sca1-biotin, and Cd11b-biotin (*see* **Note 7**).
2. Mix and incubate on ice for 10 min.
3. Wash cells by adding 1 ml MACS buffer and mix. Centrifuge cells at 350×*g* for 5 min.
4. Remove supernatant (do not use vacuum, remove slowly with the pipette). Leave a residual volume of ~100 μl and add 10 μl of streptavidin microbeads.
5. Mix and incubate on ice for 15 min.
6. Wash cells by adding 1 ml MACS buffer and mix. Centrifuge cells at 350×*g* for 5 min. Remove supernatant and resuspend the pellet in MACS buffer. At this step, all samples from the same genotype may be pooled in a final volume of 500 μl MACS buffer.
7. Place a LD column in the magnetic field of a suitable MACS separator (i.e., VarioMACS™). Rinse the column by adding 2 ml of MACS buffer (*see* **Note 8**).
8. Put a clean 15 ml conical tube under the column, and add the cell suspension onto the column.
9. Collect unlabeled cells that pass through and wash with 2 × 1 ml MACS buffer. Collect total effluent. This is the unlabeled cell fraction.
10. Centrifuge cells at 350×*g* for 5 min (*see* **Note 9**). Remove supernatant and resuspend the cell pellet in 300 μl MACS buffer. Transfer into a 1.5 ml tube.

3.3 Positive Selection of α7-Integrin+ Cells

1. Add 5 μl of α7-integrin-biotin antibody in the 1.5 ml tube (*see* **Note 10**).
2. Mix and incubate on ice for 10 min.
3. Wash cells by adding 1 ml MACS buffer and mix. Centrifuge cells at 350×*g* for 5 min.
4. Remove supernatant (do not use vacuum, remove slowly with the pipette). Leave a residual volume of ~100 μl and add 10 μl of streptavidin microbeads.
5. Mix and incubate on ice for 15 min.

6. Wash cells by adding 1 ml MACS buffer and mix. Centrifuge cells at 350×*g* for 5 min. Remove supernatant and resuspend in 500 µl MACS buffer.
7. Place a MS column in the magnetic field of a suitable MACS separator (i.e., VarioMACS™). Rinse the column by adding 500 µl of MACS buffer.
8. Put a clean 15 ml conical tube under the column, and add the cell suspension onto the column.
9. Collect unlabeled cells that pass through and wash with 3 × 500 µl MACS buffer. This is the unlabeled cell fraction that can be discarded.
10. Remove column from the magnetic field and place it in a new 15 ml collection tube.
11. Add 1 ml of MACS buffer onto the column. Immediately flush out the cells by firmly pushing the plunger into the column. This step can be repeated to increase the yield (pool the effluent).
12. Centrifuge cells at 350×*g* for 5 min. The cell pellet should not (or barely) be visible. Remove the supernatant, resuspend in 5 ml of pre-warmed myoblast growth media and put in a 60 × 15 mm collagen-coated culture dish (*see* **Note 11**).
13. Incubate in a 37 °C 5 % CO_2 incubator for at least 48 h. Change medium every 2 days and split when cells reach ~75–80 % confluency (*see* **Note 12**).

4 Notes

1. The protocol described here has been optimized for the use of muscle tissue obtained from the posterior limbs of one mouse aged between 6 and 8 weeks. Isolation of muscle tissue from younger mice may increase the number of satellite cells/myoblasts. In addition to posterior limbs, it is also possible to dissect muscle tissue from anterior limbs or diaphragm to increase the number of purified myoblasts. However, the buffer volumes and incubation times should be optimized accordingly. If using more than one mouse of the same genotype, it is recommended to treat all the individual mice independently and pool the cells at the step specified in the protocol, i.e., during the negative selection step.
2. If the dissected muscle tissue is contaminated with blood, it is recommended to perform 2–3 wash steps by transferring muscle tissue into new culture dishes filled with PBS, until most of the blood is cleared. Alternatively, it is possible to add the facultative Ter-119-biotin antibody to the lineage mix containing

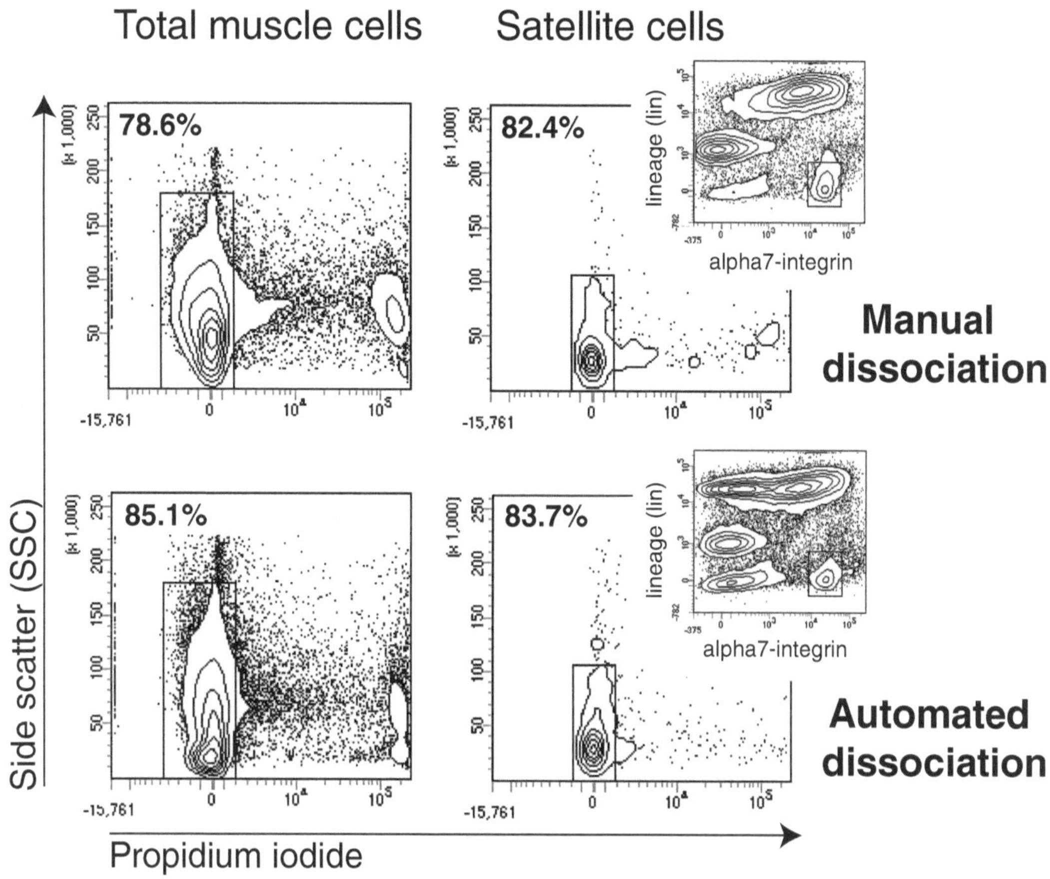

Fig. 1 Automated muscle dissociation protocol leads to increased cell viability. Muscle tissue from posterior limbs of one mouse was dissociated either manually (mincing with scissors and triturating with a pipette) or with the automated dissociation protocol. Propidium iodide was used to stain dead cells in total muscle extract (*left panel*) or specifically in lin$^-$ α7-integrin$^+$ satellite cells (*right panel*). A representative percentage of cell viability is shown in each plot

Sca1, Cd45, Cd31, and Cd11b. Ter-119 is a specific marker of red blood cells [16]. A red blood cell lysis buffer may also be employed. Red blood cells do not express α7-integrin and should be eliminated during the positive selection step, but if they are too abundant, it could influence the overall yield of myoblast purification.

3. The volume of collagenase/dispase solution is optimized for efficient dissociation of muscle tissue from the posterior limbs of one mouse per CTube.
4. The automated protocol for cell dissociation leads to increased satellite cell viability compared to the manual method using scissors/razor blades and triturating with pipette (Fig. 1).

5. The 50 ml conical tubes have a flare tip compared to the 15 ml tubes. When centrifuging in 50 ml tubes, the cell pellet can be loose and the supernatant is difficult to remove without also aspirating muscle cells. Therefore, it is recommended to centrifuge in 15 ml tubes.
6. The optimal centrifugation speed and time should be defined by the investigator. High centrifugation speed and time usually leads to better cell yield, but some variations could be observed depending of the centrifuge model.
7. The biotin-conjugated antibodies can be purchased separately. Alternatively, some satellite cell isolation kits are available. The efficiency of one given antibody could vary between manufacturers and even from one lot to the other. Each new vial of antibody should be tested and the volume needed per reaction should be optimized accordingly. The purpose of the negative selection step is to clear out lin^+ cells. The column flow-through contains satellite cells/myoblasts and must be kept. DO NOT add the α7-integrin-biotin antibody at this step.
8. The protocol described here makes use of a manual separator. Alternatively, an automated separator can be employed, as long as the column flow-through is kept. In that case, the buffer volumes and machine settings have to be optimized by the investigator.
9. At this step, the cell pellet should be smaller than at the previous step. If red blood cells have not been eliminated previously (*see* **Note 2**), the pellet should also appear as pink or red.
10. The purpose of the positive selection step is to keep α7-integrin-labeled cells. These cells are retained by the column. After discarding the flow-through, satellite cells/myoblasts need to be eluted from the column. To this end, the magnetic field should be removed before the elution step.
11. The size of the culture dish has to be defined by the investigator. The number of satellite cells/myoblasts per mouse is dependent of the genotype and the age of the mouse. It also depends on the number of mice used.
12. Freshly isolated satellite cells are very small. After one day or two in culture, they will grow in size and start to divide as primary myoblasts. Primary myoblasts grow better when plated at approximately 15–25 % confluence. If they get more than 80 % confluent, they may start to die or spontaneously differentiate into myotubes. This protocol is very efficient in eliminating fibroblast contamination from the culture mix. However, if a few fibroblasts persist, cells can be split using cold trypsin, or trypsin without EDTA, as primary myoblasts will detach easily, whereas fibroblasts will be left behind.

Acknowledgments

Y.X.W. is supported by fellowships from QEII-GSST and the Canadian Institutes of Health Research. M.A.R. holds the Canada Research Chair in Molecular Genetics. These studies were carried out with support of grants to M.A.R. from the US National Institutes for Health [R01AR044031], the Canadian Institutes for Health Research [MOP-12080, MOP-81288], E-Rare-2: Canadian Institutes of Health Research/Muscular Dystrophy Canada [ERA-132935], the Muscular Dystrophy Association, and the Stem Cell Network.

References

1. Bentzinger CF, Wang YX, Dumont NA, Rudnicki MA (2013) Cellular dynamics in the muscle satellite cell niche. EMBO Rep 14(12):1062–1072. doi:10.1038/embor.2013.182
2. Chang NC, Rudnicki MA (2014) Satellite cells: the architects of skeletal muscle. Curr Top Dev Biol 107:161–181. doi:10.1016/b978-0-12-416022-4.00006-8
3. Rando TA, Blau HM (1994) Primary mouse myoblast purification, characterization, and transplantation for cell-mediated gene therapy. J Cell Biol 125(6):1275–1287
4. Springer ML, Rando TA, Blau HM (2002) Gene delivery to muscle. Curr Protoc Hum Genet Chapter 13:Unit13.4. doi:10.1002/0471142905.hg1304s31
5. Pastoret C, Sebille A (1995) mdx mice show progressive weakness and muscle deterioration with age. J Neurol Sci 129(2):97–105
6. Stedman HH, Sweeney HL, Shrager JB, Maguire HC, Panettieri RA, Petrof B, Narusawa M, Leferovich JM, Sladky JT, Kelly AM (1991) The mdx mouse diaphragm reproduces the degenerative changes of Duchenne muscular dystrophy. Nature 352(6335):536–539. doi:10.1038/352536a0
7. Trensz F, Haroun S, Cloutier A, Richter MV, Grenier G (2010) A muscle resident cell population promotes fibrosis in hindlimb skeletal muscles of mdx mice through the Wnt canonical pathway. Am J Physiol Cell Physiol 299(5): C939–C947. doi:10.1152/ajpcell.00253.2010
8. Blanco-Bose WE, Yao CC, Kramer RH, Blau HM (2001) Purification of mouse primary myoblasts based on alpha 7 integrin expression. Exp Cell Res 265(2):212–220. doi:10.1006/excr.2001.5191
9. Irintchev A, Zeschnigk M, Starzinski-Powitz A, Wernig A (1994) Expression pattern of M-cadherin in normal, denervated, and regenerating mouse muscles. Dev Dyn 199(4):326–337. doi:10.1002/aja.1001990407
10. Beauchamp JR, Heslop L, Yu DS, Tajbakhsh S, Kelly RG, Wernig A, Buckingham ME, Partridge TA, Zammit PS (2000) Expression of CD34 and Myf5 defines the majority of quiescent adult skeletal muscle satellite cells. J Cell Biol 151(6):1221–1234
11. Lee JY, Qu-Petersen Z, Cao B, Kimura S, Jankowski R, Cummins J, Usas A, Gates C, Robbins P, Wernig A, Huard J (2000) Clonal isolation of muscle-derived cells capable of enhancing muscle regeneration and bone healing. J Cell Biol 150(5):1085–1100
12. Rosen GD, Sanes JR, LaChance R, Cunningham JM, Roman J, Dean DC (1992) Roles for the integrin VLA-4 and its counter receptor VCAM-1 in myogenesis. Cell 69(7): 1107–1119
13. Sherwood RI, Christensen JL, Conboy IM, Conboy MJ, Rando TA, Weissman IL, Wagers AJ (2004) Isolation of adult mouse myogenic progenitors: functional heterogeneity of cells within and engrafting skeletal muscle. Cell 119(4):543–554. doi:10.1016/j.cell.2004.10.021
14. Fukada S, Higuchi S, Segawa M, Koda K, Yamamoto Y, Tsujikawa K, Kohama Y, Uezumi A, Imamura M, Miyagoe-Suzuki Y, Takeda S, Yamamoto H (2004) Purification and cell-surface marker characterization of quiescent satellite cells from murine skeletal muscle by a novel monoclonal antibody. Exp Cell Res 296(2):245–255. doi:10.1016/j.yexcr.2004.02.018

15. Uezumi A, Ojima K, Fukada S, Ikemoto M, Masuda S, Miyagoe-Suzuki Y, Takeda S (2006) Functional heterogeneity of side population cells in skeletal muscle. Biochem Biophys Res Commun 341(3):864–873. doi:10.1016/j.bbrc.2006.01.037

16. Kina T, Ikuta K, Takayama E, Wada K, Majumdar AS, Weissman IL, Katsura Y (2000) The monoclonal antibody TER-119 recognizes a molecule associated with glycophorin A and specifically marks the late stages of murine erythroid lineage. Br J Haematol 109(2):280–287

Check for updates

Chapter 4

Isolation, Culture, and Immunostaining of Skeletal Muscle Myofibers from Wildtype and Nestin-GFP Mice as a Means to Analyze Satellite Cells

Pascal Stuelsatz, Paul Keire, and Zipora Yablonka-Reuveni

Abstract

Multinucleated myofibers, the functional contractile units of adult skeletal muscle, harbor mononuclear $Pax7^{+}$ myogenic progenitors on their surface between the myofiber basal lamina and plasmalemma. These progenitors, known as satellite cells, are the primary myogenic stem cells in adult muscle. This chapter describes our laboratory protocols for isolating, culturing, and immunostaining intact myofibers from mouse skeletal muscle as a means for studying satellite cell dynamics. The first protocol discusses myofiber isolation from the flexor digitorum brevis (FDB) muscle. These short myofibers are plated in dishes coated with PureCol collagen (formerly known as Vitrogen) and maintained in a mitogen-poor medium (± supplemental growth factors). Employing such conditions, satellite cells remain at the surface of the parent myofiber while synchronously undergoing a limited number of proliferative cycles and rapidly differentiate. The second protocol discusses the isolation of longer myofibers from the extensor digitorum longus (EDL) muscle. These EDL myofibers are routinely plated individually as adherent myofibers in wells coated with Matrigel and maintained in a mitogen-rich medium, conditions in which satellite cells migrate away from the parent myofiber, proliferate extensively, and generate numerous differentiating progeny. Alternatively, these EDL myofibers can be plated as non-adherent myofibers in uncoated wells and maintained in a mitogen-poor medium (± supplemental growth factors), conditions that retain satellite cell progeny at the myofiber niche similar to the FDB myofiber cultures. However, the adherent myofiber format is our preferred choice for monitoring satellite cells in freshly isolated (Time 0) myofibers. We conclude this chapter by promoting the Nestin-GFP transgenic mouse as an efficient tool for direct analysis of satellite cells in isolated myofibers. While satellite cells have been often detected by their expression of the Pax7 protein or the $Myf5^{nLacZ}$ knockin reporter (approaches that are also detailed herein), the Nestin-GFP reporter distinctively permits quantification of satellite cells in live myofibers, which enables linking initial Time 0 numbers and subsequent performance upon culturing. We additionally point out to the implementation of the Nestin-GFP transgene for monitoring other selective cell lineages as illustrated by GFP expression in capillaries, endothelial tubes and neuronal cells. Myofibers from other types of muscles, such as diaphragm, masseter, and extraocular, can also be isolated and analyzed using protocols described herein. Collectively, this chapter provides essential tools for studying satellite cells in their native position and their interplay with the parent myofiber.

Key words Skeletal muscle, Satellite cells, Isolated myofiber, Flexor digitorum brevis, Extensor digitorum longus, Diaphragm, Masseter, Extraocular muscles, Pax7, MyoD, Myogenin, Nestin-GFP, $Myf5^{nLacZ}$, MLC3F-nLacZ, 3F-nlacZ-2E

The original version of this chapter was revised. The erratum to this chapter is available at https://doi.org/10.1007/978-1-4939-6771-1_21

Eusebio Perdiguero and DDW Cornelison (eds.), *Muscle Stem Cells: Methods and Protocols*, Methods in Molecular Biology, vol. 1556,
DOI 10.1007/978-1-4939-6771-1_4, © Springer Science+Business Media LLC 2017

1 Introduction

Myofibers are the functional contractile units of skeletal muscle. While myofibers are established during embryogenesis by fusion of myoblasts into myotubes, processes involved in their growth and repair continue throughout life. These processes are supported by myogenic progenitors known as satellite cells that are located between the basal lamina and the plasmalemma of the myofiber [1–3] (Fig. 1). During postnatal growth, at least some satellite cells are proliferative and contribute progeny that fuse with the enlarging myofibers [4–6]. In mature muscle, satellite cells are typically quiescent [5]. Indeed, a recent study with an inducible Cre/loxP model of satellite cell ablation has suggested that uninjured normal muscle does not require satellite cells for myofiber maintenance [7]. However, another recent study has reported that some satellite cells may still continue to fuse with mature myofibers, possibly contributing to myofiber homeostasis [8]. In all, satellite cell activity in mature uninjured muscle is much reduced compared to growing muscle, but satellite cells can be rapidly recruited in response to muscle injury, producing progeny that fuse into existing myofibers, or form new myofibers [9–12]. In addition to generating differentiated myogenic progeny, at least some SCs can self-renew, thereby meeting the defining criteria of bona fide resident stem cells [13, 14]. Insights into the cascade of cellular and molecular events involved in satellite cell myogenesis are essential for understanding the mechanisms controlling muscle maintenance as well as for developing strategies to enhance muscle repair after trauma or in myopathies [15–18].

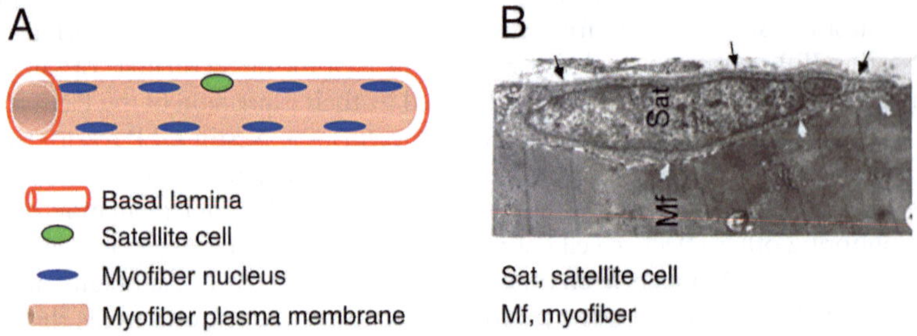

Fig. 1 A schematic (**A**) and EM micrograph (**B**) of satellite cell location. The myofiber basement and plasma membranes have been routinely detected by immunostaining with antibodies against laminin and dystrophin, respectively. In panel (**A**), myofiber nuclei (myonuclei) depicted at the myofiber periphery represent the state of healthy adult myofibers; immature myofibers present in regenerating muscles display centralized myonuclei (not shown). In panel (**B**), *black arrows* depict the basal lamina, and *white arrows* depict apposing satellite cell and myofiber plasma membranes; note the sarcomeric organization within the myofiber. A *black* and *white* version of this figure was published in [42]

Satellite cells were initially identified using electron microscopy by their location under the myofiber basal lamina [1, 19, 20] (Fig. 1). Later, it has become possible to monitor satellite cells by light microscopy based on the expression of a range of markers that can be detected by immunostaining [21]. In particular, the specific expression of the paired box transcription factor Pax7 by satellite cells [22] combined with the availability of an excellent antibody [23] for immunodetection of this protein has provided a uniform means to identify satellite cells in their native position in a range of species [24–28]. Additionally, several genetically manipulated reporter mice have permitted the detection of satellite cells based on specific expression of a fluorescent tag or of beta-galactosidase (β-gal) [29, 30]. Working with the Nestin-GFP transgenic mouse (i.e., GFP expression is driven by regulatory elements of the nestin gene), originally developed to trace neural stem cells, we have discovered that satellite cells harbor strong GFP expression in all muscle groups examined [31, 32] (*see* **Note 1** for further details on the Nestin-GFP mouse). This feature has allowed direct detection of satellite cells (GFP^+) in freshly isolated myofibers (Fig. 2), providing valuable insights on satellite cell dynamics upon physical activity, in aging, and in a number of mutant mice (MyoD-null, α7integrin-null, dystrophin-null) [13, 32–34].

Satellite cell progeny can be distinguished from their quiescent progenitors based on distinctive gene expression patterns [10, 13, 35].

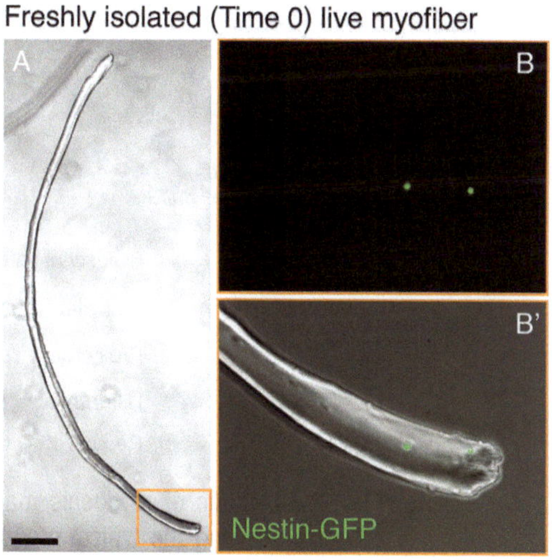

Fig. 2 Lower (**A**) and higher (**B**, **B'**) magnification images of a freshly isolated (Time 0) adherent myofiber prepared from the diaphragm muscle of an adult Nestin-GFP transgenic mouse. (**A**) A phase micrographs. (**B**, **B'**) Parallel fluorescent and merged phase/fluorescent micrographs demonstrating the expression of the Nestin-GFP transgene in satellite cells. Nestin-GFP expression is observed by direct detection of the GFP fluorescence in live myofibers. Isolated myofibers were plated individually in 24-well multiwell tissue culture dishes coated with Matrigel. Scale bar 250 μm

In particular, the myogenic regulatory factors MyoD and myogenin have been used extensively to monitor progeny of satellite cells [24, 36–38]. Proliferating progeny (myoblasts) continue to express Pax7 but, distinct from their quiescent progenitors, also express MyoD. A decline in Pax7 along with the induction of the muscle-specific transcription factor myogenin marks myoblasts that have entered into the differentiation phase and subsequently fuse into myotubes. Reemergence of quiescent cells that express Pax7, but not MyoD (reserve cells), defines a self-renewing population of satellite cells [14, 24, 35, 37–40]. Interestingly, the aforementioned Nestin-GFP transgene, which is expressed by satellite cells, is not retained in the proliferating and differentiating progeny, yet strong GFP expression reemerges in the reserve cells [13, 31, 32]. The typical progression through proliferation, differentiation, and renewal stages of the satellite cells and their progeny according to myogenic marker expression (by immunostaining) and Nestin-GFP transgene expression is summarized schematically in Fig. 3.

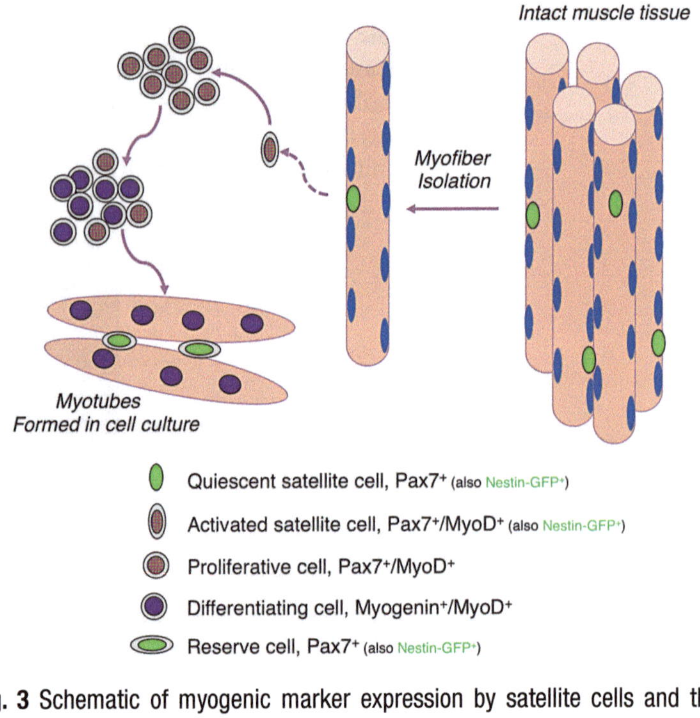

Fig. 3 Schematic of myogenic marker expression by satellite cells and their progeny, showing the typical progression through proliferation, differentiation, and renewal stages in myofiber cultures (modified from (2)). This model is based on the behavior of satellite cells in adherent myofiber cultures plated in Matrigel-coated dishes and maintained in mitogen-rich growth medium (*see* **item 6** in Subheading 2.6), which supports both proliferation and differentiation of satellite cell progeny. This model also reflects the dynamics of satellite cells when grown clonally or in primary cultures under the same conditions as for the single myofiber described here. The expression of the characteristic myogenic markers was analyzed by immunostaining with commonly used antibodies (Table 2, [24, 42]). Nestin-GFP expression was monitored by direct GFP fluorescence

Two main cell culture approaches have been employed in the study of satellite cells: (1) primary myogenic cultures prepared from mononucleated cells dissociated from the whole muscle and (2) cultures of isolated myofibers (also referred to below as "fibers") where the satellite cells remain in their in situ position underneath the myofiber basal lamina. Protocols for obtaining primary myogenic cultures involve releasing satellite cells from their niche. Steps of mincing, enzymatic digestion, and repetitive triturations of the muscle are required for breaking down both the connective tissue network and the myofibers in order to release the satellite cells from the muscle bulk [24, 41, 42]. These steps are followed by procedures for removing tissue debris and reducing the contribution of non-myogenic cells typically present in primary isolates of myogenic cells [14, 31, 32, 43–45]. In contrast, protocols for isolating individual muscle fibers result in the release of intact myofibers that retain satellite cells in their native position underneath the basal lamina [24, 31, 36, 37, 44]. These protocols allow the study of satellite cells and their progeny in their in situ position on the myofiber, and after they migrate from the parent myofiber.

This chapter describes the main protocols used in our laboratory for isolation and culture of single myofibers from mouse skeletal muscles (*see* **Note 2** for historical perspective of our laboratory contribution). One protocol, first introduced by Bekoff and Betz [46] and further developed by Bischoff [47, 48], has been adopted by us for studies of satellite cells in isolated myofibers from both rats [36, 44, 49] and mice [24, 50, 51]. In this case, single myofibers are isolated from the flexor digitorum brevis (FDB) muscle of the hind feet. Because these FDB myofibers are short and do not get tangled, typically multiple myofibers are processed and cultured together. These isolated FDB myofibers are plated in dishes coated with PureCol collagen (formerly known as Vitrogen) and maintained in serum replacement medium (± supplemental growth factors). Employing these conditions, satellite cells remain at the surface of the parent myofiber while synchronously undergoing a limited number of proliferative cycles and rapidly differentiate (Fig. 4). A second approach, first introduced by Rosenblatt and colleagues [52, 53], allows isolation of longer myofibers from a variety of muscles [24, 30, 32, 35, 40, 54]. These longer myofibers can get tangled, and therefore, when working with muscles such as the EDL, we typically process and culture these longer myofibers individually. In our laboratory, EDL myofibers are routinely plated as adherent myofibers in wells coated with Matrigel (a commercially available matrix that facilitates rapid and firm adherence of the myofibers) using a growth-promoting rich medium, conditions in which satellite cells migrate away from the parent myofiber, proliferate extensively, and generate numerous differentiating progeny (Fig. 5). Additionally, we describe herein our approach for culturing EDL myofibers in suspension (adapted from [37]) using culture conditions where the satellite cells are

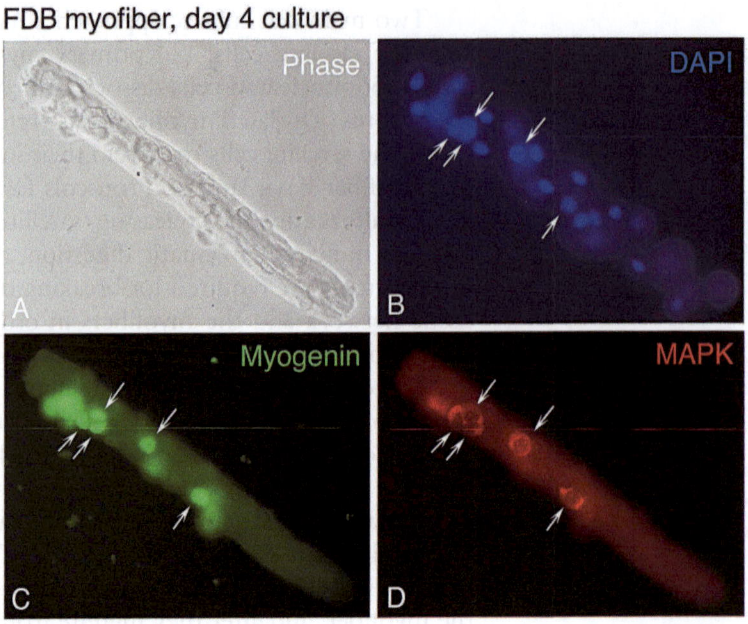

Fig. 4 Parallel phase and immunofluorescent micrographs of an isolated FDB adherent myofiber with associated satellite cells undergoing myogenesis (day 4 culture). Myofibers were isolated from an adult mouse and cultured in 35-mm tissue culture dishes coated with isotonic Vitrogen collagen in solution (now known as PureCol). Cultures were maintained for 4 days in basal medium (*see* **item 6** in Subheading 2.6) containing fibroblast growth factor 2 (FGF2, 2 ng/ml) and fixed with methanol as described in Subheading 3.3.1. (**A**, **B**) Phase and DAPI-stained images (both myofiber nuclei and satellite cell nuclei are labeled with DAPI). (**C**, **D**) Myofiber culture immunostained by double immunofluorescence for myogenin (identifies the nuclei of myogenic cells that have entered the differentiated step) and ERK1/ERK2 mitogen-activated protein kinases (MAPK) (identifies the cytoplasm of all the satellite cell progeny [51, 55]). Arrows in parallel panels point to the location of the same cell. Additional immuno-positive cells present on the myofiber are not shown, as not all positive nuclei or cells on the fibers are in the same focal plane. All micrographs were taken with a 40× objective. A *black* and *white version* of this figure was first published in [107]

retained at the myofiber surface. Similar to the FDB myofiber cultures [24, 50, 51], these non-adherent EDL myofibers, when maintained in a mitogen-poor medium, are suitable for studying the effect of supplemental growth factors on satellite cell activity [34] (Fig. 6). We also include in this chapter our protocols for monitoring satellite cells in freshly isolated (Time 0) myofibers, which enables satellite cell quantification [13, 33, 34]. Notably, for these Time 0 analyses, we recommend working with myofibers that are allowed to adhere to the dish surface before fixation rather than using non-adherent myofibers (*see* **Note 3**). Table 1 compares the different protocols of myofiber isolation from FDB and

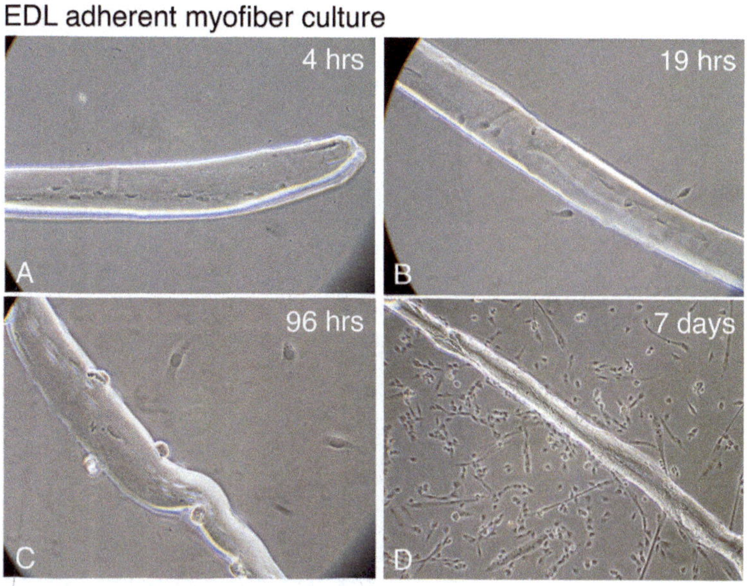

Fig. 5 Phase micrographs of EDL adherent myofibers depicting the temporal development of myogenic cultures from cells emanating from individual myofibers. Myofibers were isolated from an adult mouse and cultured individually in 24-well multiwell tissue culture dishes coated with Matrigel. Cultures were maintained in mitogen-rich growth medium and fixed with paraformaldehyde, as described in Subheading 3.3.2. Satellite cells begin to emigrate from the myofiber within the first day in culture and continue to emigrate during subsequent days. Progeny of satellite cells that have emigrated from the myofibers proliferate, differentiate, and fuse into myotubes, establishing a dense myogenic culture. (**A**) Satellite cells remained attached to the muscle fiber during the first hours after culturing. (**B**) Nineteen hours after culturing, 2–3 cells detached from the fiber but remained in close proximity to the fiber. (**C**) Four days following culturing, more cells are seen in the vicinity of the myofibers (at least four cells are visible). (**D**) By day 7, progeny of satellite cells that emigrated from the myofiber have established a culture containing mostly proliferating myoblasts and some myotubes. Micrographs in panels (**A**)–(**C**) were taken with a 40× objective to show details of the few cells that emigrated from the myofiber, while the micrograph in panel (**D**) was taken with a 10× objective to show the establishment of a dense myogenic culture. See our published study for additional details about growth of satellite cell progeny in long-term EDL myofiber cultures [24]. From [107]

EDL muscles and the specific use of each procedure by our lab. Also detailed in this chapter are our protocols for immunocytochemical analysis of cultured adherent or non-adherent myofibers and a summary of the primary antibodies used in our immunostaining studies depicted in this chapter (Table 2; *see* **Note 2** for a historical perspective of our laboratory contribution). The EDL single myofiber isolation, culturing, and immunostaining procedures described here have also been adapted in our laboratory for

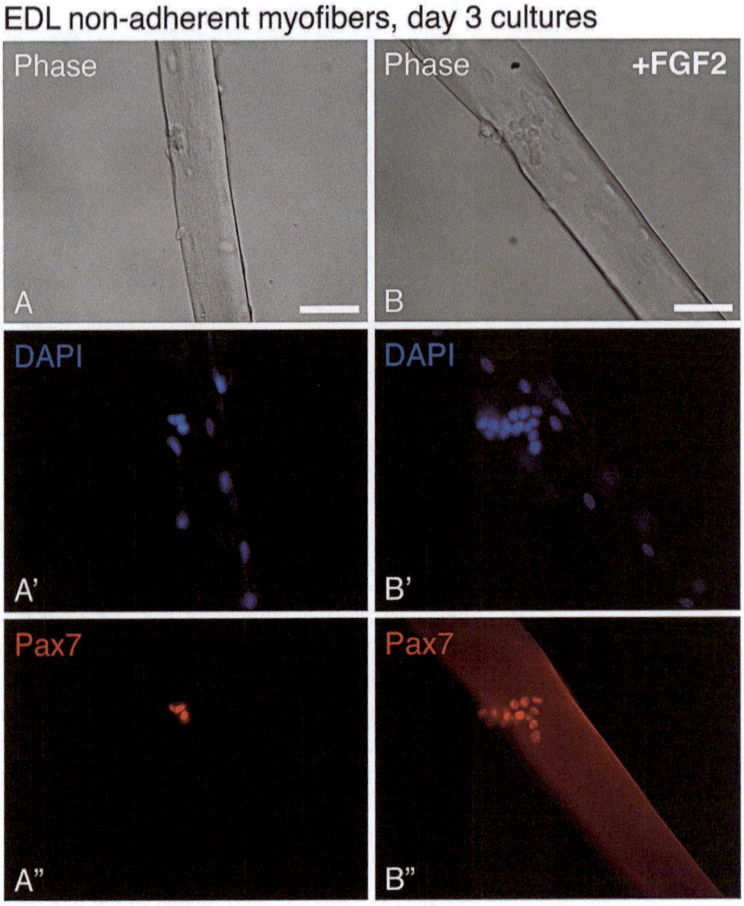

Fig. 6 Parallel phase and immunofluorescent micrographs of non-adherent EDL myofibers with associated satellite cells undergoing myogenesis (day 3 cultures, ±FGF2). Myofibers were isolated from an adult mouse and cultured individually in non-coated 24-well multiwell tissue culture dishes. Cultures were maintained for 3 days in basal medium (*see* **item 7** in Subheading 2.6) without (A-A") or with (B-B") FGF2 supplement (5 ng/ml), then fixed with paraformaldehyde, and analyzed by immunostaining for Pax7, as described in Subheading 3.3.4. (**A**, **B**) Phase images. (**A**′, **B**′) DAPI-staining (both myofiber nuclei and satellite cell nuclei are labeled with DAPI). (**A**″, **B**″) Pax7 immunostaining; the higher number of positive cells in panel B reflects the proliferative response of satellite cells to FGF2 supplementation. Scale bars 50 μm. Contributed by our former lab member, Elena Danoviz

the soleus [24, 31], diaphragm [54], extraocular [32], and masseter muscles (*see* **Note 4** for details on myofiber isolation from the diaphragm, masseter, and extraocular muscles). Regardless of the specific muscle type being processed, the isolation of intact myofibers requires that the donor muscle is handled in a delicate manner, primarily at the tendons without stressing the muscle itself upon tissue harvesting. To assist the investigator in the optimal harvesting of the FDB and the EDL muscles for fiber isolation, we

Table 1
Characteristics of myofiber cultures from flexor digitorum brevis (FDB) and extensor digitorum longus (EDL) muscles

Culture type	# of fibers per culture	Culture dish	Dish coating	Medium	Satellite cell profile after culturing	Summary
FDB myofibers Adherent (Fig. 2)	~30–50	35-mm plate	PureCol. Thick, gel-like layer of native collagen type I prepared from bovine hide (from advanced BioMatrix; product formerly known as Vitrogen) (*see* **Note 5**)	DMEM based, mitogen poor; exogenous growth factors can be added to study their effect on satellite cell activation, proliferation, and differentiation [24, 34, 36, 50, 51, 55]	Satellite cells remain at the surface of the parent myofiber as they undergo a limited number of proliferative cycles and rapidly differentiate without fusing with the parent myofiber	Cultures may model in vivo behavior of satellite cells in intact fibers during routine muscle maintenance Cultures typically are maintained short term (~1–3 days) and can be employed for studying satellite cell activation and entry into the cell cycle. Steps of proliferation and differentiation are highly synchronous [24, 36, 50, 51] Satellite cells can be monitored in freshly isolated (Time 0) FDB adherent myofibers
EDL myofibers Non-adherent (Fig. 3)	1	24-well multiwell plate	None			
EDL myofibers Adherent (Fig. 4)	1	24-well multiwell plate	Thin coating of diluted Matrigel, a basement membrane preparation isolated from a mouse tumor (from BD Biosciences) (*see* **Note 6**)	DMEM based, mitogen rich; medium can be modified to a mitogen poor to allow analysis of the role of exogenous factors on satellite cell activation [13, 24, 31, 52, 53, 56]	Satellite cells emigrate from the parent myofiber and undergo multiple rounds of proliferation, giving rise to an elaborate network of myotubes, as in primary cultures from the whole muscle	Cultures may model events after acute muscle trauma where new myofibers are formed Cultures typically are maintained long term (up to 2 weeks) and employed in studies of progeny of satellite cells that emigrate away from the parent myofiber [13, 24, 31, 52] Satellite cells can be monitored in freshly isolated (Time 0) myofibers [13, 24]. This is our preferred method for quantifying satellite cells in Time 0 myofibers (vs. non-adherent myofibers)

present here figures depicting step-by-step "real-live" images of muscle dissection and harvesting (Figs. 7 and 8). We conclude this chapter with protocols for monitoring satellite cells on isolated myofibers from the Nestin-GFP transgenic mouse (*see* **Notes 1** and **2** for mouse description and historical perspective of our use of this mouse for satellite cell detection). This can be achieved by direct GFP detection (Fig. 2, Table 3). Additionally, Nestin-GFP^{+} satellite cells can be identified in combination with (1) Pax7 immunostaining (Fig. 9) or (2) the expression of LacZ-based reporter genes of interest to the skeletal muscle field, namely, My5^{nLacZ} (which marks nuclei of satellite cells, [13, 30]) and MLC3F-nLacZ (AKA 3F-nlacZ-2E, which is specific for myofiber nuclei, [13, 30]) (Fig. 10). Of note, the Nestin-GFP transgene can additionally be useful for monitoring other selective cell types, as we describe later in the chapter, showing GFP expression in capillaries, endothelial tubes and neuronal cells (Fig. 11).

Collectively, following the procedures and protocol notes detailed in this chapter, investigators can successfully isolate, culture, and analyze myofibers from well-studied EDL and FDB muscles, as well as from other types of muscles, such as diaphragm, masseter, and extraocular muscles. These approaches provide essential tools for studying satellite cells in their native position and their interplay with the parent myofiber.

2 Materials

2.1 General Comments

1. As a general rule, only sterile materials and supplies are to be used. All solutions (apart from the Matrigel and Vitrogen preparations) are sterilized by filtering through 0.22-μm filters, all glassware and dissection tools are sterilized by autoclaving, and all cell culture procedures are performed using sterile techniques.
2. Cultures are maintained at 37 °C and 5 % CO_2 in a humidified tissue culture incubator.
3. All culture media are stored at 4 °C and used within 3 weeks from preparation.
4. Before starting the isolation procedure, tissue culture medium is pre-warmed to 37 °C and then held at room temperature throughout the procedures. Before transferring solutions/ media into the tissue culture hood, spray the glass/plastic containers with 70 % ethanol.
5. The quantities of glassware, media, and reagents as well as the time intervals for enzymatic digestion described in this chapter are appropriate for the isolation of myofibers from one adult mouse of the age and strain detailed below (*see* **Note 7**).

2.2 General Equipment

The following facilities are required for the cultures described in this chapter:

1. Standard humidified tissue culture incubator (37 °C, 5 % CO_2 in air).
2. Tissue culture hood.
3. Bunsen burner.
4. Water bath (37 °C).
5. pH meter and pH paper strips (e.g., EMD, ColorpHast strips).
6. Stereo dissecting microscope with transmitted light base (microscope is either placed inside a tissue culture hood or in an isolation box/clean area; optimally when using fixative, it can also be placed in a histological-grade fume hood).
7. Inverted phase contrast/fluorescent microscope for monitoring myofiber cultures.

2.3 Surgical Tools

1. Straight operating scissors: V. Mueller, fine-tipped, Sharp/Sharp stainless steel, 165 mm ($6^1/_2$"), for delicate cutting and fine incisions.
2. Dissecting scissors: stainless steel, 140-mm (5½") length; both blades blunt, to protect the surrounding tissue from any unwanted nicks.
3. Dressing forceps: V. Mueller, serrated, stainless steel, rounded points, and 140-mm ($5^1/_2$") length.
4. Two very fine point forceps: extra-fine tips, smooth spring action, stainless steel. Straight, 110-mm ($4^1/_4$") length.
5. Microscissors, Vannas type: 8-cm long, straight 5-mm blades, 0.1-mm tips.
6. Scalpel handle and blades: size 3 handle for blade numbers 10–15 and sterile blades (#10).
7. Two straight, 5" hemostatic forceps.
8. Dissecting board with tissue pins.

2.4 Animals

1. All mice described here were on C57BL/6 background, 3–6 months old. Mice were euthanized by cervical dislocation (*see* **Note 8** for further details).
2. Wild-type mice.
3. Nestin-GFP transgenic heterozygous reporter mouse (GFP expression is driven by regulatory elements of the nestin gene [57]) for satellite cell detection in isolated myofibers as in our previous publications [13, 32, 33]. *See* **Note 1** for further details about the Nestin-GFP mouse. This Nestin-GFP mouse was used alone or crossed with one of the following mice.

4. Myf5^{nLacZ}, knockin heterozygous (Myf5^{tm1Pas}, JAX 018626) in which one of the Myf5 alleles was modified to direct LacZ expression [58, 59], resulting in β-gal protein expressed in the nuclei of satellite cells [13, 30].
5. MLC3F-nLacZ (MLC3F = muscle-specific myosin light chain 3F), transgenic heterozygous (Tg(Myl1-lacZ)1Ibdml, JAX 018627, mouse also referred to as 3F-nlacZ-2E) [30, 60]. In this reporter, regulatory elements of MLC3F drive β-gal expression in myofiber nuclei but not in satellite cells [13, 30].
6. mdx^{4cv}, dystrophin-null mdx^{4cv} [32, 61, 62]. The mdx^{4cv} allele is one of several mdx "cv" strains that were generated by point mutations upon mutagen treatment of male mice [61–63]. The mdx^{4cv} strain has been preferred by some laboratories due to the reduced occurrence of revertants (i.e., spontaneously appearing dystrophin^{+} myofibers [64]) compared to the "standard" mdx mice (spontaneous mutation) [62, 65–68].

2.5 Plastic and Glassware for Myofiber Isolation and Culture

1. Standard 9" glass Pasteur pipettes; fire polish the ends to avoid damage to myofibers, which are transferred using these pipettes. As noted above in **item 1** in Subheading 2.1, all Pasteur pipettes are sterilized by autoclaving before use.
2. Standard 5" glass Pasteur pipettes. Prepare three gradually narrower-bore pipettes from standard 5" Pasteur pipettes. Use a file or a diamond knife to prepare a set of pipettes with bore diameter of approximately 3, 2, and 1 mm. Shake the pipette to remove any glass fragments and fire polish the sharp ends. These pipettes are used to triturate the digested muscle in order to release single myofibers.
3. Syringe filters, 0.22-μm PVDF low protein binding filters (Millipore is recommended).
4. Disposable plastic syringes, 3 or 10 ml.
5. Bottle top filters, 0.22 μm.
6. Polypropylene conical centrifuge tubes, sterile, 15 and 50 ml.
7. Three glass Corex tubes, 15 ml (Sorvall centrifuge tubes or alternatively 15-ml bicarbonate Sorvall tubes).
8. Wide-bore 100-μl micropipette tips. Trim 100-μl micropipette tips approximately 3 mm from the end. Use of these trimmed micropipettes minimizes myofiber shearing when transferring or dispensing FDB myofibers.
9. Tissue culture dishes, 35 mm.
10. Plastic Petri dishes 60 × 15 mm and 100 × 15 mm (for muscle and myofiber rinsing).
11. Twenty-four-well multiwell tissue culture dishes (*see* **Note 9**).

12. Two L-shape bent pipette spreaders prepared from standard 9" Pasteur pipettes. Use flame to first seal the distal end, and then flame about 3/4" from the sealed end until the pipette starts to bend. The bent pipettes are used to spread the coating solution on the tissue culture dishes; the length of the bent end is designed for working with the 35-mm culture dishes for FDB myofiber cultures. Spreaders should be prepared in advance and allowed to cool before use.
13. 1-ml serological glass pipettes (used for Matrigel aliquoting and dispensing; *see* **Note 6**).
14. Cryogenic vials sealed with O-rings (used for storing Matrigel aliquots; *see* **Note 6**).

2.6 Media, Enzymes, and Cell Culture Reagents

1. DMEM/high glucose: Dulbecco's Modified Eagle Medium with 4500 mg/L glucose, 4.0 mM L-glutamine, and 110 mg/L sodium pyruvate. Supplemented with 100 U/ml penicillin and 100 μg/ml streptomycin.
2. Horse serum (HS): standard, not heat inactivated (*see* **Note 10**). Original bottles are stored at −80 °C; once thawed and aliquoted, store at −20 °C.
3. Fetal bovine serum (FBS): standard, not heat inactivated (*see* **Note 11**). Original bottles are stored at −80 °C; once thawed and aliquoted, store at −20 °C.
4. Controlled process serum replacement (Sigma-Aldrich). Stored at −80 °C; once thawed and aliquoted, store at −20 °C. *See* **Note 12** for product composition and availability.
5. Chicken embryo extract (CEE; *see* **Notes 13** and **14**). Stored at −80 °C for long term or −20 °C when aliquoted.
6. FDB myofiber culture medium: DMEM/high glucose (supplemented with antibiotics), 20 % controlled process serum replacement, and 1 % HS.
7. Culture medium for EDL adherent myofibers: DMEM with antibiotics, 20 % FBS, 10 % HS, and 1 % CEE.
8. Culture medium for non-adherent EDL myofibers: DMEM with antibiotics, 10 % HS.
9. PureCol collagen (Advanced BioMatrix) for coating 35-mm tissue culture dishes (*see* **Note 5**). This collagen in solution was formerly known as Vitrogen when sold by Cohesion Technologies).
10. Concentrated (7×) DMEM solution made from powder DMEM (Sigma-Aldrich). This is used to prepare isotonic PureCol collagen (*see* **Note 5**).
11. Matrigel (BD Biosciences; *see* **Note 6**) for coating 24-well multiwell dishes. We typically dispense Matrigel into aliquots of 100–200 μl and freeze back at −20 °C. *See* **Note 6** for all handling details.

12. Collagenase (type I, Sigma-Aldrich). The final working solution is prepared as described in **step 3** in Subheading 3.1.1.
13. 100 ml of DMEM containing 10 % HS. HS is freshly filtered on the day of use through a 0.22-μm filter. This DMEM-10 % HS medium is used for FDB myofiber purification as detailed in Subheading 3.1.5. Also, all Pasteur pipettes and micropipette tips are pre-flushed with this DMEM-10 % HS medium to prevent sticking of myofibers during manipulation.
14. HS, 10 ml, freshly filtered on day of use with 0.22-μm syringe filter. Used to coat Petri dishes and pre-flush Pasteur pipettes to minimize potential sticking of myofibers during isolation procedure.

2.7 Reagents and Solutions for Fixing and Immunostaining Isolated Myofibers

1. Pre-fixation rinse solution: DMEM as in **item 1** in Subheading 2.6.
2. Fixative for FDB myofibers: ice-cold 100 % methanol (*see* **Note 15**).
3. Fixative for EDL myofibers: 4 % paraformaldehyde in a sodium phosphate buffer containing 0.03 M sucrose (*see* **Notes 16** and **17** for specific buffer details and preparation). Store at 4 °C, and pre-warm to room temperature before use. To maintain quality and effectiveness of fixative, pre-warm only the volume that is required for immediate use.
4. Tris-buffered saline (TBS) rinse solution: 0.05 M Tris, 0.15 M NaCl, and pH 7.4 (*see* **Note 18**).
5. Detergents: Triton X–100 and Tween 20.
6. Detergent solutions: TBS containing 0.05 % Tween 20 (TBS-TW20) and TBS containing 0.5 % Triton X–100 (TBS-TRX100).
7. Blocking reagent: normal goat serum. Can be stored at −80 °C; once thawed and aliquoted, store at −20 °C.
8. Blocking solution: TBS containing 1 % normal goat serum (TBS-NGS).
9. Primary and secondary antibodies: *see* Table 2.
10. DAPI solution (4′,6-diamidino-2-phenylindole, dihydrochloride): stock concentration 10 mg/ml and a working concentration of 1 μg/ml diluted in TBS-NGS prior to use (*see* **Note 19**).
11. Mounting medium: Vectashield (Vector Laboratories); store at 4 °C.
12. Sterile glycerol solution: 25 % glycerol in TBS, store at 4 °C.
13. Cover glass, 22 mm^2.

Table 2
Primary and secondary antibodies used in the current manuscript

Primary antibodies				
Antigen	**Source**	**Dilution**	**Refs.**	**Used in**
Mouse monoclonal antibodies (IgG1)				
MyoD (clone 5.8A)	BD Biosciences	1:800	[24, 51, 69]	Fig. 3
Myogenin (clone F5D, supernatant)	Developmental Studies Hybridoma Bank (DSHB)	1:5	[24, 36, 51, 70, 71]	Fig. 3 and 4
Pax7 (bioreactor supernatant)	Developmental Studies Hybridoma Bank (DSHB)	1:100–1:200	[23, 24, 31]	Figs. 3, 6, 9
Rabbit polyclonal antibodies				
MyoD	Santa Cruz Biotechnology, sc760	1:400	[31, 72]	Fig. 3
MyoD	A gift from Dr. Stefano Alemá	1:50	[36, 73]	Fig. 3
MAPK (ERK1/ERK2)	A gift from Drs. Rony Seger and Edwin G. Krebs	1:1000	[24, 50, 51, 55, 74, 75]	Fig. 4
GFP	Abcam, ab6556	1:10,000	[13, 31]	Fig. 10A'
Secondary antibodies				
Antigen	Conjugate	Source	Dilution	Used in
Goat anti-mouse IgG	Fluorescein	Organon Teknika	1:100	Fig. 4
Goat anti-rabbit IgG	Rhodamine	Organon Teknika	1:100	Fig. 4
Goat anti-mouse IgG1	Alexa Fluor® 568	Molecular Probes	1:1,000	Figs. 6 and 9
Goat anti-rabbit IgG	Alexa Fluor® 568	Molecular Probes	1:1,000	Fig. 10A

2.8 Reagents and Solutions for X-gal Staining

1. Fixative solution: buffered paraformaldehyde, as in **item 3** in Subheading 2.7.
2. Rinse solution: Dulbecco's phosphate-buffered saline (PBS).
3. X-gal dilution buffer: 5 mM $K_3Fe(CN)_6$, 5 mM $K_4Fe(CN)_6{\cdot}3H_2O$, and 2 mM $MgCl_2$ in PBS, stored at 4 °C.
4. X-gal substrate: X-gal is constituted at 40 mg/ml in *N,N*-dimethylformamide and aliquots are stored at −20 °C.
5. DAPI working solution and Vectashield as in **items 10** and **11** in Subheading 2.7.

3 Methods

3.1 Isolation of Single Myofibers from the Flexor Digitorum Brevis Muscle

The information in this introductory section is provided to assist in the identification of the flexor digitorum brevis (FDB) muscles. The FDB is a superficial, multipennate, broad, and thin muscle of the foot and paw [48, 76]; it arises from the tendon of the plantaris as three slender muscles converging into long tendons. At the base of the first phalanx, it divides into two, passes around the tendon of the flexor hallucis longus obliquely across the dorsum of the foot, and ends as the tendons insert into the second phalanx of the second through the fifth digits. As the FDB contracts, digits 2–5 are flexed. For additional details about the anatomy of the FDB muscle, *see* **Note 20**.

For uniformity, we typically use only the hind limb muscles in our studies. Figure 7 depicts "real-live" images of steps in FDB muscle isolation that emphasize the location of the specific tendons that are handled during the process. It is of utmost importance to delicately manipulate the muscle of interest only at the tendons during its excision and further processing.

3.1.1 Initial Steps Prior to Harvesting the Muscle and Preparation of Digestive Enzyme

1. Add 3 ml of DMEM to six 35-mm tissue culture dishes and place the dishes in the tissue culture incubator until muscle dissection begins.
2. Add 3 ml of DMEM containing 10 % HS to three 35-mm tissue culture dishes and place them in the tissue culture incubator until needed for the isolated single myofibers.
3. Add 6 mg of collagenase type I to 3 ml of DMEM in order to prepare 0.2 % (w/v) collagenase type I solution. Use a 0.22-μm syringe filter attached to a 3- or 10-ml syringe to filter the collagenase solution into a 35-mm tissue culture dish (*see* **Note 21**). We prepare this solution fresh for each experiment.

3.1.2 Dissection of FDB Muscle

1. Euthanize one mouse according to institute regulations.
2. Spray the hind foot (Fig. 7A) lightly with 70 % ethanol.
3. All the following steps, until the muscle is dissected out, are carried in an enclosure dedicated for this procedure in order to limit contamination.
4. Secure the mouse, lying on its back, to the dissecting board by pinning down the forelimb diagonally across from the limb being dissected.
5. Use a scalpel to carefully cut the skin circumferentially just above the ankle joint, so that the skin above and below the cut site are completely separated (after this circular cut, the skin below resembles a sock).
6. Using scissors, cut the skin in a straight line along the center of the ventral part of the foot almost all the way to the digits (the cut as viewed from the front of the foot should resemble a "T") (Fig. 7B).
7. Using a hemostat, grasp one of the upper corners of the cut tissue (at the junction of the circular and longitudinal cuts) and reflect the skin away from the foot (Fig. 7C).
8. Hold the scalpel with its blade parallel to the longitudinal axis of the partially exposed muscle and carefully separate the skin from the connective tissue. Be especially careful not to cut into

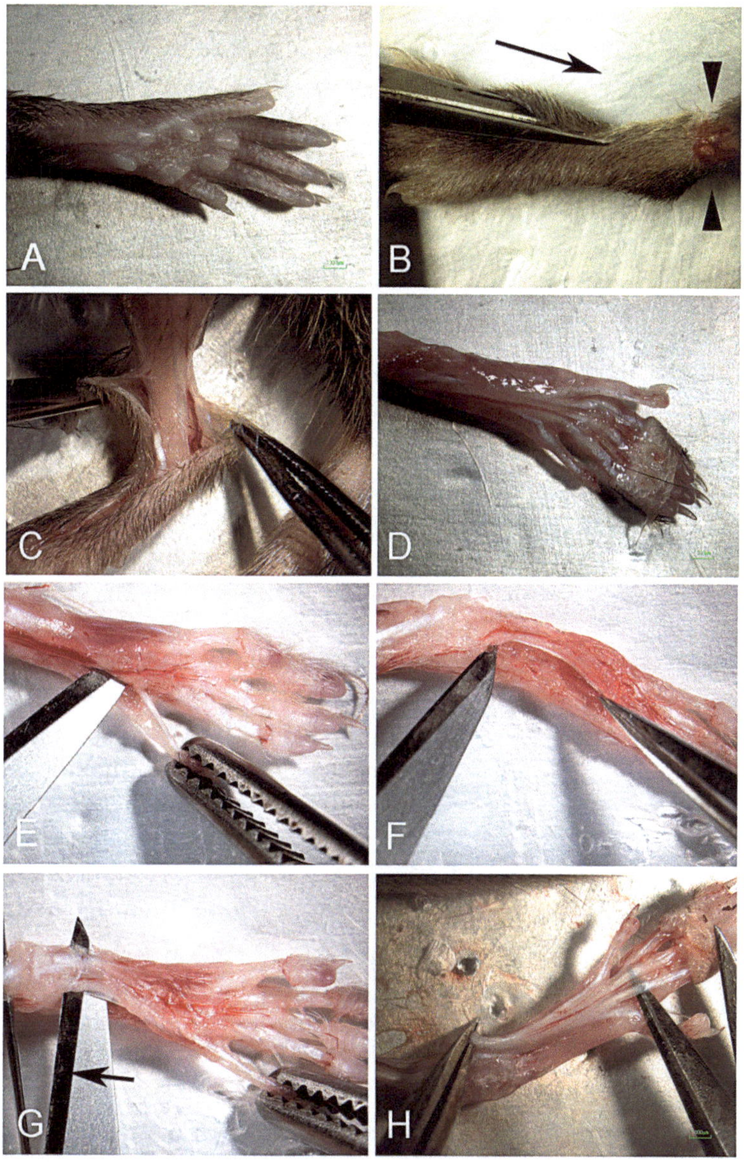

Fig. 7 Dissection of FDB muscle from the rear foot of adult mouse. (**A**) Rear foot before dissection. (**B**) Cutting of "T" toward the ankle; *arrow* shows the direction of the longitudinal cut and *arrowheads* identify the circumferential cut at the ankle. (**C**) Peeling the skin back from the ankle exposing the muscles and tendons. (**D**) Digit tendons of the FDB exposed on the sole of the foot. (**E**) Cutting the connective tissue under the FDB toward the heel of the foot. (**F**) Freeing the FDB from the underlying connective tissue. (**G**) Cutting the FDB at the heal origin; *arrow* indicates direction of cutting. (**H**) Preparing the release of the FDB from its tendon insertion points at the digits. From [86]

the muscle tissue at the back of the leg, as the FDB is the most superficial muscle of the back of the foot.

9. Clamp a second hemostat to the other corner of the cut tissue and repeat **step 8**.
10. When the skin is completely cut away from the foot, the FDB should be exposed all the way to the tendons reaching the digits (Fig. 7D).
11. Turn the mouse over so that it lies on its stomach, and identify the FDB. During the next steps of the dissection, be careful not to injure the small medial plantar artery that supplies blood to the FDB to limit blood cell contamination of the myofiber preparation. This artery passes along the medial part of the sole of the foot and branches into the digits.
12. Carefully run the tip of the scalpel along each side of the FDB to dissect the connective tissue holding the muscle in place (Fig. 7E).
13. When the FDB is separated from the surrounding muscles, carefully lift the FDB by inserting one arm of your smooth forceps or a fine blunt probe underneath the FDB so that the flat side of the scalpel may be inserted horizontally underneath it.
14. With the blade of the scalpel underneath the muscle, running horizontal and parallel to the muscle, cut away the underlying connective tissue (Fig. 7F). It is best to cut toward the heel and only lift that portion of the muscle directly over the scalpel.
15. Cut underneath the tendon to separate the muscle and a large portion of its tendon from the heel bone (Fig. 7G).
16. Grasp the freed tendon as far as possible from the muscle tissue with a hemostat and gently lift the FDB away from the leg. While lifting the FDB, use the scalpel, running parallel to the muscle, to cut through the connective tissue while holding the foot down.
17. Continue cutting through the connective tissue until the tendons that connect the FDB muscle to the digits have been exposed (Fig. 7H). When about half the length of the three tendons has been exposed, cut the tendons and release the entire muscle from the leg. The fourth small lateral tendon (attached to the fifth digit) and its attached myofibers can be trimmed off.
18. Retrieve from the incubator three 35-mm tissue culture dishes containing DMEM and place them close to the dissection area.
19. Place the harvested FDB in one of the 35-mm tissue culture dishes.
20. For harvesting the FDB from the other hind foot, repeat **steps 4–18**, and place the muscle in a second 35-mm tissue culture dish.

21. Place the 35-mm tissue culture dishes, one at a time, under the stereo dissecting microscope.
22. Use fine point forceps to pull the connective tissue perpendicular to the line of the muscle and use fine dissection scissors to cut it off.
23. Once the muscle is clean, shorten the tendons but do not cut all of them off.
24. Use a wide-bore Pasteur pipette to transfer the cleaned muscle to another 35-mm tissue culture dish containing DMEM.
25. Repeat **steps 21–24** to clean the second FDB muscle.

3.1.3 Enzymatic Digestion

1. Working in the tissue culture hood, transfer the two cleaned FDB muscles to a 35-mm tissue culture dish containing 1.5 ml of the 0.2 % collagenase I solution.
2. Place this 35-mm culture dish inside the tissue culture incubator for 2.5 h (*see* **Notes** 7 and **21**). Gently swirl the dish every 15–20 min during digestion, or, if available, one can use a low-speed agitator placed inside the tissue culture incubator. In the latter case, the speed should be adjusted to the lowest possible speed for minimal agitation, to avoid damage to myofibers.
3. At the end of the digestion period, transfer each muscle to a 35-mm tissue culture dish containing 10 % HS.

3.1.4 Separation of the Three Tendons and Release of FDB Myofibers

1. Pre-flush all Pasteur pipettes with 10 % HS, prepared as described in **item 14** in Subheading 2.6.
2. Place one muscle at a time under the stereo dissecting microscope.
3. Identify the two grooves running between the three tendons separating the middle from the two lateral tendons.
4. Being careful not to touch the muscle, insert the tip of a pair of forceps into one of the grooves and hold the muscle in place by securing the connective tissue between the tendons to the dish.
5. Use another pair of forceps to gently pull away the connective tissue that holds the tendons and their attached muscle tissue together.
6. Continue removing the connective tissue until the lateral tendons are separated from the middle tendon and its attached myofibers.
7. Holding the muscle only at its tendons, transfer the muscle preparation to a 35-mm dish containing 3 ml of DMEM containing 10 % HS.
8. While grasping one end of the middle tendon with a pair of forceps, use a second pair of forceps to grip its surrounding connective tissue sheath and pull gently. If the sheath does not

come off easily, use fine point forceps to pull the connective tissue perpendicular to the line of the muscle and cut it off.

9. Repeat **steps 1–7** with the second FDB muscle until all six tendons and their attached myofibers are in the 35-mm tissue culture dish containing 10 % HS.
10. For one tendon at a time: hold one end of the tendon with a pair of forceps and with the tip of a second pair gently separate the myofibers from the tendon. The liberation of the myofibers from the two lateral tendons should be easy, while the middle tendon requires patience since the myofibers are attached to it more firmly.
11. Use a wide-bore Pasteur pipette to gently triturate the clumps of myofibers until they disengage into single myofibers. The number of trituration rounds can vary, but it may take at least five times. Excessive trituration can lead to fiber damage (*see* **Note 7**).
12. Remaining clumps should be transferred to another 35-mm tissue culture dish containing 10 % HS and further triturated until disengaged into single myofibers.
13. Set the stereo dissecting microscope magnification so that the small pieces of connective tissue floating around in the suspension are visible, and use fine forceps (or standard narrow-bore Pasteur pipette, fire polished) to pick them out. Continue until the myofiber suspension is clean of any connective tissue debris.
14. Triturate the myofiber suspension ten more times using a 9" Pasteur pipette with a fire-polished tip to further separate small clumps of myofibers.

3.1.5 Further Purification of FDB Myofibers

1. Add 10 ml of DMEM containing 10 % HS to each of the three glass Corex tubes.
2. Using the trimmed 100-μl pipette tip, transfer the myofiber suspension to the top of the 10 % HS column in the first Corex tube. Allow the myofibers to settle (at 1× *g*) through the HS column for 15 min at room temperature (*see* **Note 22**). This step is important for purifying the myofibers from free mononucleated cells, debris, and occasional damaged myofibers.
3. As soon as the myofibers are settled, aspirate about 11 ml of the supernatant (leaving about 1–1.5 ml). Triturate the myofiber suspension gently with a 5" fire-polished Pasteur pipette, and transfer the suspension to the next Corex tube as described in **step 2**.
4. Allow myofibers to settle and transfer the myofiber suspension to the third Corex tube as in **steps 2** and **3**.
5. Allow myofibers to settle and harvest the final myofiber suspension. Following the third purification, the residual volume

of medium to be left with the myofiber suspension depends on the number of culture dishes and the desired myofiber number per dish. Typically in our studies the volume of the final myofiber suspension is 300 μl, which is sufficient for culturing 4–6 dishes.

3.1.6 Preparation of Isotonic PureCol Collagen and Coating Culture Dishes

Isotonic PureCol collagen can be prepared during the settling of myofibers. The isotonic mixture should be kept on ice. Stock PureCol is an acidic solution, and when made isotonic, it gels rapidly if not maintained at 4 °C (*see* **Note 5**):

1. Place PureCol collagen stock bottle, 7× DMEM, and one 15-ml conical tube on ice.
2. On ice: Add 1 volume of 7× DMEM and 6 volumes of PureCol to the 15-ml conical tube and mix gently. Calculate the volume of stock PureCol needed for the experiment based on using 120 μl isotonic PureCol collagen to coat each 35-mm tissue culture dish. Use pH paper strips to ensure a neutral pH of the PureCol collagen in DMEM solution. The pH of this solution rises slightly after coating the culture dish. If the pH remains acidic after coating a test dish, add one to two drops of 1 *M* NaOH to the PureCol collagen in DMEM solution.
3. On ice: Transfer 120 μl of isotonic PureCol collagen to the center of a 35-mm culture dish and immediately use the L-shape spreader to coat the dish evenly. The coated culture plates need to be kept on ice until used as detailed below, to avoid premature matrix gelling.

3.1.7 FDB Myofiber Culturing

1. Gently swirl the myofiber suspension (in the 15-ml tube) for even distribution of myofibers throughout the residual medium.
2. Remove one culture dish at a time from ice to allow rapid warming to room temperature.
3. Use a wide-bore, 100-μl micropipette tip to dispense about 50 μl of the myofiber suspension per each culture dish.
4. Gently swirl the culture dish to allow even distribution of the myofibers.
5. Repeat **steps 2–5**, one dish at a time, for additional culture dishes.
6. Transfer the culture dishes to the tissue culture incubator for a minimum of 20–30 min to allow the formation of PureCol collagen matrix and the adherence of the myofibers to the matrix.
7. Remove dishes from the incubator. Gently add 1 ml of myofiber culture medium (*see* **item 6** in Subheading 2.6) to each dish without agitating the myofibers, and return dishes to the incubator.

When the effect of growth factors on satellite cell proliferation/differentiation is investigated, parallel FDB cultures are maintained in culture medium with/without additives, and the medium is replaced every 24 h (residual medium should be left when performing medium change to avoid drying of the myofibers) to ensure that growth factors do not become rate limiting. These cultures can be used for monitoring satellite cells and their progeny in live cultures and for fixed/immunostained cultures as detailed in Fig. 2 and in Subheading 3.3.

3.2 Isolation of Single Myofibers from the Extensor Digitorum Longus Muscle

3.2.1 Preparation of Matrigel Working Mixture and Coating 24-Well Tissue Culture Dishes with Matrigel (For Adherent Myofibers Only)

1. Matrigel solution preparation and plate coating (**steps 1–6**) are done on ice. Thaw the required amount of Matrigel by placing frozen aliquot(s) on ice for at least 30 min and as much as 1.5 h to allow the Matrigel stock to completely liquefy for subsequent dilution to the working solution (*see* **Note 6**).
2. Prechill a 15-ml conical tube on ice. Add about 400–900 μl of ice-cold DMEM to the thawed Matrigel and gently mix by several repetitive drawings through a 1-ml glass pipette. Transfer the diluted Matrigel into the 15-ml conical tube and add ice-cold DMEM to dilute the Matrigel to a final concentration of 1 mg/ml. Gently mix the Matrigel and DMEM by several repetitive drawings through a 1-ml glass pipette. An optimal Matrigel stock is at ~10 mg/ml protein concentration, further diluted at 1:10 for the working Matrigel solution. Stock protein concentration can vary greatly from lot to lot and should be monitored. Allow the diluted Matrigel solution to cool on ice for at least 15 min.
3. After 15 min, use a chilled 1-ml glass pipette to draw up the diluted Matrigel solution and coat wells with an appropriate volume (250 to 300 μl per well for a 24-well plate). In our experience, 2 ml of working Matrigel solution can be used to coat an entire 24-well plate; we typically coat 6–8 wells at a time as detailed next.
4. Per each series of wells, leave the culture plate coated with the Matrigel working solution on ice for 7 min, then use the same pipette as before (held cooled in a tube on ice) to remove the Matrigel solution, and place it back in the 15-ml conical tube that is kept on ice. This will leave a thin coat of Matrigel at the bottom of the wells.
5. Once all of Matrigel solution has been placed back in the tube, use the same pipette to coat the next set of wells. Leave the diluted Matrigel in each well for 7 min.
6. Once the desired number of wells per one tray has been coated, tilt the tray and use a 20-μl pipette tip to carefully remove residual Matrigel, and place it back in the 15-ml conical tube that is kept on ice (*see* **Note 23**).

7. Incubate the Matrigel-coated multiwell dishes in the tissue culture incubator for at least 1 h.
8. About 10 min before culturing myofibers, take the Matrigel-coated, 24-well dish out of the incubator to the tissue culture hood and open the lid. This will allow evaporation of water that otherwise will condense on the underside of the lid when moving the dish from the warm incubator to room temperature. If allowed to form, the condensation will drip into the well, disturbing the Matrigel coating.

3.2.2 Coating Glassware and Plasticware Dishes with Horse Serum (HS)

This is done to minimize adherence of myofibers to plasticware and glassware used during the isolation process:

1. For each EDL muscle being processed, coat six plastic 100-mm Petri dishes and one to two 60-mm Petri dishes with filtered horse serum (HS). Successively transfer a volume of HS to each Petri dish that is sufficient to cover the bottom of the plate, and then swirl the dish to coat evenly. Allow each dish to sit with HS for about 2–3 min at room temperature and then remove HS and apply it to the next dish. After all dishes have been coated, add 9–12 ml of DMEM to each 100-mm Petri dish and 3–5 ml to each 60-ml Petri dish. One may consider processing a pair of EDLs together (which reduces usage of materials and supplies), but we typically process each EDL alone to allow for better separation of fibers with less debris.
2. Incubate the Petri dishes in the tissue culture incubator until needed following muscle digestion.
3. Coat the fire-polished Pasteur pipettes by flushing HS solution through the pipettes several times. The coated pipettes are then placed vertically in sterile plastic tubes (e.g., 5-ml Falcon tubes) to maintain sterility and also for reflushing HS through the pipettes to refresh the coating.

3.2.3 Preparation of the Digesting Enzyme Solution and Post-digestion Rinse Plates

1. Prepare 0.4 % collagenase type I solution by dissolving 0.012 g of collagenase in 3 ml of DMEM. Use a 0.22-μm syringe filter attached to a 3- or 10-ml syringe to filter the collagenase solution into a 35-mm tissue culture dish. We prepare this solution fresh for each experiment (*see* **Note 21**).
2. Fill three 100-mm Petri dishes with 9 ml DMEM and place in tissue culture incubator to warm dishes for later use as rinse dishes.

3.2.4 Dissection of EDL Muscle

The information in this introductory section is provided to assist in the identification of the extensor digitorum longus (EDL) muscles. The EDL muscle is situated at the ventral-lateral aspect of the hind limb, running from the knee to the ankle, extending to the second to fifth digits [76]. The EDL actually consists of four combined muscle bellies and their tendons; the bellies arise from the lateral condyle of the tibia and the front edge of the fibula (two tendons at the origin of the muscle). The tendons lie close to each other and appear as one glistening white tendon that continues down to the surface of the ankle. At the ankle joint, it separates to four tendons, each attached to one of the second to fifth digits. As the EDL contracts, the four digits are extended. For additional details about the anatomy of the EDL muscle, *see* **Note 20**.

As detailed in Subheading 3.1, we typically use only the hind limb muscles in our studies. Figure 8 depicts "real-live" images of the steps in EDL muscle isolation with emphasis on the location of the specific tendons that are handled during the process. It is of utmost importance to delicately manipulate the muscle of interest only at the tendons during its excision and further processing:

1. Euthanize one mouse according to institute regulations.
2. Spray the hind limbs with 70 % ethanol.
3. Secure the mouse, lying on its back, to the dissecting board by pinning down the forelimb diagonal to the hind limb to be dissected.
4. Use straight rounded-tip scissors to cut through the skin, opening a small incision above the knee.
5. Holding the skin with fine forceps, insert the rounded-tip scissors beneath the skin and carefully open the scissors to loosen the skin from the underlying muscles.
6. Extend the incision to a point just in front of the digits.
7. Loosen the skin as you go, being careful not to cut the underlying muscles or blood vessels.
8. Cut and remove the skin from the knee to the paw (Fig. 8A) and cut the fascia (thin connective tissue layer that covers the muscles) that overlays the EDL and TA muscles (Fig. 8B). This will facilitate the identification of the tendons.
9. Identify the four tendons in the foot at the insertion of the EDL, each extending to one of the digits but not the large toe.
10. Use the microscissors to cut all four tendons (Fig. 8C).
11. Using fine forceps, gently isolate and pull the portion of the tendon before its division (into four tendons) at the ankle up from the paw (Fig. 8D, E); the tendon should slide up and out from under the connective tissue sheath at the ankle, with the four divisions trailing behind it. Carefully work the tendon of the EDL out from underneath the TA tendon and pull the tendon out of the ankle with the four divisions trailing behind it (Fig. 8E).
12. Identify the two tendons that are located by the knee cap, facing the lateral part of the leg (i.e., opposite to the midline of the body).

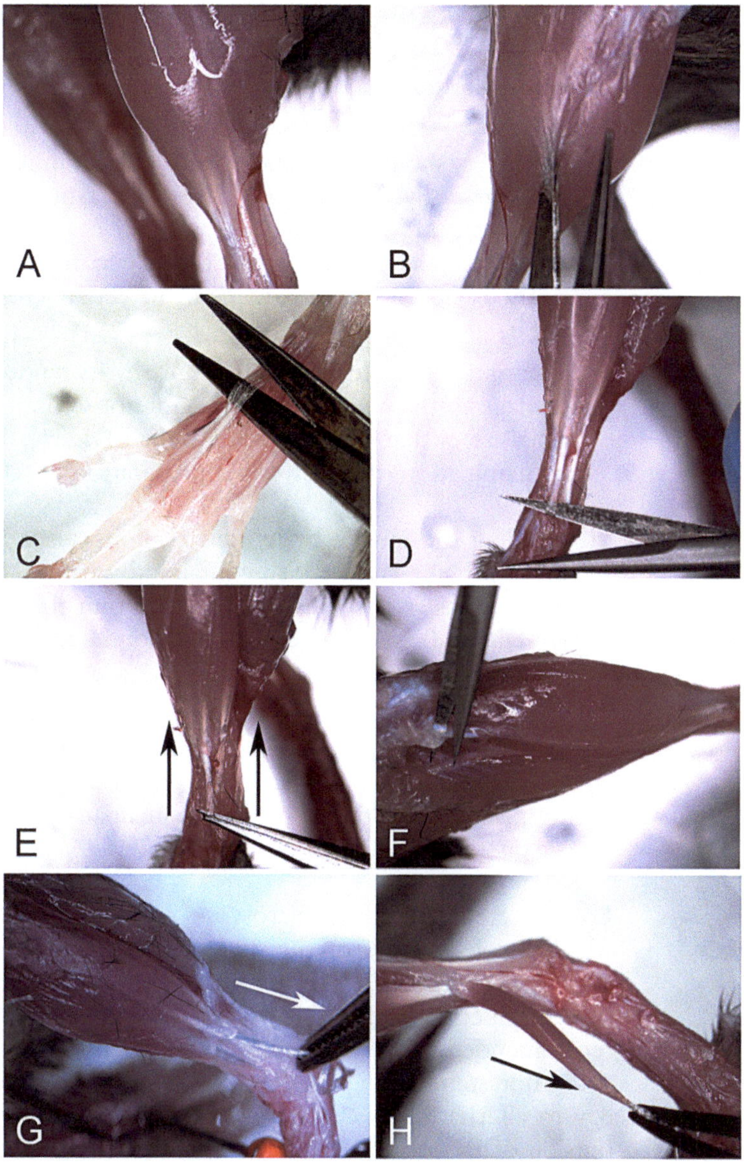

Fig. 8 Dissection of EDL muscle from the hind limb of adult mouse. (**A**) Anterior lower hind limb with the skin removed. (**B**) The fascia covering the anterior lower hind limb muscles is removed to allow access to tendons. (**C**) The four foot tendon insertion points of the EDL are isolated and cut. (**D**) The common tendon of the EDL is carefully exposed and isolated at the ankle. (**E**) Once isolated and foot insertions are cut, the EDL tendons are pulled proximally up from the foot; *arrows* indicate the direction of pulling. The tendons should easily slide underneath the connective tissue sheath at the ankle up from the foot. If the tendons do not easily slide out, then reexamine the foot tendons to ensure that they have been cut. (**F**) Origin of the EDL is exposed and then cut at the lateral surface of the tibia condyle head. (**G**) Grasping only the EDL tendon (do not grasp the muscle as it can easily be damaged), carefully pull distally toward the toes to remove the EDL muscle; *arrow* indicates the direction of pulling. (**H**) The EDL should slide underneath the TA muscle and should pull out easily. It is important to pull gently and there should be little resistance; if the muscle does not slide out easily, one or both tendons at the muscle origin may still be attached to the bone. In this case identify the attached tendon and cut it. From [86]

13. Use microscissors to cut these tendons as far as possible from the muscle itself (Fig. 8F).
14. Grasp the four tendons and carefully pull distally toward the toes to remove the EDL muscle.
15. The EDL should slide underneath the TA muscle and should pull out easily (Fig. 8G, H). It is important not to apply too much force. If the muscle does not slide out easily, one or both tendons at the muscle origin at the knee may still be attached to the bone. In this case, identify the yet attached tendon and cut it.
16. The muscle should be handled only by its tendons to prevent damage to the myofibers. Be careful not to injure the anterior tibial artery that supplies blood to the EDL, to avoid blood cell contamination of the myofiber preparation.

3.2.5 Enzymatic Digestion

1. Holding the muscle by its four tendons, place the EDL in a 35-mm tissue culture dish containing warm DMEM to rinse. Next, transfer the muscle to the 35-mm tissue culture dish containing 0.4 % collagenase I solution. A pair of EDLs can be digested in the same dish.
2. Place the dish inside the tissue culture incubator for 45–60 min (*see* **Notes 7** and **21**). Gently swirl the dish every 15–20 min during digestion (alternatively, one can use a low-speed agitator placed inside the tissue culture incubator) to facilitate muscle dissociation.

3.2.6 Liberation of Single Myofibers from Muscle Bulk

Use a stereo dissecting microscope throughout the procedure, which involves rinses of the digested muscle bulk and a 3-step sequence of muscle bulk trituration to release myofibers. All Pasteur pipettes used in this process should be fire polished. It is recommended to spend no more than 5–7 min at a time per each trituration step. When processing multiple EDLs, it is a good strategy to alternate between muscle bulks so that only one EDL is outside of the incubator at a time in order to minimize muscle cooling. Additionally, the recommended number of rinses of the digested muscle and of individual myofibers as detailed in this section should not be overlooked. The myofiber rinses are essential for minimizing the contribution of non-myogenic cells that are released from the muscle bulk during the enzymatic digestion. Unless myofibers are well rinsed, such non-myogenic cells will be co-isolated with the myofibers and eventually produce many progeny in the rich culture conditions:

1. Inspect the muscle under the stereo dissecting microscope to make sure that the myofibers are loosened from the muscle bulk; the muscle should look like a loose skein of yarn. If the

myofibers are not loosened, continue enzymatic digestion for another 10 min and check again.

2. Retrieve from the incubator the three 100-mm Petri dishes containing 9-ml DMEM (rinse plates). Use the widest-bore Pasteur pipette to transfer the muscle bulk from the collagenase solution to the first DMEM rinse plate to wash away the collagenase and debris that might have dissociated from the muscle during digestion. Transfer the muscle to the second then third Petri dish for further dilution of any possible collagenase that may remain. These rinses must be performed with great care; limit the amount of mechanical manipulation of the muscle or swirling of the dish until the trituration step is reached.
3. Retrieve from the incubator one of the six 100-mm Petri dishes that were pre-coated with HS and filled with DMEM (this will be the holding dish for the muscle bulk and will be used in several of the steps described below). Transfer the rinsed muscle bulk into the holding dish. Place the dish in the incubator for approximately 10 min to allow the tissue to warm up.
4. Retrieve from the incubator a second HS-coated DMEM containing 100-mm dish. Using the same widest-bore pipette, transfer the muscle bulk to this dish (first trituration dish). Return the holding dish to the incubator to warm.
5. Use another HS-coated Pasteur pipette (tip diameter: approximately 3–4 mm) to triturate the muscle along its length. This orientation of the EDL muscle during triturations is critical to prevent damage to the myofibers.
6. When single myofibers are liberated from the muscle, its diameter decreases. Therefore, use a narrower-bore pipette for subsequent triturations.
7. When 10–15 viable single myofibers are released, retrieve from the incubator the holding plate, transfer the muscle bulk into it, and place it back in the incubator. Additionally, place the dish with the single myofibers in the tissue culture incubator to keep the fibers warm (typically the fibers from this first trituration round are not used, but save the plate in case it is needed). Allow the holding plate with muscle bulk to warm up for at least 5–10 min in the incubator before the next round of trituration.
8. Retrieve from the incubator a third HS-coated DMEM containing 100-mm dish (second trituration dish) and the holding plate with muscle bulk. Transfer the muscle bulk to the second trituration dish using the same widest-bore pipette used in the first trituration dish. Using a wide-bore pipette with a smaller diameter, triturate the tissue until 30–50 myofibers are obtained (but do not triturate the tissue for more than

5–7 min). Transfer the muscle bulk back to the holding dish and place both the holding dish and the dish with released myofibers back in the incubator.

9. Follow the pattern of moving the muscle bulk as described above in **steps** 7 and **8**, and create a third trituration dish; triturate the muscle bulk until approximately 100 myofibers have been released. When the third trituration step is complete, transfer the tissue back into the holding dish, and place both the holding dish and the dish with released myofibers back into the incubator. Typically, three rounds of triturations are sufficient to dissociate the muscle bulk entirely.
10. Using a HS-coated 9" pipette (standard bore size), begin to transfer individual fibers from the second and third trituration plates to the remaining two HS-coated DMEM containing 100-mm Petri dishes (collection plates). Refresh the HS coating of the pipette before each fiber transfer so that fibers do not adhere to the glass. Alternate between (at least) two collection plates to minimize cooling of the myofibers. As a general scheme:
    (a) Transfer ten fibers from the second trituration dish to one of the collection plates, and then move both plates back to the incubator.
    (b) Remove the third trituration dish from the incubator and transfer ten fibers to a second collection plate. Try to avoid using the first trituration dish as the fibers from this trituration are much more fragile and often have more non-myogenic cells attached to them.
    (c) Repeat this process when triturating the second EDL, alternating with the first EDL throughout the processing. If using only one EDL, always allow the plates to rest for 10 min in the incubator before repeating the process. Collect those fibers that are relatively straight and are not partially contracted.
11. Once a large enough number of fibers have been collected (generally 20–30 per collection plate), begin selecting fibers that will be used for analysis. Remove the 100-mm collection plates one at a time and visually inspect the fibers under the highest magnification available. Avoid fibers that have visible associated debris, and also avoid those that are kinked or partially contracted (*see* **Note** 7). Transfer fibers that pass these criteria to another HS-coated 60-mm dish (final dish). Try not to place too many fibers into one single plate as the fibers may become entangled with each other or associated with debris that may have been carried over in the transfer (as an approximation, no more than 3 fibers/1 cm^2 of dish surface area). Although including this step of fiber selection and transfer to

the final plate requires extra time, it allows for another wash step to remove non-myogenic cells that may have been carried over during trituration, thereby ensuring a more optimal fiber preparation.

3.2.7 Plating Freshly Isolated EDL Adherent Myofibers for Time 0 Satellite Cell Analysis

As noted in the introduction, for Time 0 analyses, we recommend working with myofibers that have been allowed to adhere to the dish surface before fixation rather than using non-adherent myofibers (*see* **Note 3**):

1. Transfer a Matrigel-coated, 24-well multiwell dish from the incubator to the tissue culture hood and open its lid to allow moisture (generated during the incubation period) to evaporate.
2. Bring the 60-mm Petri dish containing single fibers (final dish) to the dissection microscope along with the 24-well plate.
3. Under the dissection microscope, use a fire-polished, HS-coated 9" Pasteur pipette to select fibers that are free of associated tissue debris. Transfer one fiber at a time with residual DMEM (~150 μl) into the center of a Matrigel-coated well (one fiber per well). The fiber should be sitting in a droplet of DMEM, on top of the Matrigel to ensure that it does not dry out. After myofibers are dispensed to the desired number of wells, check again under the stereo dissecting microscope to ensure that indeed there is a myofiber in each well. This step is necessary since occasionally myofibers adhere to the Pasteur pipette and are not released into the well or the fiber becomes damaged in the transfer. Avoid excessive agitation of the fibers.
4. If needed, add a myofiber to empty well(s) or replace with an intact myofiber. Minimize the length of time the 60-mm final plate and the multiwell dish are held at room temperature; transfer dishes back to the incubator after 10 min to warm while continuing dispensing fibers.
5. After the desired number of fibers has been dispensed (1 per well), place the plate back in the incubator for 3 h to allow the fibers to adhere more firmly to the Matrigel. Minimize the amount of time that the fibers remain outside of the incubator and do not subject the plate to sudden motion. Extra special care should be exercised when handling such early time points for microscopic examination or immunostaining because the fibers are only loosely adhered and too much manipulation can damage the fibers and cause contraction.

3.2.8 Culturing Single Adherent EDL Myofibers

1. Transfer a Matrigel-coated, 24-well multiwell dish from the incubator to the tissue culture hood and open its lid to allow moisture, generated during the incubation period, to evaporate. Add 500 μl of pre-warmed, culture medium (*see* **item 7** in Subheading 2.6) to each well.

2. Bring the 60-mm Petri dish containing single fibers (final dish) to the dissection microscope along with the 24-well plate.
3. Follow **steps 3** and **4** in Subheading 3.2.7.
4. When the desired number of fibers has been plated, place the 24-well multiwell dish in the tissue culture incubator. Avoid handling the plate (i.e., to inspect fibers) for a minimum of 18 h (overnight).
5. After the fibers have been in culture for three days, gently add an additional 500 μl of complete media to the fibers. After three more culture days, replace the entire old medium with 500 μl of fresh growth medium. Continue changing the media every three days. We typically maintain myofiber cultures for 10–14 days without any apparent decline in culture quality. Depending on the goal of the project, we also have maintained fiber cultures for up to 3 weeks, but Matrigel may be partially degraded by then, and myotubes may detach from the plate. Moreover, the medium change schedule may need to be more frequent for longer culture periods.

3.2.9 Culturing Single Non-adherent EDL Myofibers

1. Add 500 μl of pre-warmed, culture medium (*see* **item 8** in Subheading 2.6) to each well of a non-coated 24-well plate.
2. Bring the 60-mm Petri dish containing single fibers (final dish) to the dissection microscope along with the 24-well plate.
3. Follow **steps 3** and **4** in Subheading 3.2.7.
4. When the desired number of fibers has been plated, place the 24-well multiwell dish in the tissue culture incubator. When the effect of growth factors on satellite cell proliferation/differentiation is investigated, parallel specimens are maintained in 0.3 ml of culture medium with/without additives, and the replenishment of the medium is ensured by adding fresh medium (0.2 ml) on culture day 1 and performing partial medium change (0.25 ml) on culture day 2. This approach ensures that growth factors do not become rate limiting and that myofibers are not disturbed during medium change.

3.3 Fixing and Immunofluorescent Labeling of FDB and EDL Myofibers

This section details current protocols used in our laboratory to fix myofiber cultures for immunofluorescent studies of satellite cells and their progeny. Extra care should be taken when working with Time 0 myofibers as these are loosely attached to the plate compared to myofibers in later time point cultures. Use of a stereo dissecting microscope throughout the procedure to ensure that the fibers are not lost or become damaged is recommended. All wash steps should be performed using a glass fire-polished Pasteur pipette to minimize myofiber damage.

Freshly plated (Time 0) and cultured FDB myofibers are typically fixed with ice-cold methanol (the preferred fixative when working with dishes coated with PureCol collagen), whereas freshly plated (Time 0) and cultured EDL myofibers are typically fixed with paraformaldehyde that is pre-warmed to room temperature; methanol fixation is not optimal for the longer EDL myofibers as it can lead to their hypercontraction. Overall, when analyzing single myofibers via immunofluorescence, fixatives should be optimized for both preserving the myofibers and the antigens being analyzed. Table 2 summarizes primary and secondary antibodies used in our immunostaining studies depicted in this chapter. Protocols described in this section are also appropriate for detecting proliferating satellite cells in single myofibers by autoradiography following labeling with ^{3}H-thymidine [47, 49] or when analyzing proliferation using bromodeoxyuridine [31, 35, 56, 77]. All steps are done in a sterile manner. Handling antibodies strictly in the tissue culture hood minimizes possible contamination and helps maintain antibody stocks for years.

3.3.1 Freshly Isolated (Time 0) and Cultured FDB Adherent Myofibers

1. Warm DMEM in a water bath set at 37 °C.
2. Rinse cultures with 500 μl warm DMEM three times. Following the final rinse, add 1 ml ice-cold 100 % methanol to each 35-mm tissue culture dish, and transfer the dishes to 4 °C for 10 min.
3. Return dishes to room temperature, aspirate the methanol, and allow the dishes to air-dry for 10–15 min in the tissue culture hood (*see* **Note 24**).
4. Add 1.5 ml of blocking solution (TBS-NGS) to each culture dish, to block nonspecific antibody binding.
5. Cultures are then kept at 4 °C for at least overnight and up to 2 weeks. Bring cultures to room temperature when ready to start antibody labeling.
6. Dilute the appropriate primary antibody in the NGS-TBS blocking solution. If not otherwise published, before diluting your antibody, test a range of dilutions to determine the lowest concentration of antibody that gives a clear specific signal without nonspecific background.
7. Rinse the cultures three times with 500 μl TBS-TW20.
8. Remove the final TBS-TW20 rinse and add 100 μl of the primary antibody solution. Incubate for 1 h at room temperature followed by an overnight incubation at 4 °C in a humidified chamber (*see* **Notes 25** and **26**). Primary and secondary antibodies are applied at the center of the dish followed by a light swirling on a flat surface to ensure optimal spreading of the antibody across the dish. This approach allows using just 100 μl antibody solution, which is beneficial for conserving antibody stocks.
9. Dilute the appropriate secondary antibody in the NGS-TBS blocking solution. Secondary antibodies are often diluted at

1:1000 or greater, but the researcher needs to determine the optimal dilutions for their specific study.

10. Rinse cultures with 500 μl TBS-TW20 three times.
11. Remove the final TBS-TW20 rinse and add 100 μl of the diluted secondary antibody. Incubate for 1–2 h at room temperature.
12. Remove the secondary antibody and wash three times with 500 μl TBS-TW20.
13. For nuclear visualization, add at least 100 μl of DAPI working solution (1 μg/ml, diluted in TBS-NGS prior to use; *see* **item 10** in Subheading 2.7) and incubate for 30 min at room temperature.
14. Rinse the cultures twice with 500 μl TBS-TW20 followed by a final rinse with 500 μl TBS.
15. Remove the TBS and mount in Vectashield mounting medium. Add one drop at the center of each culture dish, and cover gently with a cover slip, avoiding generating air bubbles. Cultures should be viewed as soon as possible but, if not, then stored at 4 °C sealed in Parafilm, covered with aluminum foil to protect from light, and viewed within a week after immunostaining to avoid fading.

3.3.2 Freshly Isolated (Time 0) Adherent EDL Myofibers

1. Use a fire-polished Pasteur pipette to slowly add the 4 % paraformaldehyde fixative solution (pre-warmed to room temperature) until the droplet containing the fiber (*see* Subheading 3.2.7) has approximately doubled in volume. Allow the fiber to sit in the fixative solution for 10 min at room temperature, then carefully remove (using a Pasteur pipette) the DMEM-paraformaldehyde fixative mixture, and rinse each well three times with 500 μl TBS.
2. Add 500 μl of TBS-TRX100 for 5 min at room temperature. Alternatively, Triton X−100 can be omitted as some antigens may be more optimally detected if Triton X−100 has not been used.
3. Add 500 μl of blocking solution (TBS-NGS) to each of the 24 wells, to block nonspecific antibody binding.
4. Follow **steps 5–14** as described in Subheading 3.3.1. However, when exposed to antibodies, the 24-well multiwell trays should be continuously and gently swirled as described in **Note 26**, as uneven antibody staining can otherwise occur.
5. Remove the final TBS rinse, and add one drop of Vectashield mounting medium as in **step 15**, Subheading 3.3.1. We prefer not to use cover slips when working with 24-well multiwell trays. Instead, we add 300 μl of the glycerol mounting solution (25 % glycerol in TBS) following the initial drop of

Vectashield to allow sufficient mounting medium coverage of individual wells in 24-multiwell trays.

6. Observe and analyze the fibers under the microscope, then seal the multiwell tray with Parafilm, and store at 4 °C and in the dark (e.g., can be stored wrapped with aluminum foil) when not in use. Trays should be viewed as soon as possible. If they cannot be viewed immediately, they should be stored at 4 °C, sealed in Parafilm, and covered with aluminum foil to protect from light. Typically, we aim to complete analyses within a week after immunostaining to avoid fading.

3.3.3 EDL Adherent Myofiber Cultures

1. While observing each myofiber under the stereo dissecting microscope, use a Pipetman to gently, without agitating the culture or touching the myofiber, add an equal volume (500 μl) of 4 % paraformaldehyde fixative solution (pre-warmed to room temperature) to the culture medium in each well in the 24-well dish. Allow 10 min at room temperature for the fixation, then carefully remove (by aspiration or using a pipette) the culture medium-paraformaldehyde fixative mixture, and rinse each well three times with 500 μl TBS.
2. Follow **steps 2–6** as described in Subheading 3.3.2.

3.3.4 EDL Non-adherent Myofiber Cultures

1. Slowly add to the culture medium an equal volume of 4 % paraformaldehyde fixative solution (pre-warmed to room temperature) while observing each myofiber under the stereo dissecting microscope. Allow the fiber to sit in the fixative solution for 10 min at room temperature, then carefully remove (using a Pasteur pipette) the culture medium-paraformaldehyde fixative mixture, and rinse each well three times with 500 μl TBS.
2. Follow **steps 5–14** as described in Subheading 3.3.1, except that extra care had to be taken due to the non-adherent nature of the myofibers.
3. Follow **step 5** as described in Subheading 3.3.2 for mounting the specimen.
4. Observe and analyze the fibers under the microscope, then seal the multiwell tray with Parafilm, and store at 4 °C and in the dark (e.g., can be stored wrapped with aluminum foil) when not in use. Typically, we aim to complete analyses within a week after immunostaining to avoid fading.

3.4 Detection of Satellite Cells in Freshly Isolated Myofibers Based on Nestin-GFP Transgene Expression

As detailed in the introduction, the Nestin-GFP transgenic mouse provides an efficient tool for monitoring satellite cells on freshly isolated live (or fixed) myofibers by direct detection of GFP fluorescence (Fig. 2), which enables linking initial Time 0 numbers and subsequent performance upon culturing [31]. The Nestin-GFP transgene also facilitates the isolation of satellite cells using

Table 3
Satellite cell numbers according to Nestin-GFP detection on freshly isolated live myofibers prepared from EDL, diaphragm (DIA), and masseter (MAS) muscles of 6-month-old mice[a]

Muscle type	Number of mice	Number of fibers	Nestin-GFP$^+$ cells per fiber (mean ± SEM)	Agreement between Nestin-GFP$^+$ cells and	
				Pax7$^+$ cells[b] (%)	Myf5^{nLacZ+} cells[c] (%)
EDL	11	275	8.51 ± 0.19	98.8	98.7
DIA	8	161	7.18 ± 0.23	98.3	96.8
MAS[d]	7	176	6.10 ± 0.27	99.3	98.4

[a]Analyses are done with adherent myofibers (*see* **Note 3**)
[b]Based on 13 EDL, 12 DIA, and 16 MAS fibers isolated from three mice
[c]Based on 58 EDL, 22 DIA, and 83 MAS fibers isolated from three mice
[d]*See* **Note 3**

fluorescent-activated cell sorting and subsequent studies of purified populations [13, 32]. Table 3 illustrates satellite cell quantification in three different muscles based on direct detection of the Nestin-GFP transgene on live myofibers isolated, compared with satellite cells detection by Pax7 (Fig. 9) and Myf5^{nLacZ} (Fig. 10) expression.

3.4.1 Co-detection of Nestin-GFP and Pax7 on Fixed Myofibers

Myofibers are fixed and immunostained following steps described in Subheading 3.3.2. When performing Pax7 immunostaining, a direct detection of GFP fluorescence is typically satisfactory; however, if warranted, an enhancement of the GFP signal can be achieved by GFP immunostaining (Table 2).

3.4.2 Detection of Nestin-GFP in Combination with X-gal Staining for lacZ Reporters

As detailed in the introduction, we have typically crossed the Nestin-GFP transgene with Myf5^{nLacZ} knockin mice, providing dual marking for satellite cells, and MLC3F-nLacZ transgenic mice, which distinguishes myofiber nuclei but is not expressed in satellite cells (Fig. 10). The detection of the Nestin-GFP transgene in satellite cells is possible in combination with X-gal staining for lacZ reporters following the procedure described below:

1. Fibers are plated as in Subheading 3.2.7.
2. Pre-warm the X-gal dilution buffer at 37 °C.
3. Fix myofibers. Use a fire-polished Pasteur pipette to slowly add the 4 % paraformaldehyde fixative solution (pre-warmed to room temperature) until the droplet containing the fiber has approximately doubled in volume. Allow the fiber to sit in the fixative solution for 10 min at room temperature, then carefully remove (using a fire-polished Pasteur pipette) the DMEM-paraformaldehyde fixative mixture, and rinse each well three times with 500 μl PBS (note the use of PBS here versus the use of TBS rinses when performing immunostaining).

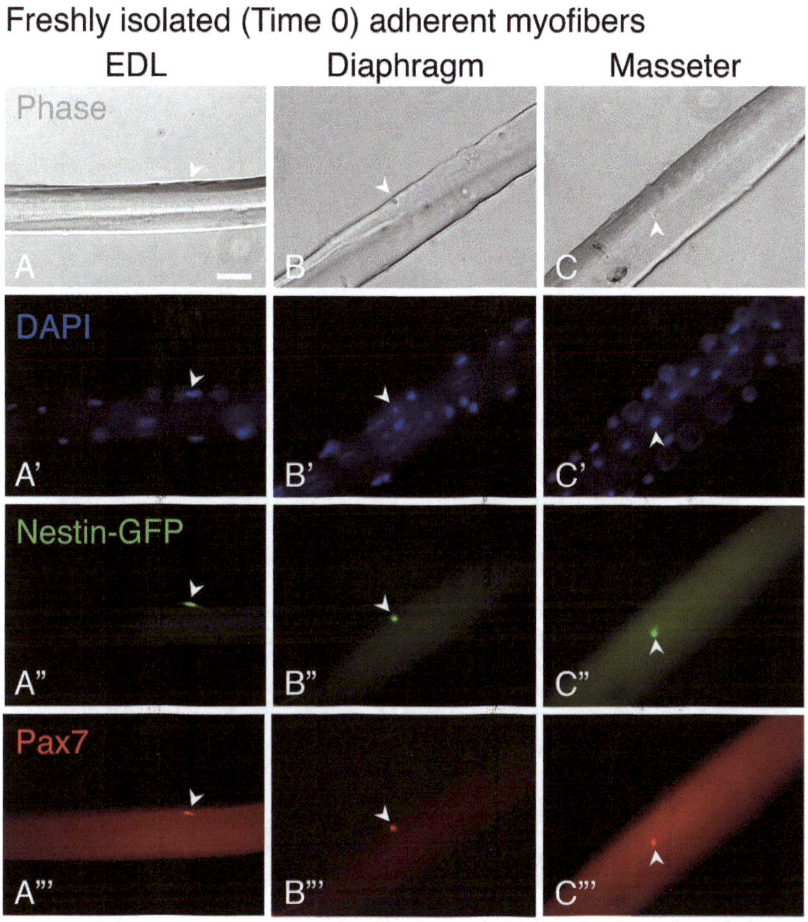

Fig. 9 Parallel phase and fluorescent micrographs of Time 0 adherent myofibers isolated from EDL (**A**–**A**‴), diaphragm (**B**–**B**‴), and masseter (**C**–**C**‴) muscles, demonstrating co-expression of the Nestin-GFP transgene and the satellite cell marker Pax7. Myofibers were isolated from adult Nestin-GFP mice and plated individually in 24-well multiwell tissue culture dishes coated with Matrigel. Myofibers were fixed with paraformaldehyde and analyzed by immunostaining as described in Subheading 3.3.2. (**A**, **B**, **C**) Phase images. (**A**′, **B**′, **C**′) DAPI-staining (both myofiber nuclei and satellite cell nuclei are labeled with DAPI). (**A**″, **B**″, **C**″) Nestin-GFP expression in satellite cells is observed by direct detection of the GFP fluorescence. (**A**‴, **B**‴, **C**‴) Pax7 immunofluorescence expression in Nestin-GFP^{+} satellite cells. Arrowheads depict Nestin-GFP^{+}/Pax7^{+} satellite cells. Scale bar 50 μm

4. Dilute the X-gal substrate in the pre-warmed X-gal dilution buffer at a 1:40 ratio, and then dispense 500 μl of this X-gal staining solution to each of the desired wells in the 24-well multiwell dish.

5. Incubated myofibers in X-gal staining solution using a 37 °C chamber (we typically use the humidified tissue culture incubator). In our laboratory, myofibers are incubated for up to 16 h when working with Myf5^{nLacZ} reporter or up to 2 h for MLC3F-nLacZ reporter. However, conditions may vary; therefore, myofibers should be monitored regularly during incubation and the X-gal staining stopped when positive nuclei

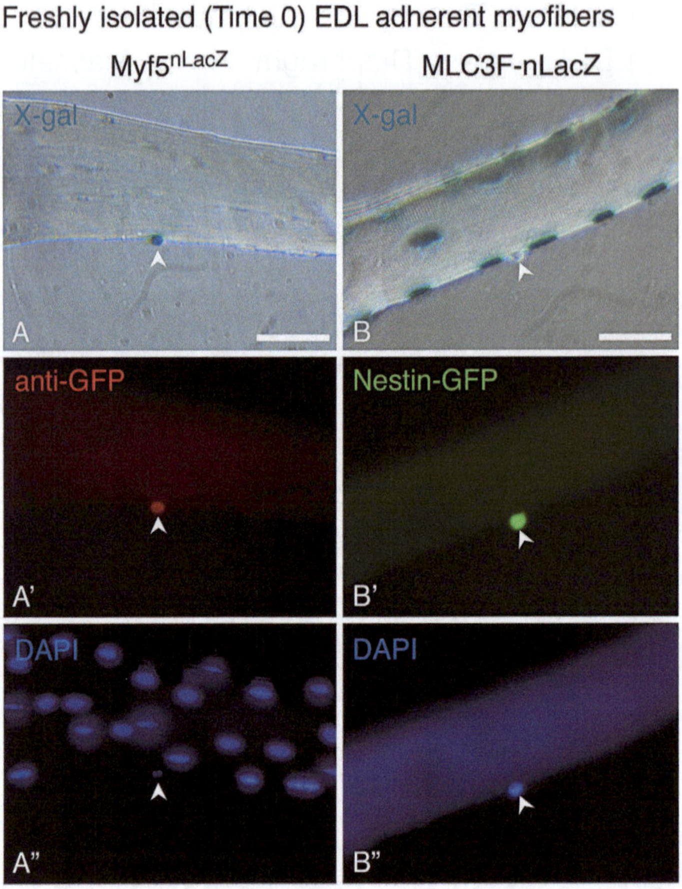

Fig. 10 Parallel phase and fluorescent micrographs depicting Nestin-GFP and LacZ reporter detection in Time 0 adherent EDL myofibers isolated from (**A–A′**) an adult Nestin-GFP/Myf5^{nLacZ} mouse and (**B–B″**) an adult Nestin-GFP/MLC3F-nLacZ mouse. Myofibers were plated individually in 24-well multiwell tissue culture dishes coated with Matrigel and fixed with paraformaldehyde and analyzed by X-gal staining as described in Subheading 3.4.2. (**A–A″**) For optimal detection of the Myf5^{nLacZ} reporter in satellite cells, myofibers were incubated for 16 h in X-gal staining solution, which also resulted in a strong green autofluorescence of the entire myofiber together with quenching of direct Nestin-GFP fluorescent signal in the LacZ^{+} satellite cells. Therefore, to recover a clear Nestin-GFP signal in satellite cells, the myofiber was immunostained with an anti-GFP combined with a red fluorescent secondary antibody. The strong X-gal staining of the satellite cell nucleus also typically extinguishes DAPI fluorescent signal as shown in panel **A″**. (**B-B"**) Optimal detection of the MLC3F-nLacZ reporter in myonuclei is typically achieved by as little as 15 min incubation in the X-gal staining solution, with lack of LacZ expression in GFP^{+} satellite cells; satellite cells remained negative for X-gal staining even after overnight incubation. The extinguished DAPI fluorescent signal in myonuclei by strong X-gal staining also allows identification of satellite cell nuclei that are LacZ negative but DAPI^{+}/GFP^{+}. *Arrowheads* depict a GFP^{+}/LacZ^{+} satellite cell that is DAPI^{-} (**A–A″**) and a GFP^{+}/LacZ^{-} satellite cell that is DAPI^{+} (**B–B″**). Images in (**B–B″**) were previously published in [13]. Scale bars 50 μm

can be clearly distinguished. A particular attention should be exerted with the Myf5^{nLacZ} reporter, as a too short exposure will only stain some satellite cells, while a too long exposure may lead to myofiber nuclei staining as well.

6. Rinse each well three times with 500 μl TBS.
7. If immunostaining is desired, follow **steps 3–5** as described in Subheading 3.3.2. When working with myofibers from Nestin-GFP/Myf5^{nLacZ} mice, the use of an anti-GFP antibody (combined with a red fluorescent secondary antibody) is necessary to enable specific GFP signal detection above the overall green fluorescent background produced by the prolonged incubation of myofibers in X-gal staining solution (Fig. 10). Differently, when working with myofibers from Nestin-GFP/MLC3F-nLacZ mice, a 15-min incubation in the X-gal staining solution is typically sufficient and direct GFP signal is preserved (Fig. 10).
8. For nuclear visualization, add to the rinsed well at least 100 μl of DAPI working solution (1 μg/ml, diluted in TBS-NGS prior to use; *see* **item 10** in Subheading 2.7), and incubate for 30 min at room temperature. Then, rinse twice with 500 μl TBS-TW20 followed by a final rinse with 500 μl TBS.
9. Remove the final rinse, and add one drop of Vectashield plus 300 μl 25 % glycerol-TBS (as in **step 5** Subheading 3.3.2)
10. Observe and analyze the fibers under the microscope, then seal the multiwell tray with Parafilm, and store at 4 °C and in the dark (e.g., can be stored wrapped with aluminum foil) when not in use. Typically, we aim to complete analyses within a week following fiber harvesting.

4 Notes

1. The Nestin-GFP transgenic mouse used in our studies has been developed originally to study neural stem cells [57]. Indeed, nestin, an intermediate filament protein, was originally identified as a marker of neural progenitors. However, nestin protein expression was subsequently detected in other cell types [78, 79], including myogenic cells [31, 80–82]. Furthermore, Nestin-GFP transgenic expression was identified by us in satellite cells in all muscle groups examined [31, 32, 54] and in a number of other progenitor lineages [83, 84]. While in the context of isolated myofibers, this Nestin-GFP transgene can be used to specifically detect satellite cells [31, 32]; if myofibers are isolated with remnants of associated vasculature or neuromuscular junction, non-myogenic cell types that express the Nestin-GFP reporter can at times also develop in long-term cultures (Fig. 11A–C). This feature has provided

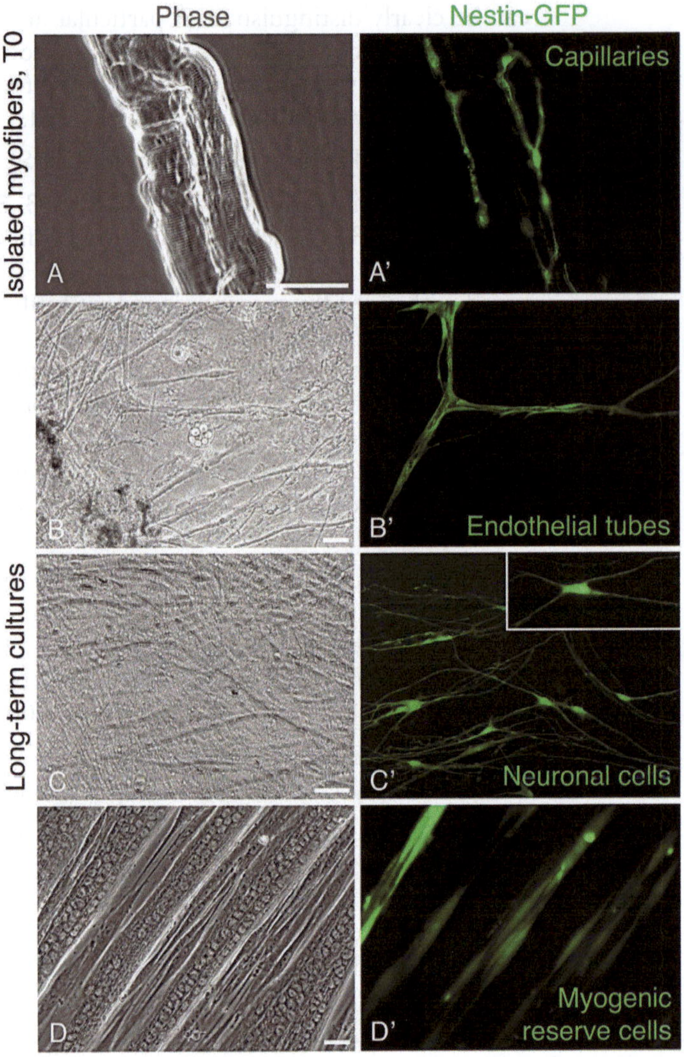

Fig. 11 Parallel phase and direct GFP fluorescent micrographs depicting expression of the Nestin-GFP transgene in endothelial, neuronal, and myogenic reserve cells. (**A**, **A′**) Two associated Time 0 EDL myofibers isolated from adult Nestin-GFP/mdx^{4cv} mice, demonstrating the presence of remnants of a capillary bed whose endothelial cells express the Nestin-GFP transgene as we previously published [31]. Notably, such fibers isolated with associated connective tissue are prevalent when working with adult and old mdx (dystrophin-null) mice even if thorough rinses are performed during the liberation of single myofibers (Subheading 3.2.6). (**B**–**D′**) Long-term/high-density cultures prepared from the skeletal muscle of adult Nestin-GFP mice and grown in Matrigel-coated plates maintained in mitogen-rich growth medium (*see* **item 7** in Subheading 2.6), demonstrating Nestin-GFP expression in (B, B') endothelial tubes (also immunostained for the endothelial marker CD31, not shown) [31], (**C**, **C'**) neuronal cells (also immunostained for the marker Tuj1, not shown) [108] (we postulated that these neuronal cells originate from neuromuscular junctions that can occasionally be detected at the surface of isolated myofibers) [10], and (**D**, **D'**) myogenic reserve cells (also immunostained for Pax7, Fig. 2) [31, 32]. The myogenic reserve cell represents the only Nestin-GFP$^+$ cell type that develops in pure myogenic cultures derived solely from satellite cells, typically residing adjacent to myotubes. Notably, all the three Nestin-GFP$^+$ cell types discussed above emerge only in high-density cultures, and these cultures are clearly void of Nestin-GFP expressing cells before such cells appeared. Scale bars 50 μm. Images in Panel (**A**) were contributed by our former lab member, Andrew Shearer

an additional tool to evaluate the level of myogenic purity of the isolated myofiber and the resulting cultures. Indeed, adipogenic cells that have been occasionally observed in myofiber cultures and initially considered to derive from satellite cells undergoing fate change [80, 85] could be eliminated by further rinses of isolated fibers before culturing ([13, 86]; current manuscript) and were eventually proven not of satellite cell origin, but rather derived from fibro-/adipogenic cells co-isolated with myofibers ([87, 88]; additional unpublished studies from our lab tracing lineage fate with myogenic reporter mice). The aforementioned Nestin-GFP$^+$ non-myogenic cells (Fig. 11B, C) should not be mistaken for the Nestin-GFP$^+$ myogenic reserve cells (Fig. 11D) that represent the only Nestin-GFP$^+$ cell type developing in long-term pure myogenic cultures and like satellite cells are Pax7$^+$, MyoD$^-$, and quiescent [13, 31].

Notably, it is unknown if other Nestin-GFP transgenic mice that have been reported in the literature (apart from the one used in our studies [31, 57]) are of relevance for satellite cell detection. Both the regulatory elements used in the transgene cassette and the genomic insertion site of the transgene can contribute to the specific expression of Nestin-GFP by satellite cells as seen in our studies.

2. Isolation and analysis of single myofibers can be a challenging and time-intensive task. Just by the look of this chapter, it is clear that there are numerous details involved. So why did we get into all this? Following our initial publication on satellite cell isolation and culture [43], it appeared essential to progress toward investigating satellite cells at their native niche for pursuing studies on satellite cell dynamics and the role of growth factors (while bypassing the complexity of the whole tissue). For this, we turned in the late 1980s to the isolated myofiber model, adapting the approach of Richard Bischoff of single myofiber isolation, especially focusing on details provided in his 1986 seminal articles about studying satellite cells in myofibers isolated from the hind feet of the rat flexor digitorum brevis (FDB) muscle [47, 89]. Once we had the FDB myofiber culture system in place, it became clear that studying satellite cell proliferative response in isolated myofibers solely based on tritiated thymidine uptake would require a lot of wait time for the autoradiographies to develop after each experiment. At that stage, we elected to try a novel approach for tracing proliferating satellite cells in isolated myofibers, developing an immunostaining protocol for the proliferating cell nuclear antigen [36]. Indeed, we demonstrated the same pattern of satellite cell proliferation when comparing cell tracing in FDB myofibers by PCNA labeling versus tritiated thymidine autoradiography [49]. Satellite cell proliferation peaked early in these

myofiber cultures (days 1.5–2) and then declined to a baseline level by culture day 4, and it was unclear where the cells headed once proliferation declined. We then learned that several labs had developed antibodies against the myogenic regulatory factors MyoD and myogenin and were able to receive aliquots of these antibodies for further immunostaining studies of isolated FDB myofibers—we were pleased to find out that only nuclei of satellite cells (or their progeny) were positive for MyoD and myogenin proteins, while myofiber nuclei were not, enabling studies on satellite cell progression from activation to differentiation [36, 49, 51]. We are grateful to the developers of these antibodies for their generous antibody gifts long before these antibodies became widely available to the research community: Stefano Alemo, rabbit anti-MyoD; Peter Dias, mouse anti-MyoD clone 5.8A; and Woodring Wright, mouse anti-myogenin clone F5D. Collectively, our immunostaining approach established that the satellite cells were rapidly transiting through proliferation and differentiation in the FDB myofiber cultures. A year following our first lab publication on the analysis of satellite cells in isolated myofibers [36], David Rosenblatt and colleagues reported on a different type of isolated myofiber cultures (mostly from the extensor digitorum longus "EDL" muscle) in which the satellite cells emigrated from the parent myofiber and gave rise to a large progeny pool, similar to satellite cell behavior in primary cultures [52]. Indeed, in the early 2000s, we began adapting the Rosenblatt myofiber model to compare the growth capacity of satellite cells from young and old mice [24]. Nevertheless, we have continued using myofiber culture models where satellite cells are retained at the myofiber surface for studies where we wish to model closely the in vivo state of muscle homeostasis (which does not involve major muscle trauma and robust activation of satellite cells for tissue repair) [24, 34, 44, 50, 51]. Indeed, the association of the satellite cell with its parent myofiber has been shown to regulate the satellite cell self-renewal versus commitment to proliferate and differentiate [90, 91].

In this chapter, we are also providing details regarding the utility of the Nestin-GFP mouse (described in *see* **Note 1**) to trace satellite cells in live (and fixed) myofibers. We extend our gratitude to Grigori Enikolopov for making this mouse available to us in the mid-2000s [31, 57]. Initially, we have introduced the Nestin-GFP transgenic mouse to our lab based on our long-standing interest in the expression of the intermediate filament proteins desmin, vimentin, and nestin in myogenic and vascular smooth muscle cells [80, 92–94]. While we were considering the possibility that the Nestin-GFP transgene would permit a direct means for tracing proliferating progeny of satellite cells, its expression in quiescent satellite cells was discovered by serendipity. It certainly was an astonishing surprise and a fluorescence magic when we observed for the first time the satellite cell expressing the green fluorescence protein while the rest of the freshly isolated myofiber was void of fluorescence when our team member, Gabi Shefer, got first look of EDL myofibers from this transgenic mice. This magic feeling comes back each time we look at a freshly isolated myofiber from the Nestin-GFP mouse, regardless of muscle group being processed. Since then, we have been extensively utilizing the Nestin-GFP transgenic mouse to trace, analyze, and isolate satellite cells.

Overall, the isolation and analysis of single myofibers can turn out to be an extremely detailed process, and one should approach this task with a passion or one should not start this route at all. It can take many attempts before things all fit in place. But the reward eventually is enormous, being able to look at the myofiber and its satellite cells and follow these magnificent cells as they perform their myogenic task at the myofiber niche.

3. Working with freshly isolated EDL myofibers from Nestin-GFP adult mice, we compared satellite cell numbers before and after fixation of adherent and non-adherent fibers and concluded that for the non-adherent Time 0 fibers, some satellite cells can be lost during the fixation process. This may explain the differences in average satellite cell numbers obtained for masseter myofibers from adult mice between our data (adherent myofibers, Table 3, MAS) and a previously published study (non-adherent myofibers, [95]), whereas satellite cell numbers in adherent EDL myofibers are in agreement between the latter group [96] and our data (Table 3, [24, 31]).

4. Isolation of myofibers from masseter, diaphragm, and extraocular muscles. Details in this section are provided in brief and focus mainly on muscle harvesting and dissociation. All reagents are as described for EDL myofiber isolation and culture in earlier sections of this chapter. For all muscles, tissue is dissociation with 0.4 % collagenase (type I, source as listed **item 12**, Subheading 2.6, and preparation as in **step 1**, Subheading 3.2.5).

   The diaphragm muscle consists of two portions; the costal muscle, radially arrayed from a central non-contractile tendon (the central tendon), and the crural diaphragm through which the aorta, thoracic duct, and esophagus are transmitted. When isolating single myofibers from the diaphragm [54], we have found that at the risk of contaminating the preparation with a greater amount of debris and shorter intercostal fibers of the ribcage, removal of the diaphragm in whole with its immediate supporting ribcage structure provides the greatest fiber yield.

After removing the diaphragm, rinse the muscle in a dish containing pre-warmed DMEM. Observe the diaphragm under the dissection scope, and without touching the muscle, carefully remove any obvious fat or connective tissue; otherwise, this material can foul the prep. Other steps that ensure higher and purer yield of fibers are the addition of a wash step post-enzymatic digest and not "over-digesting" the muscle. Digesting for 45–60 min in 0.4 % collagenase is recommended. The muscle is triturated centrally from the position of the central tendon and ends of the ribcage using the largest bore, fire-polished pipette. Expect no more than 50 ideal fibers (undamaged and without associated debris) per diaphragm.

The masseter muscle of the jaw (like the diaphragm) requires harvesting some of the supporting skeletal structure along with the muscle for the highest fiber yields. The masseter muscle is a multilayered muscle with complex and extensive investment geometry. We have found that including the origin and investment surfaces in the enzymatic digest, rather than risking damage to the muscle at the tendon-bone interface, ensures more intact muscle fibers. Once removing the masseter muscles with the associated bones, rinse and remove debris in a pre-warmed 100-mm Petri dish containing DMEM. Digest the masseter muscle en bulk with jaw and skull bone attachments in 0.4 % collagenase for 45 min followed by careful washes and triturations. This preparation typically yields 30–100 fibers.

EOM are a unique set of muscles that control eye movements. There are 6 EOMs per eye accessed by first bisecting the skull from the top of the head and removing the brain to expose the bone at top of the eye socket. Once exposed, the eye socket is broken along its suture lines to expose the eye. Following removal of the lacrimal gland, the entire eye with its associated muscles is then removed "en bulk" by first cutting the optic nerve at a point just behind the annulus of Zinn (the point where 5 of the 6 EOM muscles meet). The isolation is completed by cutting the remaining soft tissue from around the eye, freeing the eye from the socket. For a comprehensive protocol of EOMs isolation please refer to our recently published method chapter [97]. The eye, with the muscles still attached, is placed in 0.4 % collagenase digest for 90 min at 37 °C and then transferred to fresh 0.4 % collagenase for an additional 45–60-min digestion. Gentle swirling of the digest every 15 min will help increase the yield. The rinsing and trituration steps that follow are similar to those described for the EDL, except with less vigor. This procedure yields an array of short and long, thick and thin EOM myofibers for analysis; 30–100 fibers are typically derived per preparation.

5. PureCol collagen (Advanced BioMatrix, product was formerly known as Vitrogen) is a sterile solution of purified, pepsin-solubilized, bovine hide collagen (97 % type I, 3 % type III) dissolved in 0.01 *N* HCl and stored at 4 °C until used. In our studies, PureCol collagen is made isotonic by mixing 6 volumes of stock PureCol collagen with 1 volume of 7× DMEM. The isotonic solution is prepared just prior to coating dishes because it gels rapidly at room temperature. To obtain consistent coating, the culture dishes should be precooled and coated on ice. When removed from the ice, these dishes warm up rapidly and are ready for myofiber addition. Preparations of collagen type I from other sources (e.g., Sigma-Aldrich) have been used by some laboratories as an alternative to PureCol collagen. The use of alternative sources would require prescreening to ensure compatibility; we only have experience with the bovine-derived product.
6. Matrigel (BD Biosciences) is a solubilized basement membrane preparation extracted from the Engelbreth-Holm-Swarm mouse sarcoma, a tumor rich in extracellular matrix proteins. Its major component is laminin, followed by collagen IV, entactin, and heparan sulfate proteoglycan [98]. Matrigel is shipped on dry ice and is stored at −20 °C until aliquoted. Matrigel should be thawed on ice; never use at a warmer temperature, as it will prematurely gel. To ensure Matrigel stability, we follow the manufacturer's handling instructions, thawing the product on ice (overnight in an ice bucket placed at 4 °C). Once liquefied, Matrigel is aliquoted with prechilled, 1-ml serological glass pipettes into tubes chilled on ice. Typically, we aliquot 200 μl each into 2-ml cryogenic vials sealed with O-rings. These aliquots are stored at −20 °C. We have observed some batch-to-batch variation in the time it takes to thaw the aliquots for final dish coating; therefore, for consistency, we typically allow Matrigel aliquots to thaw for 1.5 h. Matrigel can be purchased in its standard format or in its growth factor reduced format. We have typically used the growth factor reduced format but more recently have begun using the standard format for routine studies in rich growth medium. Life Technologies carries Matrigel-like products that might be useful as an alternative to Matrigel (e.g., Geltrex); however, we do not have sufficient experience with the latter product for detailed recommendations.
7. Aged mice (up to 33 months old) and other mouse strains have also been used in our studies following the same myofiber isolation procedures (e.g., [13, 24, 34]. Adjustments, such as concentration of collagenase, length of muscle digestion, and extent of muscle trituration for releasing myofibers, may be needed when isolating myofibers from younger/older mice,

other mouse strains, mutant mice, or other rodents such as rats. Prolonged digestion and extensive trituration of the muscle bulk will result in poor yields of intact myofibers. Myofibers that are damaged in the course of the isolation can be distinguished from the intact myofibers since they typically hypercontract. Bent myofibers are also damaged to some degree and should not be collected when preparing myofiber cultures.

8. When harvesting muscles for fiber preparation, we prefer cervical dislocation for euthanizing mice as this method is more rapid and minimizes muscle stiffening that occurs after death. Muscle stiffening can make the isolation of single fibers more difficult and decrease overall fiber yield.
9. Falcon Primaria 24-well multiwell dishes (BD Biosciences) were initially used for single myofiber cultures; however, we find that the standard, less expensive, Falcon 24-well multiwell dishes are as good.
10. Horse serum (HS) is used for tissue culture medium and for coating plastic and glassware. HS used for tissue culture media should be pre-characterized by comparing sera from various suppliers (e.g., over years of studies, our preferred serum lots came typically from Gibco-Life Technologies, HyClone, or Sigma-Aldrich). We select HS based on its capacity to support proliferation and differentiation of primary chicken myoblasts cultured at standard and clonal densities [99]. One may consider replacing HS with bovine serum albumin (BSA) for coating plastic and glassware to further minimize any possible activation of satellite cells during myofiber isolation. However, attention should be given to the purity of the BSA as some lots may contain growth-promoting factors.
11. Fetal bovine serum (FBS) should be pre-characterized by comparing sera from several suppliers (e.g., over years of studies, our preferred serum lots came from Gibco-Life Technologies, HyClone, or Sigma-Aldrich). We select FBS based on the capacity of the serum to support proliferation and differentiation of mouse primary myoblasts cultured at various cell concentrations. Only sera able to support growth and differentiation over a wide range of concentrations, down to a clonal density, are employed in our studies. Primary myogenic cultures are prepared according to our published procedures [41, 44].
12. The controlled processed serum replacement 2 (CPSR-2, Sigma-Aldrich) that had been routinely used in our myofiber culture studies [24, 36, 44, 49–51, 55] has been discontinued. The source of this discontinued CPSR-2 was dialyzed bovine plasma. This product was further processed in a manner that also reduced lipids. Another alternative serum replacement product, serum replacement 2 (50×) (Sigma-Aldrich),

contains highly purified bovine serum albumin, insulin, and transferrin, and its use for mouse myofiber cultures has been previously described [77].

13. Chicken embryo extract (CEE) is available commercially from several sources with which we have no experience. We prepare CEE in our laboratory using 10-day-old White Leghorn embryos [100]. The procedure is similar to a previously described method [101] but uses the entire embryo. We recommend this approach over purchasing CEE if the investigator can obtain embryonated chicken eggs, as the quality is thought to be higher and the cost lower than that of purchased CEE.
14. Preparation of chicken embryo extract:
    - Embryonated chicken eggs (eight dozen, White Leghorn; from Charles River) are maintained in a standard egg incubator (incubation conditions: a dry temperature of 38 °C, a wet temperature of 30 °C, and relative humidity of 56 %). The following egg incubator is well suited for basic research use: Marsh Automatic Incubator (Lyon Technologies, Chula Vista, CA).
    - After 10 days, batches of 15–30 eggs are removed from the incubator and transferred into the tissue culture hood. All steps from here on are performed in a sterile manner.
    - Place the eggs lengthwise in the rack and spray with 70 % ethanol to sterilize. Wait for several minutes until the ethanol evaporates.
    - Crack open one egg at a time into a 150-mm Petri dish.
    - Remove the embryo from surrounding membranes by piercing it with fine forceps. Rinse the embryo by transferring it through three 150-mm Petri dishes containing DMEM supplemented with antibiotics (*see* **item 1** in Subheading 2.6). Swirl embryo a few times in each dish for a good rinse.
    - Empty the egg remains from the initial 150-mm dish (described in **step d**) into a waste beaker and repeat **steps d–f** until the final rinse dish contains about 30 embryos.
    - The embryos are transferred with fine forceps into a 60-ml disposable syringe, forced through with the syringe plunger, and the suspension is collected into a 500-ml sterile glass bottle.
    - The extract is diluted with approximately an equal volume of DMEM (supplemented with antibiotics as detailed in **item 1** in Subheading 2.6) and gently agitated for 2 h at room temperature. To ensure good agitation, keep the maximum volume to one-half bottle capacity and place the bottle at a 45° angle during the agitation.

- The extract is frozen at −80 °C for a minimum of 48 h. It is then thawed, dispensed into 50-ml conical tubes, and centrifuged at approximately 500 × *g* for 10 min to remove residual tissue.
- The supernatant is pooled, divided into 40-ml aliquots, and kept frozen at −80 °C for long-term storage. For short-term storage, we typically prepare aliquots of 2.5 ml that are kept frozen at −20 °C.
- Prior to use, the CEE-thawed aliquot should again be centrifuged at about 800–1000 × *g* for 10 min to remove aggregates. The supernatant is then collected and added to the DMEM-based medium to prepare the rich growth medium for EDL myofiber cultures. The growth medium is then passed through a sterile 0.22-μm filter (to clear remaining particles and sterilize). All details of supplies for generating the medium are in Subheading 2.6. To ensure optimal cell growth conditions, we typically prepare only 250-ml medium each time, and use it up within a few weeks.

15. Methanol is a colorless flammable liquid with an alcohol-like odor. Use nitrile gloves, safety goggles, and a fume hood when handling. It is important to refer to the MSDS instructions and institutional regulations for further information regarding storage, handling, and first aid.
16. Paraformaldehyde is a white powder with a formaldehyde-like odor. It is a rapid fixative and a potential carcinogen. When handling paraformaldehyde, wear gloves, a mask, and goggles. It is important to refer to the MSDS instructions and institutional regulations for further information regarding storage, handling, and first aid.
17. Preparation of 100 ml of 4 % paraformaldehyde with 0.03 *M* sucrose: In a fume hood, mix 4 g of paraformaldehyde powder and 80 ml of deionized water in a glass beaker. Warm the solution to 60 °C with continuous stirring to dissolve the powder. Allow the solution to cool to room temperature. Add one to four drops of 1 *N* NaOH, until the opaque color of the solution clears. Add 10-ml 1 *M* sodium phosphate. Adjust the pH to 7.2–7.4 using concentrated HCl and color pH strips. Add 1.026 g of sucrose. Bring the volume to 100 ml and filter through a 0.22 μm disposable filter unit (Millipore) into a bottle. Store at 4 °C in an aluminum foil-wrapped bottle for no more than 1 month.
18. Preparation 1 L of 10× Tris-buffered saline (TBS): Weigh 60.5 g of Tris-Base into a beaker and add 700-ml deionized water. Stir on a magnetic stirrer until the powder has dissolved and adjust the pH to 7.4. Add deionized water to bring the volume up to 1 L, mix well, then autoclave or sterilize by passing it through a 0.22-μm filter, and store at 4 °C. To make 1 L

of 1×TBS, weigh 8.766 g NaCl into a beaker and add 100 ml of 10× TB. Mix vigorously until the powder has dissolved. Add deionized water to bring the volume up to 1 L, mix well, then sterile filter, and store at 4 °C.

19. DAPI is potentially harmful. Avoid prolonged or repeated exposure. We typically dissolve the entire powder in its original container and generate a concentrated stock solution. Alternatively, a ready-made DAPI reagent is available from Molecular Probes. It is important to refer to the MSDS instructions and institutional regulations for further information regarding storage and handling.
20. For additional details about the FDB muscle anatomy, refer to [102, 103]. For additional details about the EDL muscle anatomy, refer to [104–106]. The cited references are web links, recommended as good resources for anatomical description, and schematic images of the muscles although they refer to human muscles.
21. Collagenase concentration, as well as the optimal time for enzymatic digestion, should be adjusted for younger or older mice and for other muscle groups. The enzyme sold by Sigma-Aldrich tends to have consistent specific activity between batches, but attention should be given to the specific activity with each batch. The volume needed for the preparation should be evaluated based on the size of the tissue, so that the tissue will be fully covered by the collagenase solution (e.g., 1.5 ml is sufficient to cover an EDL muscle, but more will be needed to cover a TA).
22. The time required for the myofiber suspension to settle (at 1 × *g*) through 10 ml of 10 % HS can vary between 5 and 15 min, and the investigator should adjust this time accordingly. A prolonged settling period results in a preparation with more debris and residual single-cell carryover (not necessarily myofiber associated) released from the digested tissue. Depending on mouse age, the number of rounds of myofiber settling in the 15-ml glass Corex tubes, as well as the amount of medium in the tube, may also need to be adjusted.
23. The working Matrigel solution can be used to coat additional trays after completing the first tray coating. Matrigel that has been used to coat too many dishes, however, is less effective in supporting myofiber adhesion. We typically limit reuse of diluted Matrigel to three rounds of coating. Also, we only use Matrigel that has been diluted the day of the fiber isolation to maintain consistency.
24. The tissue culture dishes are dry when the bottom appears opaque white.
25. For some antibodies the cultures may be blocked for just 2–4 h at room temperature if overnight blocking is not desired.

26. For even and continuous distribution of the antibodies (both primary and secondary), it is recommended to place the dishes on a gyrating platform rotator (e.g., Lab-Line Maxi Rotator) when staining cultures in 24-well multiwell dishes; without this agitation, the antibody solution tends to rapidly accumulate at the well periphery, leading to uneven staining across the culture.

Acknowledgments

The methods and results summarized in the current chapter reflect the culmination of continuous updates during many years of research. Our current research is supported by grants to Z.Y.R. from the National Institutes of Health (AG035377, NS090051, NS088804). Pascal Stuelsatz is supported by an AFM-telethon fellowship (#18574). The authors are additionally grateful to the granting agencies (MDA, AHA, USDA-NRI, NIH) that have funded this research in the past and to our former members of our laboratory (Antony Rivera, Gabi Shefer, Andrew Shearer, and Elena Danoviz) for their valuable contributions.

References

1. Mauro A (1961) Satellite cell of skeletal muscle fibers. J Biophys Biochem Cytol 9:493–495
2. Yablonka-Reuveni Z (2011) The skeletal muscle satellite cell: still young and fascinating at 50. J Histochem Cytochem 59:1041–1059
3. Montarras D, L'Honore A, Buckingham M (2013) Lying low but ready for action: the quiescent muscle satellite cell. FEBS J 280: 4036–4050
4. Moss FP, Leblond CP (1971) Satellite cells as the source of nuclei in muscles of growing rats. Anat Rec 170:421–435
5. Schultz E, Gibson MC, Champion T (1978) Satellite cells are mitotically quiescent in mature mouse muscle: an EM and radioautographic study. J Exp Zool 206:451–456
6. White RB, Bierinx AS, Gnocchi VF, Zammit PS (2010) Dynamics of muscle fibre growth during postnatal mouse development. BMC Dev Biol 10:21
7. Fry CS, Lee JD, Mula J, Kirby TJ, Jackson JR, Liu F, Yang L, Mendias CL, Dupont-Versteegden EE, McCarthy JJ, Peterson CA (2015) Inducible depletion of satellite cells in adult, sedentary mice impairs muscle regenerative capacity without affecting sarcopenia. Nat Med 21:76–80
8. Keefe AC, Lawson JA, Flygare SD, Fox ZD, Colasanto MP, Mathew SJ, Yandell M, Kardon G (2015) Muscle stem cells contribute to myofibres in sedentary adult mice. Nat Commun 6:7087
9. Hawke TJ, Garry DJ (2001) Myogenic satellite cells: physiology to molecular biology. J Appl Physiol 91:534–551
10. Zammit PS, Partridge TA, Yablonka-Reuveni Z (2006) The skeletal muscle satellite cell: the stem cell that came in from the cold. J Histochem Cytochem 54:1177–1191
11. Lepper C, Partridge TA, Fan CM (2011) An absolute requirement for Pax7-positive satellite cells in acute injury-induced skeletal muscle regeneration. Development 138: 3639–3646
12. Grounds MD, Yablonka-Reuveni Z (1993) Molecular and cell biology of skeletal muscle regeneration. Mol Cell Biol Hum Dis Ser 3:210–256
13. Day K, Shefer G, Shearer A, Yablonka-Reuveni Z (2010) The depletion of skeletal muscle satellite cells with age is concomitant with reduced capacity of single progenitors to produce reserve progeny. Dev Biol 340: 330–343
14. Sacco A, Doyonnas R, Kraft P, Vitorovic S, Blau HM (2008) Self-renewal and expansion of single transplanted muscle stem cells. Nature 456:502–506

15. Dumont NA, Wang YX, Rudnicki MA (2015) Intrinsic and extrinsic mechanisms regulating satellite cell function. Development 142: 1572–1581
16. Shefer G, Yablonka-Reuveni Z (2008) Ins and outs of satellite cell myogenesis: the role of the ruling growth factors. In: Schiaffino S, Partridge T (eds) Skeletal muscle repair and regeneration, Advances in muscle research, vol 3. Springer, Dordrecht, Netherlands, pp 107–144
17. Morgan JE, Zammit PS (2010) Direct effects of the pathogenic mutation on satellite cell function in muscular dystrophy. Exp Cell Res 316:3100–3108
18. Yablonka-Reuveni Z, Day K (2010) Skeletal muscle stem cells in the spotlight: the satellite cell. In: Cohen I, Gaudette G (eds) Regenerating the heart: stem cells and the cardiovascular system, Stem cell biology and regenerative medicine series. Humana Press, Springer, pp 173–200
19. Muir AR, Kanji AH, Allbrook D (1965) The structure of the satellite cells in skeletal muscle. J Anat 99:435–444
20. Yablonka-Reuveni Z (1995) Development and postnatal regulation of adult myoblasts. Microsc Res Tech 30:366–380
21. Boldrin L, Muntoni F, Morgan JE (2010) Are human and mouse satellite cells really the same? J Histochem Cytochem 58:941–955
22. Seale P, Sabourin LA, Girgis-Gabardo A, Mansouri A, Gruss P, Rudnicki MA (2000) Pax7 is required for the specification of myogenic satellite cells. Cell 102:777–786
23. Kawakami A, Kimura-Kawakami M, Nomura T, Fujisawa H (1997) Distributions of PAX6 and PAX7 proteins suggest their involvement in both early and late phases of chick brain development. Mech Dev 66:119–130
24. Shefer G, Van de Mark DP, Richardson JB, Yablonka-Reuveni Z (2006) Satellite-cell pool size does matter: defining the myogenic potency of aging skeletal muscle. Dev Biol 294:50–66
25. Shefer G, Rauner G, Yablonka-Reuveni Z, Benayahu D (2010) Reduced satellite cell numbers and myogenic capacity in aging can be alleviated by endurance exercise. PLoS One 5:e13307
26. Allouh MZ, Yablonka-Reuveni Z, Rosser BW (2008) Pax7 reveals a greater frequency and concentration of satellite cells at the ends of growing skeletal muscle fibers. J Histochem Cytochem 56:77–87
27. Lindstrom M, Thornell LE (2009) New multiple labelling method for improved satellite cell identification in human muscle: application to a cohort of power-lifters and sedentary men. Histochem Cell Biol 132:141–157
28. Reimann J, Brimah K, Schroder R, Wernig A, Beauchamp JR, Partridge TA (2004) Pax7 distribution in human skeletal muscle biopsies and myogenic tissue cultures. Cell Tissue Res 315:233–242
29. Montarras D, Morgan J, Collins C, Relaix F, Zaffran S, Cumano A, Partridge T, Buckingham M (2005) Direct isolation of satellite cells for skeletal muscle regeneration. Science 309:2064–2067
30. Beauchamp JR, Heslop L, Yu DS, Tajbakhsh S, Kelly RG, Wernig A, Buckingham ME, Partridge TA, Zammit PS (2000) Expression of CD34 and Myf5 defines the majority of quiescent adult skeletal muscle satellite cells. J Cell Biol 151:1221–1234
31. Day K, Shefer G, Richardson JB, Enikolopov G, Yablonka-Reuveni Z (2007) Nestin-GFP reporter expression defines the quiescent state of skeletal muscle satellite cells. Dev Biol 304:246–259
32. Stuelsatz P, Shearer A, Li Y, Muir LA, Ieronimakis N, Shen QW, Kirillova I, Yablonka-Reuveni Z (2015) Extraocular muscle satellite cells are high performance myo-engines retaining efficient regenerative capacity in dystrophin deficiency. Dev Biol 397:31–44
33. Shefer G, Rauner G, Stuelsatz P, Benayahu D, Yablonka-Reuveni Z (2013) Moderate-intensity treadmill running promotes expansion of the satellite cell pool in young and old mice. FEBS J 280:4063–4073
34. Yablonka-Reuveni Z, Danoviz ME, Phelps M, Stuelsatz P (2015) Myogenic-specific ablation of Fgfr1 impairs FGF2-mediated proliferation of satellite cells at the myofiber niche but does not abolish the capacity for muscle regeneration. Front Aging Neurosci 7:85
35. Yablonka-Reuveni Z, Day K, Vine A, Shefer G (2008) Defining the transcriptional signature of skeletal muscle stem cells. J Anim Sci 86:E207–E216
36. Yablonka-Reuveni Z, Rivera AJ (1994) Temporal expression of regulatory and structural muscle proteins during myogenesis of satellite cells on isolated adult rat fibers. Dev Biol 164:588–603
37. Zammit PS, Golding JP, Nagata Y, Hudon V, Partridge TA, Beauchamp JR (2004) Muscle satellite cells adopt divergent fates: a mechanism for self-renewal? J Cell Biol 166:347–357
38. Day K, Paterson B, Yablonka-Reuveni Z (2009) A distinct profile of myogenic regulatory factor detection within Pax7+ cells at S phase supports a unique role of Myf5 during posthatch chicken myogenesis. Dev Dyn 238:1001–1009
39. Halevy O, Piestun Y, Allouh MZ, Rosser BW, Rinkevich Y, Reshef R, Rozenboim I, Wleklinski-Lee M, Yablonka-Reuveni Z (2004)

Pattern of Pax7 expression during myogenesis in the posthatch chicken establishes a model for satellite cell differentiation and renewal. Dev Dyn 231:489–502
40. Collins CA, Olsen I, Zammit PS, Heslop L, Petrie A, Partridge TA, Morgan JE (2005) Stem cell function, self-renewal, and behavioral heterogeneity of cells from the adult muscle satellite cell niche. Cell 122:289–301
41. Yablonka-Reuveni Z (2004) Isolation and culture of myogenic stem cells. In: Lanza R, Blau D, Melton D et al (eds) Handbook of Stem Cells—Vol 2: Adult and Fetal Stem Cells. Elsevier, San Diego
42. Danoviz ME, Yablonka-Reuveni Z (2012) Skeletal muscle satellite cells: background and methods for isolation and analysis in a primary culture system. Methods Mol Biol 798:21–52
43. Yablonka-Reuveni Z, Quinn LS, Nameroff M (1987) Isolation and clonal analysis of satellite cells from chicken pectoralis muscle. Dev Biol 119:252–259
44. Kastner S, Elias MC, Rivera AJ, Yablonka-Reuveni Z (2000) Gene expression patterns of the fibroblast growth factors and their receptors during myogenesis of rat satellite cells. J Histochem Cytochem 48:1079–1096
45. Ieronimakis N, Balasundaram G, Rainey S, Srirangam K, Yablonka-Reuveni Z, Reyes M (2010) Absence of CD34 on murine skeletal muscle satellite cells marks a reversible state of activation during acute injury. PLoS One 5:e10920
46. Bekoff A, Betz W (1977) Properties of isolated adult rat muscle fibres maintained in tissue culture. J Physiol 271:537–547
47. Bischoff R (1986) Proliferation of muscle satellite cells on intact myofibers in culture. Dev Biol 115:129–139
48. Bischoff R (1989) Analysis of muscle regeneration using single myofibers in culture. Med Sci Sports Exerc 21:S164–S172
49. Yablonka-Reuveni Z, Rivera AJ (1997) Proliferative dynamics and the role of FGF2 during myogenesis of rat satellite cells on isolated fibers. Basic Appl Myol 7:189–202
50. Yablonka-Reuveni Z, Anderson JE (2006) Satellite cells from dystrophic (mdx) mice display accelerated differentiation in primary cultures and in isolated myofibers. Dev Dyn 235:203–212
51. Yablonka-Reuveni Z, Rudnicki MA, Rivera AJ, Primig M, Anderson JE, Natanson P (1999) The transition from proliferation to differentiation is delayed in satellite cells from mice lacking MyoD. Dev Biol 210:440–455
52. Rosenblatt JD, Lunt AI, Parry DJ, Partridge TA (1995) Culturing satellite cells from living single muscle fiber explants. In Vitro Cell Dev Biol Anim 31:773–779
53. Rosenblatt JD, Parry DJ, Partridge TA (1996) Phenotype of adult mouse muscle myoblasts reflects their fiber type of origin. Differentiation 60:39–45
54. Stuelsatz P, Keire P, Almuly R, Yablonka-Reuveni Z (2012) A contemporary atlas of the mouse diaphragm: myogenicity, vascularity, and the Pax3 connection. J Histochem Cytochem 60:638–657
55. Yablonka-Reuveni Z, Seger R, Rivera AJ (1999) Fibroblast growth factor promotes recruitment of skeletal muscle satellite cells in young and old rats. J Histochem Cytochem 47:23–42
56. Shefer G, Partridge TA, Heslop L, Gross JG, Oron U, Halevy O (2002) Low-energy laser irradiation promotes the survival and cell cycle entry of skeletal muscle satellite cells. J Cell Sci 115:1461–1469
57. Mignone JL, Kukekov V, Chiang AS, Steindler D, Enikolopov G (2004) Neural stem and progenitor cells in nestin-GFP transgenic mice. J Comp Neurol 469:311–324
58. Tajbakhsh S, Rocancourt D, Buckingham M (1996) Muscle progenitor cells failing to respond to positional cues adopt non-myogenic fates in myf-5 null mice. Nature 384:266–270
59. Tajbakhsh S, Rocancourt D, Cossu G, Buckingham M (1997) Redefining the genetic hierarchies controlling skeletal myogenesis: Pax-3 and Myf-5 act upstream of MyoD. Cell 89:127–138
60. Kelly R, Alonso S, Tajbakhsh S, Cossu G, Buckingham M (1995) Myosin light chain 3F regulatory sequences confer regionalized cardiac and skeletal muscle expression in transgenic mice. J Cell Biol 129:383–396
61. Chapman VM, Miller DR, Armstrong D, Caskey CT (1989) Recovery of induced mutations for X chromosome-linked muscular dystrophy in mice. Proc Natl Acad Sci U S A 86:1292–1296
62. Im WB, Phelps SF, Copen EH, Adams EG, Slightom JL, Chamberlain JS (1996) Differential expression of dystrophin isoforms in strains of mdx mice with different mutations. Hum Mol Genet 5:1149–1153
63. Banks GB, Combs AC, Chamberlain JS (2010) Sequencing protocols to genotype mdx, mdx(4cv), and mdx(5cv) mice. Muscle Nerve 42:268–270
64. Lu QL, Morris GE, Wilton SD, Ly T, Artem'yeva OV, Strong P, Partridge TA (2000) Massive idiosyncratic exon skipping corrects the nonsense mutation in dystrophic mouse muscle and produces functional revertant fibers by clonal expansion. J Cell Biol 148:985–996
65. Arpke RW, Darabi R, Mader TL, Zhang Y, Toyama A, Lonetree CL, Nash N, Lowe DA, Perlingeiro RC, Kyba M (2013) A new

immuno-, dystrophin-deficient model, the NSG-mdx(4Cv) mouse, provides evidence for functional improvement following allogeneic satellite cell transplantation. Stem Cells 31:1611–1620
66. Danko I, Chapman V, Wolff JA (1992) The frequency of revertants in mdx mouse genetic models for Duchenne muscular dystrophy. Pediatr Res 32:128–131
67. Decrouy A, Renaud JM, Davis HL, Lunde JA, Dickson G, Jasmin BJ (1997) Mini-dystrophin gene transfer in mdx4cv diaphragm muscle fibers increases sarcolemmal stability. Gene Ther 4:401–408
68. Judge LM, Haraguchiln M, Chamberlain JS (2006) Dissecting the signaling and mechanical functions of the dystrophin-glycoprotein complex. J Cell Sci 119:1537–1546
69. Dias P, Parham DM, Shapiro DN, Tapscott SJ, Houghton PJ (1992) Monoclonal antibodies to the myogenic regulatory protein MyoD1: epitope mapping and diagnostic utility. Cancer Res 52:6431–6439
70. Wright WE, Binder M, Funk W (1991) Cyclic amplification and selection of targets (CASTing) for the myogenin consensus binding site. Mol Cell Biol 11:4104–4110
71. Wright WE, Dac-Korytko I, Farmer K (1996) Monoclonal antimyogenin antibodies define epitopes outside the bHLH domain where binding interferes with protein-protein and protein-DNA interactions. Dev Genet 19: 131–138
72. Shefer G, Wleklinski-Lee M, Yablonka-Reuveni Z (2004) Skeletal muscle satellite cells can spontaneously enter an alternative mesenchymal pathway. J Cell Sci 117: 5393–5404
73. Anderson JE, McIntosh LM, Moor AN, Yablonka-Reuveni Z (1998) Levels of MyoD protein expression following injury of mdx and normal limb muscle are modified by thyroid hormone. J Histochem Cytochem 46:59–67
74. Gause KC, Homma MK, Licciardi KA, Seger R, Ahn NG, Peterson MJ, Krebs EG, Meier KE (1993) Effects of phorbol ester on mitogen-activated protein kinase kinase activity in wild-type and phorbol ester-resistant EL4 thymoma cells. J Biol Chem 268:16124–16129
75. Seger R, Seger D, Reszka AA, Munar ES, Eldar-Finkelman H, Dobrowolska G, Jensen AM, Campbell JS, Fischer EH, Krebs EG (1994) Overexpression of mitogen-activated protein kinase kinase (MAPKK) and its mutants in NIH 3T3 cells. Evidence that MAPKK involvement in cellular proliferation is regulated by phosphorylation of serine residues in its kinase subdomains VII and VIII. J Biol Chem 269:25699–25709
76. Greene EC (1963) Anatomy of the rat. Hafner Publishing Company, New York, NY
77. Wozniak AC, Pilipowicz O, Yablonka-Reuveni Z, Greenway S, Craven S, Scott E, Anderson JE (2003) C-Met expression and mechanical activation of satellite cells on cultured muscle fibers. J Histochem Cytochem 51:1437–1445
78. Michalczyk K, Ziman M (2005) Nestin structure and predicted function in cellular cytoskeletal organisation. Histol Histopathol 20:665–671
79. Wiese C, Rolletschek A, Kania G, Blyszczuk P, Tarasov KV, Tarasova Y, Wersto RP, Boheler KR, Wobus AM (2004) Nestin expression—a property of multi-lineage progenitor cells? Cell Mol Life Sci 61:2510–2522
80. Shefer G, Wleklinski-Lee M, Yablonka-Reuveni Z (2004) Skeletal muscle satellite cells can spontaneously enter an alternative mesenchymal pathway. J Cell Sci 117: 5393–5404
81. Kania G, Blyszczuk P, Czyz J, Navarrete-Santos A, Wobus AM (2003) Differentiation of mouse embryonic stem cells into pancreatic and hepatic cells. Methods Enzymol 365:287–303
82. Vogel W, Grunebach F, Messam CA, Kanz L, Brugger W, Buhring HJ (2003) Heterogeneity among human bone marrow-derived mesenchymal stem cells and neural progenitor cells. Haematologica 88:126–133
83. Amoh Y, Li L, Katsuoka K, Penman S, Hoffman RM (2005) Multipotent nestin-positive, keratin-negative hair-follicle bulge stem cells can form neurons. Proc Natl Acad Sci U S A 102:5530–5534
84. Davidoff MS, Middendorff R, Enikolopov G, Riethmacher D, Holstein AF, Muller D (2004) Progenitor cells of the testosterone-producing Leydig cells revealed. J Cell Biol 167:935–944
85. Asakura A, Komaki M, Rudnicki M (2001) Muscle satellite cells are multipotential stem cells that exhibit myogenic, osteogenic, and adipogenic differentiation. Differentiation 68:245–253
86. Keire P, Shearer A, Shefer G, Yablonka-Reuveni Z (2013) Isolation and culture of skeletal muscle myofibers as a means to analyze satellite cells. Methods Mol Biol 946: 431–468
87. Shefer G, Yablonka-Reuveni Z (2007) Reflections on lineage potential of skeletal muscle satellite cells: do they sometimes go MAD? Crit Rev Eukaryot Gene Expr 17:13–29
88. Starkey JD, Yamamoto M, Yamamoto S, Goldhamer DJ (2011) Skeletal muscle satellite cells are committed to myogenesis and do not spontaneously adopt nonmyogenic fates. J Histochem Cytochem 59:33–46

89. Bischoff R (1986) A satellite cell mitogen from crushed adult muscle. Dev Biol 115: 140–147
90. Kuang S, Kuroda K, Le Grand F, Rudnicki MA (2007) Asymmetric self-renewal and commitment of satellite stem cells in muscle. Cell 129:999–1010
91. Cossu G, Tajbakhsh S (2007) Oriented cell divisions and muscle satellite cell heterogeneity. Cell 129:859–861
92. Yablonka-Reuveni Z, Christ B, Benson JM (1998) Transitions in cell organization and in expression of contractile and extracellular matrix proteins during development of chicken aortic smooth muscle: evidence for a complex spatial and temporal differentiation program. Anat Embryol (Berl) 197:421–437
93. Yablonka-Reuveni Z, Nameroff M (1990) Temporal differences in desmin expression between myoblasts from embryonic and adult chicken skeletal muscle. Differentiation 45:21–28
94. Yablonka-Reuveni Z, Schwartz SM, Christ B (1995) Development of chicken aortic smooth muscle: expression of cytoskeletal and basement membrane proteins defines two distinct cell phenotypes emerging from a common lineage. Cell Mol Biol Res 41:241–249
95. Ono Y, Boldrin L, Knopp P, Morgan JE, Zammit PS (2010) Muscle satellite cells are a functionally heterogeneous population in both somite-derived and branchiomeric muscles. Dev Biol 337:29–41
96. Zammit PS, Heslop L, Hudon V, Rosenblatt JD, Tajbakhsh S, Buckingham ME, Beauchamp JR, Partridge TA (2002) Kinetics of myoblast proliferation show that resident satellite cells are competent to fully regenerate skeletal muscle fibers. Exp Cell Res 281:39–49
97. Stuelsatz P, Yablonka-Reuveni Z (2016) Isolation of mouse periocular tissue for histological and immunostaining analyses of the extraocular muscles and their satellite cells. Methods Mol Biol 1460:101-127.
98. Kleinman HK, McGarvey ML, Liotta LA, Robey PG, Tryggvason K, Martin GR (1982) Isolation and characterization of type IV procollagen, laminin, and heparan sulfate proteoglycan from the EHS sarcoma. Biochemistry 21:6188–6193
99. Yablonka-Reuveni Z, Seifert RA (1993) Proliferation of chicken myoblasts is regulated by specific isoforms of platelet-derived growth factor: evidence for differences between myoblasts from mid and late stages of embryogenesis. Dev Biol 156:307–318
100. Yablonka-Reuveni Z (1995) Myogenesis in the chicken: the onset of differentiation of adult myoblasts is influenced by tissue factors. Basic Appl Myol 5:33–42
101. O'Neill MC, Stockdale FE (1972) A kinetic analysis of myogenesis in vitro. J Cell Biol 52:52–65
102. Gray, H. (1918) The muscles and fasciæ of the foot. In: Anatomy of the human body. Available via Bartleby.com. http://www.bartleby.com/107/131.html. Accessed 17 Nov 2016
103. Gray, H. (1918) Fig. 443. In: Anatomy of the human body. Available via Bartleby.com. http://www.bartleby.com/107/illus443.html. Accessed 17 Nov 2016
104. Gray, H. (1918) Fig. 437. In: Anatomy of the human body. Available via Bartleby.com. http://www.bartleby.com/107/illus437.html. Accessed 17 Nov 2016
105. Gray, H. (1918) Fig. 441. In: Anatomy of the Human Body. Available via Bartleby.com. http://www.bartleby.com/107/illus441.html. Accessed 17 Nov 2016
106. Gray, H. (1918) The muscles and fasciæ of the leg. In: Anatomy of the Human Body. Available via Bartleby.com. http://www.bartleby.com/107/129.html. Accessed 17 Nov 2016
107. Shefer G, Yablonka-Reuveni Z (2005) Isolation and culture of skeletal muscle myofibers as a means to analyze satellite cells. Methods Mol Biol 290:281–304
108. Birbrair A, Wang ZM, Messi ML, Enikolopov GN, Delbono O (2011) Nestin-GFP transgene reveals neural precursor cells in adult skeletal muscle. PLoS One 6:e16816

Chapter 5

Characterization of *Drosophila* Muscle Stem Cell-Like Adult Muscle Precursors

Guillaume Lavergne, Cedric Soler, Monika Zmojdzian, and Krzysztof Jagla

Abstract

Uncovering how muscle stem cells behave in quiescent and activated states is central to understand the basic rules governing normal muscle development and regeneration in pathological conditions. Specification of mesodermal lineages including muscle stemlike adult muscle precursors (AMPs) has been extensively studied in *Drosophila* providing an attractive framework for investigating muscle stem cell properties. Restricted number of AMP cells, relative ease in following their behavior, and large number of genetic tools available make fruit fly an attractive model system for studying muscle stem cells. In this chapter, we describe the recently developed tools to visualize and target the body wall and imaginal AMPs.

Key words *Drosophila*, Adult muscle precursors (AMPs), Muscle stem cell, Muscle development, Imaging

1 Introduction

The development of *Drosophila* muscles is a complex process leading in fine to the formation of adult fly musculature. Two distinct waves of myogenesis take place during the life cycle of *Drosophila*. The first embryonic wave leads to the formation of larval body wall muscles and to specification of a small population of myoblasts called adult muscle precursors (aPs/AMPs) that keep an undifferentiated state and share several features with vertebrate muscle stem cells [1, 2].

Two populations of AMPs can be distinguished: abdominal AMPs and thoracic AMPs. They differ by their anatomical position and the adult muscle type they give rise to. In the thoracic segments, the AMPs gather in cluster of six to seven ventrally located cells associated with the primordia of imaginal discs. In each abdominal hemisegment, six AMPs are present in a stereotyped pattern of one dorsal AMP, two dorsolateral AMPs, two lateral AMPs, and one ventral AMP (Fig. 1). The specified abdominal AMPs stay in

Eusebio Perdiguero and DDW Cornelison (eds.), *Muscle Stem Cells: Methods and Protocols*, Methods in Molecular Biology, vol. 1556, DOI 10.1007/978-1-4939-6771-1_5, © Springer Science+Business Media LLC 2017

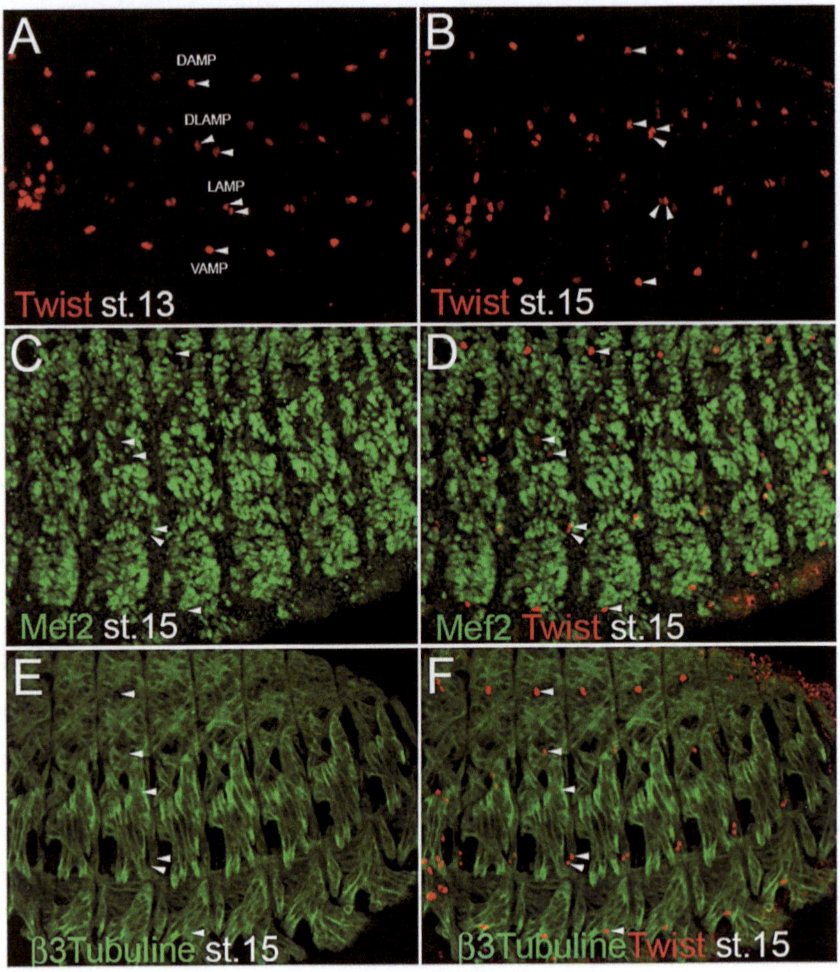

Fig. 1 Non-differentiated, muscle-committed fate of AMPs. (**a**) Lateral views of stage 13 and (**b**) stage 15 embryos stained for myogenic factor Twi. *Arrowheads* point to Twi-positive abdominal adult muscle precursors (six per hemisegment). DAMP, dorsal AMP; DLAMP, dorsolateral AMPs; LAMP, lateral AMP; and VAMP, ventral AMP. Twi-expressing AMPs are Mef2 negative (**c**, **d**) and β3-tubulin negative (*arrowheads*) indicating their non-differentiated fate

close vicinity of specific muscle fibers and are characterized by the persistent expression of the basic helix-loop-helix transcription factor Twist (Twi) [3] and the activation of the Notch pathway. The cooperation between Notch and Twist results in the expression of two known repressors of muscle differentiation: Holes in muscle (Him) [4] and Zinc finger homeodomain 1 (Zfh1) [5].

We took advantage from the fact that the Notch pathway is active in AMPs by using the Notch-responsive element of *E(spl) M6* gene to drive membrane-targeted GFP and to follow AMP cells in vivo. We observed that the embryonic AMPs send long cellular processes, which make AMPs interconnected [6] but also short filopodia that make them connected to the surrounding

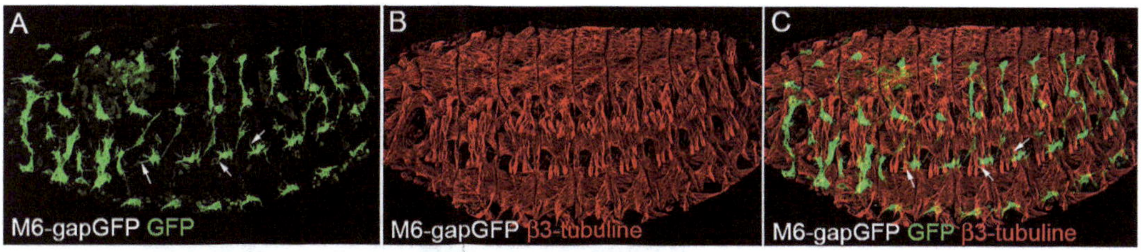

Fig. 2 AMPs send filopodia to contact neighboring muscles. (**a–c**) Lateral view of stage 15 M6-gapGFP–AMP-sensor embryo stained for GFP (**a**) to reveal AMP cell shapes and for β3-tubulin (**b**) to reveal body wall muscle network. (**c**) A merged view. *Arrows* point to filopodia that extend from lateral AMPs to contact lateral muscles

muscles (Fig. 2). The capacity to produce cellular protrusions and make connections has also been described for quiescent satellite cells sitting on muscle fibers [7]. Thus, studying AMPs in embryos could provide insights into how muscle stem cells recognize their muscle niche and how stem cell homing process is regulated.

On the other hand, in middle of the second larval instar, the quiescent AMPs get reactivated and start to proliferate to generate a pool of myoblasts required for the second wave of myogenesis and the formation of adult fly muscles. Accordingly, abdominal AMPs are at the origin of all the adult abdominal muscles, whereas thoracic AMPs associated with wing and leg imaginal discs give rise to flight and leg muscles [3]. Studying reactivation of AMPs and their exit from the quiescent state could help in understanding how homologous processes are regulated in vertebrate muscle stem cells. Here we provide some tips to detect AMPs at the onset of their reactivation in the second instar larva (Fig. 3).

As mentioned above, in the thoracic segments, most AMPs are associated with wing and leg imaginal discs. These cells, known as imaginal myoblasts or adepithelial cells, give rise to flight and leg muscles [3]. In both types of discs, imaginal myoblasts reside at precise positions [3, 8]. For instance, in wing imaginal discs, they are positioned in the notum part of the disc since the second instar. Most of the thoracic muscles, including leg muscles, direct flight muscles (DFM), and dorsoventral muscles (DVMs), are formed de novo. However, a subset of indirect flight muscles (IFMs), the dorsal longitudinal muscles (DLMs), uses larval templates for development [9, 10]. The DLMs are formed by a regeneration-like process consisting of fusion of imaginal AMPs from the wing disc with persisting larval oblique muscles (LOMs). This leads to remodeling of the three LOMs into six adult DLMs. Thus, like mammalian satellite cells, the AMPs contribute to the regeneration and remodeling of preexisting muscles.

Various genes are expressed in myoblasts associated with imaginal discs, sometimes long before they start to differentiate, and could thus be used as their markers. Several genes are expressed in all adepithelial cells. This is the case of Twist, maintained in all

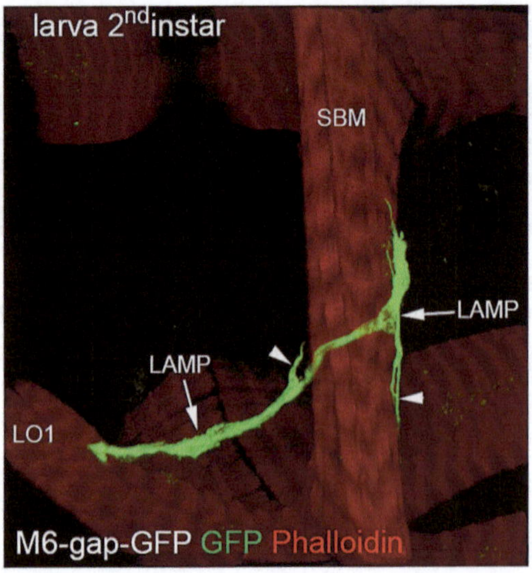

Fig. 3 Second larval instar AMPs keep filopodia-based connections with muscles. Two lateral AMPs revealed by M6-gapGFP sensor are shown (*arrows*) associated with neighboring segment border muscle (SBM) and lateral oblique 1 (LO1) muscle. *Arrowheads* point to filopodia extending from the AMP cell bodies and connecting the SBM

AMPs from embryonic stages, and Mef2 [3, 11]. Mef2 is expressed from the late third instar, but its activity is restrained until several hours after pupae formation (APF) by its inhibitor Him [4, 12]. Many genes, such as *cut*, *vestigial*, *scalloped*, or *ladybird*, involved in muscle specification in the embryo are used again to specify adult muscle precursors [13–16]. These genes are differentially expressed in myoblasts associated to leg and wing discs and mark the distinct subsets of thoracic AMPs (Fig. 4).

Taken together, the described characteristics of AMPs illustrate the diversity of myogenic processes to which they contribute through *Drosophila* development. Below we provide a set of tools that allow to visualize and to target the AMPs during embryogenesis and during larval stages.

2 Materials

2.1 *Drosophila* Stocks

The embryonic/larval AMP sensor line M6-gapGFP and the driver M6-GAL4 were generated in our lab [6]. Note that many useful *Drosophila* lines are already available at Bloomington Drosophila Stock Center.

2.2 Embryo Collection and Dechorionation

1. Egg collection cages: 100 ml plastic beakers with pierced holes on the top attached to freshly prepared small plastic Petri dishes plated with grape juice (50 % grape juice, 2.5 % agar)

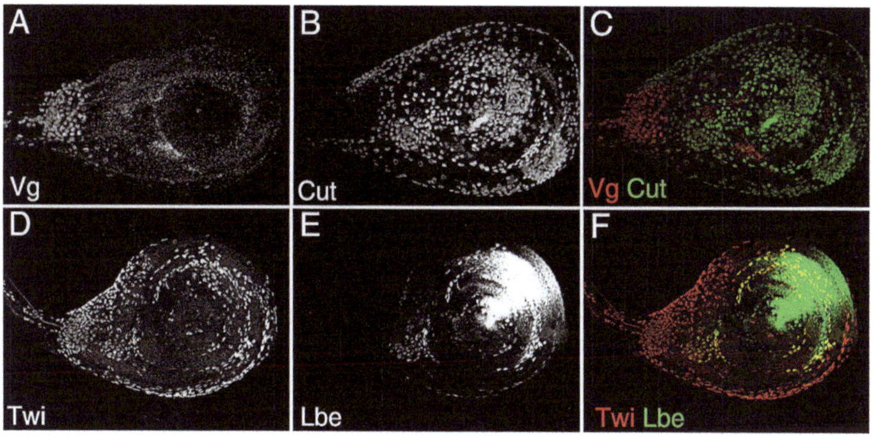

Fig. 4 Markers of AMPs associated with leg imaginal discs. (**a–c**) Two distinct subsets of leg disc AMPs are detected by Vg and Cut transcription factors. Vg labels AMPs in the stalk region of the disc, whereas Cut-positive AMPs are associated with the leg disc proper. (**d–f**) Twist marks all imaginal AMPs, a subset of which expresses homeodomain transcription factor Lbe

and covered with yeast paste (dry baker's yeast mixed with water till obtaining a paste consistency).

2. Plastic baskets prepared from cutoff 15 ml Falcon tubes with Nitex® mesh screwed by a cap cut in the center to allow embryo collection.
3. 3 % commercial bleach.
4. Paint brushes and distilled water in squirt bottle.

2.3 In Vivo Imaging of Embryonic AMPs

1. Dissecting microscope.
2. Agar (50 % grape juice, 5 % agar) and needles with handle.
3. Technau glue: heptane/glue mix.
4. 22 × 40 mm coverslips.
5. Halocarbon oil (Voltalef 10S).

2.4 Larvae Dissection

1. Dissecting microscope.
2. Sylgard® dish and three well glass dish.
3. Dissection solution: phosphate-buffered saline (PBS 1×, pH 7.4) with 25 mM EDTA.
4. Insect pins® 0.2 mm diameter (Fine Science Tools).
5. Sharp forceps (Dumont #55) and scissors (Fine Science Tools).

2.5 Immunostaining of Embryonic and Larval AMPs

1. Fixative solutions:

   4 % formaldehyde, 80 mM PIPES pH 7.4, 1.6 mM EGTA, and 0.8 mM $MgSO_4$ (modified 4 %) FA.

   4 % formaldehyde in PBS 1× (4 % FA).

   4 % paraformaldehyde in PBS 1× (4 % PFA).

2. Heptane, methanol, and ethanol.
3. Rocking platform.
4. PBS 1×, pH 7.4.
5. 0.01 % Triton-X-100 PBS 1× (0.01 % PBT).
6. 0.1 % Triton-X-100 PBS 1× (0.1 % PBT).
7. 0.5 % Triton-X-100 PBS 1× (0.5 % PBT).
8. Blocking solution: 10 % horse serum in PBT.
9. Fluoromount-G® mounting medium.

2.6 Detection and Markers of AMPs

2.6.1 Expression of Fluorescent Proteins

A common way for expressing a variety of fluorescent proteins in a tissue-specific manner is to use the GAL4/UAS system [17]. For visualization of AMPs, we dispose several tools relying on this system. We developed a new M6-GAL4 line [6] driving expression of UAS reporters specifically in all embryonic AMPs till the end of the third larval instar. While M6-GAL4 expression is decreasing, other lines can be used to follow larval AMPs. The 1151-GAL4 line is the most common line used for driving expression of fluorescent protein in imaginal myoblasts [8, 18]. As this driver does not drive expression during embryogenesis, it is also useful for expressing any molecule without affecting embryonic development. Other known Gal4 driver lines can be used such as 24B-Gal4 and Mef2-Gal4 [19, 20] to follow myogenic progression of myoblasts derived from AMPs. Fluorescent proteins expressed in imaginal myoblasts can be visualized on living or fixed discs without performing any immunostaining. On the contrary, an immunostaining is required to visualize fluorescent proteins in fixed preparation of embryos due to the alcohol-based fixation method.

2.6.2 Markers of Embryonic AMPs

Expression of several genes can be used to selectively mark the various types of adult muscle precursors at each stage of development (listed in Table 1). During embryogenesis, AMPs are characterized by the expression of Twist (Twi), which is the most common marker used to visualize them. Shared markers of embryonic AMPs also include the transcription factors Cut [6] and Zfh1 [21] as well as the Mef2 inhibitor Him [4]. However, AMPs are a heterogeneous population as illustrated by the expression of different identity genes in specific AMP subpopulations. Within the abdominal segments, the AMPs display differential expression of identity markers depending on their dorsoventral position. The lateral AMPs express the transcription factors Ladybird (Lb) [15] and Krüppel (Kr) [6, 22], whereas the ventral AMPs express Slouch (Slou) and Pox meso [23, 24]. Starting from the embryonic stage 12, a subpopulation of thoracic AMPs that expresses Vestigial (Vg) can also be distinguished [13].

Table 1
General markers of embryonic and imaginal disc associated with AMP cells

Genes	Embryonic AMP expression	Adepithelial cell expression	Vertebrate orthologue	Antibody	Reference/source
Twist	All AMPs	All AMPs of wing and leg discs	Twi	Rabbit (1/500)	Figeac et al. 2010 [6]
Mef2	Not expressed	All AMPs of wing and leg discs	Mef2	Rabbit (1/1000)	Lilly et al. 1995 [28]
Vg	Subset of thoracic AMPs	In wing disc, large group of proximal AMPs; in leg disc, small group of most proximal AMPs	VGLL	Rabbit (1/200)	Kim et al. 1996 [22]
Cut	All AMPs	In wing disc, small distal group of AMPs; in leg disc, most AMPs	Cux2	Mouse (1/100)	DHSB
Lb	Lateral AMPs	All AMPs associated to leg discs (except most proximal region); ventral myoblasts express a higher level than dorsal ones	Lbx	Mouse (1/200)	Jagla et al. 1997 [29]

2.6.3 Markers of Imaginal Myoblasts Associated with Leg and Wing Discs

Twist expression is maintained in all AMPs, including imaginal myoblasts, and is detectable in both leg and wing discs associated with myoblasts until several hours after pupae formation [3]. Mef2, a key factor of differentiation in embryo and adult myogenesis, is first expressed from the very late third instar and is maintained until adult stages [11, 25, 26]. Embryonic identity genes are also required in adult muscle diversification and can be used as specific markers to detect different populations of AMPs associated with imaginal discs. In wing disc, two populations of AMPs are distinguishable, Vestigial (Vg), and low levels of Cut are expressed in myoblasts giving rise to the IFMs, whereas myoblasts required for DFM development show a high level of Cut but no Vg [13]. In leg disc, Vg is expressed in a small subset of myoblasts associated with the dorsal proximal region and very likely participating to thoracic flight muscle. Other leg myoblasts expressing a high level of Cut [14]; (Fig. 4a–c) are at the origin of all leg muscles. We also found that Ladybird early (Lbe) [15], an orthologue of Lbx1, a key regulator of appendicular myogenesis in vertebrates [27], is expressed in leg myoblasts. Lbe is initially detected at the early third instar in a small subset of leg myoblasts, and then its expression extends progressively to all myoblasts (except most proximal myoblasts expressing Vg). However, dorsal myoblasts are characterized by a lower level of Lbe expression compared to ventral myoblasts [14] (Fig. 4d–f).

3 Methods

3.1 Embryo Collection and Dechorionation

1. Set up the cross in fly food vial with about 40 females and 10–20 males for 2 days at 25 °C, and then transfer the flies to the egg collection cage.
2. Collect embryos after overnight egg laying.
3. Transfer embryos into a plastic basket using distilled water and a paintbrush. Then wash with distilled water to remove yeast paste residuals.
4. Proceed to dechorionation by transferring the basket-containing embryos into a 3 % bleach solution for 3 min. Ensure that embryos are well immerse during this step. Dechorionated embryos should float on the surface (*see* **Note 1**).
5. Thoroughly rinse the embryos with distilled water. Place the basket on a paper towel to absorb water prior to perform in vivo confocal microscopy (*see* Subheading 3.3) or immunostaining (*see* Subheading 3.5).

3.2 Second and Third Instar Larva Collection

Proper staging of larvae is critical for the accuracy of following experiments, especially when studying reactivation of AMPs. Here we describe an easy way to ensure the recovery of staged larvae:

1. Set up the cross as mentioned in Subheading 3.1, **step 1** (e.g., for the weekend).
2. The day of egg collection: synchronize the eggs laying by changing the plates 2–3 times with 2 h interval.
3. Collect the plates with embryos after 2 h of laying (T0–T2).
4. Allow the embryos to develop at 25 °C.
5. To ensure proper staging of larvae, at 22 h from T0, remove all hatched larvae.
6. At 24 h from T0, collect all hatched larvae (first larval instar, first larval day), and put on a fresh plate enriched in yeast/fly food*.
7. Allow the larvae to develop for 66 h at 25 °C.
8. After 72–96 h (96–120 h from T0), collect the larvae (third instar) for imaginal disc staining (*see* Subheading 3.5) or larva body wall staining (*see* Subheading 3.6).
   *For second instar larvae:

7 Allow the larvae to develop for 26 h at 25 °C.

8 After 26 h (50 h from T0), collect the larva (beginning of the second instar, second larval day) for dissection.

3.3 In Vivo Confocal Microscopy of Embryonic AMPs

One of the advantages of *Drosophila* model is the possibility to easily follow embryonic development by live imaging. Nevertheless, special care should be taken with the choice of fluorescent proteins and settings when performing in vivo confocal microscopy (*see* **Note 2**):

1. Transfer the dechorionated embryos with a paintbrush onto a slice of agar laid on a glass slide.
2. Under a dissecting microscope, use a needle with handle to carefully align embryos of the desired stage (*see* **Note 3**).
3. Spread a drop of heptane/glue on the surface of a 22 × 40 mm coverslip, and let it dry. Transfer the embryos on the coverslip by delicately pressing the coverslip on the slice of agar.
4. Cover the embryos with Voltalef 10S oil.
5. Embryos are ready for live imaging.

3.4 Embryo Immunostaining

The protocol presented below is routinely used in our lab for most of immunostaining procedure. Please note that in some specific cases, some steps should be adapted (*see* **Notes 4** and **5**).

1. Prepare a mix of 600 μL of heptane and 600 μL of fixative solution (modified 4 % FA) in a 1.5 ml Eppendorf tube (*see* **Note 4**).
2. Transfer the dechorionated embryos with a paintbrush in the heptane/fixative solution.
3. Incubate for 20 min at room temperature on a rocking device.
4. After fixation, carefully remove the lower phase leaving only the embryos in the heptane phase. Add 600 μL of methanol, and shake vigorously during 4 min (*see* **Note 5**).
5. Let the embryos settle, and carefully remove the heptane and methanol phases. Embryos which stay between the two phases did not popped out their vitelline membrane and should be discarded.
6. Wash two times with methanol by letting embryos sink at the bottom of the tube, and discard the methanol.
7. Wash one time in ethanol 100 %. At this stage, dehydrated embryos can be stored in ethanol 100 % for several months at −20 °C.
8. Prior to immunostaining, embryos need to be rehydrated using ethanol/0.1 % PBT series. Incubate embryos in ethanol 70 %, then in ethanol 50 %, and finally in ethanol 30 % for 5 min each, at room temperature on a rocking device.
9. Rinse three times for 5 min with 0.1 % PBT at room temperature on a rocking device.
10. Remove the 0.1 % PBT, and block nonspecific protein-binding sites by adding the blocking solution (10 % horse serum in 0.1 % PBT) for at least 1 h at room temperature on a rocking device.

11. Remove the blocking solution, and add primary antibodies at recommended concentration in blocking solution (*see* **Note 6**).
12. Incubate overnight at 4 °C on a rocking device.
13. Wash with 0.1 % PBT quickly two times and then three times for 10 min on a rocking device.
14. Remove the 0.1 % PBT, and add appropriate secondary antibodies at recommended concentration in blocking solution. If secondary antibody is conjugated to light sensitive compound, keep the sample in the dark from this step.
15. Incubate 2 h at room temperature on a rocking device.
16. Repeat **step 13**, and then transfer embryos on slide in a drop of Fluoromount-G® mounting medium using a cutoff pipet cone.

3.5 Imaginal Disc Immunostaining

The following protocol describes the immunostaining of imaginal discs of the third instar larvae; for pupal stage discs, some precautions should be taken (*see* **Note 8**). All steps of fixation, blocking, washing, and antibody incubations are performed on a rocking device.

1. Collect the third instar larvae with a paintbrush (*see* **Note 8**).
2. Transfer collected larvae to a three-well clear glass dish filled with PBS 1×.
3. Pre-dissection: using forceps cut the larvae in half and invert the inside–outside tissue so that associated with the head region imaginal discs are outside, and remove the trachea and fat body.
4. Transfer inverted sections into 1.5 ml Eppendorf tube filled with fixative solution (4 % PFA), and incubate for 20–30 min at room temperature on a rocking device (*see* **Note 8**).
5. Rinse three times for 10 min with PBT 0.5 % at room temperature.
6. Remove the 0.5 % PBT, and add the blocking solution (10 % horse serum in 0.5 % PBT) for at least 1 h at room temperature.
7. Remove the blocking solution, and add primary antibodies at recommended concentration.
8. Incubate overnight at 4 °C.
9. Rinse three times with 0.5 % PBT for 10 min.
10. Add appropriate secondary antibodies diluted in blocking solution.
11. Incubate 2 h at room temperature.
12. Rinse three times for 10 min with PBS 1×.

13. Transfer immunostained sections back into three-well glass dish filled with PBS 1×.
14. Carefully dissect each disc with fine forceps, and transfer them on microscope slide in a drop of Fluoromount-G® mounting medium.

3.6 Second and Third Instar Larvae Immunostaining

To detect/visualize AMP cells in the second and third instar larvae, we use M6-gapGFP individuals which are dissected as described below and then stained with anti-GFP and/or anti-Twist antibodies. Muscle pattern is revealed using phalloidin staining (*see* Fig. 3).

1. Put the larva of the desired stage on a Sylgard® dish.
2. Using the forceps, pin the larva with the small pins in the anterior end and posterior end keeping the dorsal side up.
3. Pour the larva with a drop of dissection solution (PBS containing 25 mM EDTA), and with sharp scissors, incise the body wall close to the posterior end, and open the larva to the anterior end.
4. Using the forceps, gently remove all internal tissue (intestine, fat body, salivary gland, etc.).
5. Incise finely the body wall close to the pins, and using the forceps, pin the body wall making it flat.
6. Remove the brain and remaining tissues.
7. Rinse the larva in dissection solution.
8. Remove the remaining dissection solution, and add the fixative solution (4 % FA) for 20 min at room temperature.
9. Remove the fixation solution and rinse in 0.01 % PBT.
10. Transfer the dissected larva into 1.5 ml Eppendorf tube. Up to five dissected larvae can be pooled at this step.
11. Rinse three times for 5 min at room temperature.
12. Incubate the tissue in blocking solution (10 % horse serum in 0.01 % PBT) for 30 min at room temperature.
13. Incubate in primary antibody diluted in blocking solution at 4 °C overnight.
14. Rinse three times with 0.01 % PBT for 10min at room temperature.
15. Incubate in secondary antibody diluted with 0.01 % PBT for 2 h at room temperature.
16. Repeat **step 12**.
17. If needed to visualize body wall muscles, incubate with phalloidin, diluted in 0.01 % PBT for 20 min at room temperature in the dark.
18. Repeat **step 12**, and then mount on a slide in Fluoromount-G® mounting medium (*see* **Note 7**).

4 Notes

1. Proper removal of the chorion can be observed under a dissecting microscope. Exceeding 3 min of bleach exposure can affect normal architecture of the embryos.
2. Photobleaching during in vivo confocal imaging can occur when applying high-laser power, high-resolution, and low-speed scanning settings.
3. Experimented manipulator can easily recognize the different embryo stages under a dissecting microscope. To facilitate alignment of a defined stage embryo, it is also possible to set up short egg-laying window as mentioned in Subheading 3.2, **steps 2** and **3**, and then let embryos develop until the desired stage is reach.
4. Alternative fixative methods can be applied depending on the antigen targeted. For example, for immunostaining of the cytoskeleton, the use of 4 % paraformaldehyde, either in PBS or phosphate buffer fixative solutions, is recommended.
5. For some staining, like phalloidin, the devitellinization step requires to avoid the use of methanol. In this case, replace the methanol with 90 % ethanol and vortex for 1 min.
6. It is recommended to use a minimal volume of 200 μL when performing immunostaining in 1.5 ml Eppendorf tube.
7. When mounting the slide, take care to ensure that the body wall muscles are well oriented for proper visualization (e.g., on the upper side for inverted microscope).
8. After pupae formation (APF), dissection of imaginal discs is more delicate. As pupal case is a hard tissue and inside pressure increases with pupae age, it is preferable to open the pupae progressively and delicately from its center. After 2 h of APF, imaginal discs are not anymore strongly connected to surrounding tissues (e.g., brain, cuticle, etc.), and they tend to disperse; it is recommended to not transfer the samples into an Eppendorf but rather to perform all steps in three-well clear glass.

Acknowledgments

We wish to thank all members of the Jagla lab for stimulating discussions. We are grateful to former Jagla lab members, N. Figeac, for initial characterization of AMP markers and to R. Aradhya for generating M6-GAL4 and M6-gapGFP lines. This work is supported by the ANR grant ID-CELL-SPE to KJ, the "Equipe" FRM grant to K.J., and the AFM-Téléthon grant to G.L.

References

1. Ruiz Gomez M, Bate M (1997) Segregation of myogenic lineages in Drosophila requires numb. Development 124(23):4857–4866
2. Figeac N, Daczewska M, Marcelle C, Jagla K (2007) Muscle stem cells and model systems for their investigation. Dev Dyn 236(12):3332–3342. doi:10.1002/dvdy.21345
3. Bate M, Rushton E, Currie DA (1991) Cells with persistent twist expression are the embryonic precursors of adult muscles in Drosophila. Development 113(1):79–89
4. Liotta D, Han J, Elgar S, Garvey C, Han Z, Taylor MV (2007) The Him gene reveals a balance of inputs controlling muscle differentiation in Drosophila. Curr Biol 17(16):1409–1413. doi:10.1016/j.cub.2007.07.039
5. Postigo AA, Ward E, Skeath JB, Dean DC (1999) zfh-1, the Drosophila homologue of ZEB, is a transcriptional repressor that regulates somatic myogenesis. Mol Cell Biol 19(10):7255–7263
6. Figeac N, Jagla T, Aradhya R, Da Ponte JP, Jagla K (2010) Drosophila adult muscle precursors form a network of interconnected cells and are specified by the rhomboid-triggered EGF pathway. Development 137(12):1965–1973. doi:10.1242/dev.049080
7. Tavi P, Korhonen T, Hanninen SL, Bruton JD, Loof S, Simon A, Westerblad H (2010) Myogenic skeletal muscle satellite cells communicate by tunnelling nanotubes. J Cell Physiol 223(2):376–383. doi:10.1002/jcp.22044
8. Soler C, Daczewska M, Da Ponte JP, Dastugue B, Jagla K (2004) Coordinated development of muscles and tendons of the Drosophila leg. Development 131(24):6041–6051. doi:10.1242/dev.01527
9. Fernandes J, Bate M, Vijayraghavan K (1991) Development of the indirect flight muscles of Drosophila. Development 113(1):67–77
10. Fernandes JJ, Keshishian H (1996) Patterning the dorsal longitudinal flight muscles (DLM) of Drosophila: insights from the ablation of larval scaffolds. Development 122(12):3755–3763
11. Ranganayakulu G, Zhao B, Dokidis A, Molkentin JD, Olson EN, Schulz RA (1995) A series of mutations in the D-MEF2 transcription factor reveal multiple functions in larval and adult myogenesis in Drosophila. Dev Biol 171(1):169–181. doi:10.1006/dbio.1995.1269
12. Soler C, Taylor MV (2009) The Him gene inhibits the development of Drosophila flight muscles during metamorphosis. Mech Dev 126(7):595–603. doi:10.1016/j.mod.2009.03.003
13. Sudarsan V, Anant S, Guptan P, VijayRaghavan K, Skaer H (2001) Myoblast diversification and ectodermal signaling in Drosophila. Dev Cell 1(6):829–839
14. Maqbool T, Soler C, Jagla T, Daczewska M, Lodha N, Palliyil S, VijayRaghavan K, Jagla K (2006) Shaping leg muscles in Drosophila: role of ladybird, a conserved regulator of appendicular myogenesis. PLoS One 1:e122. doi:10.1371/journal.pone.0000122
15. Bernard F, Dutriaux A, Silber J, Lalouette A (2006) Notch pathway repression by vestigial is required to promote indirect flight muscle differentiation in Drosophila melanogaster. Dev Biol 295(1):164–177. doi:10.1016/j.ydbio.2006.03.022
16. Deng H, Hughes SC, Bell JB, Simmonds AJ (2009) Alternative requirements for Vestigial, Scalloped, and Dmef2 during muscle differentiation in Drosophila melanogaster. Mol Biol Cell 20(1):256–269. doi:10.1091/mbc.E08-03-0288
17. Brand AH, Perrimon N (1993) Targeted gene expression as a means of altering cell fates and generating dominant phenotypes. Development 118(2):401–415
18. Anant S, Roy S, VijayRaghavan K (1998) Twist and Notch negatively regulate adult muscle differentiation in Drosophila. Development 125(8):1361–1369
19. Fyrberg C, Becker J, Barthmaier P, Mahaffey J, Fyrberg E (1997) A Drosophila muscle-specific gene related to the mouse quaking locus. Gene 197(1-2):315–323
20. Ranganayakulu G, Elliott DA, Harvey RP, Olson EN (1998) Divergent roles for NK-2 class homeobox genes in cardiogenesis in flies and mice. Development 125(16):3037–3048
21. Sellin J, Drechsler M, Nguyen HT, Paululat A (2009) Antagonistic function of Lmd and Zfh1 fine tunes cell fate decisions in the Twi and Tin positive mesoderm of Drosophila melanogaster. Dev Biol 326(2):444–455. doi:10.1016/j.ydbio.2008.10.041
22. Kim J, Sebring A, Esch JJ, Kraus ME, Vorwerk K, Magee J, Carroll SB (1996) Integration of positional signals and regulation of wing formation and identity by Drosophila vestigial gene. Nature 382(6587):133–138. doi:10.1038/382133a0
23. Knirr S, Azpiazu N, Frasch M (1999) The role of the NK-homeobox gene slouch (S59) in somatic muscle patterning. Development 126(20):4525–4535
24. Duan H, Zhang C, Chen J, Sink H, Frei E, Noll M (2007) A key role of Pox meso in

somatic myogenesis of Drosophila. Development 134(22):3985–3997. doi:10.1242/dev.008821

25. Cripps RM, Olson EN (1998) Twist is required for muscle template splitting during adult Drosophila myogenesis. Dev Biol 203(1):106–115. doi:10.1006/dbio.1998.9040
26. Soler C, Han J, Taylor MV (2012) The conserved transcription factor Mef2 has multiple roles in adult Drosophila musculature formation. Development 139(7):1270–1275. doi:10.1242/dev.077875
27. Buckingham M, Bajard L, Chang T, Daubas P, Hadchouel J, Meilhac S, Montarras D, Rocancourt D, Relaix F (2003) The formation of skeletal muscle: from somite to limb. J Anat 202(1):59–68
28. Lilly B, Zhao B, Ranganayakulu G, Paterson BM, Schulz RA, Olson EN (1995) Requirement of MADS domain transcription factor D-MEF2 for muscle formation in Drosophila. Science 267(5198):688–693
29. Jagla K, Jagla T, Heitzler P, Dretzen G, Bellard F, Bellard M (1997) ladybird, a tandem of homeobox genes that maintain late wingless expression in terminal and dorsal epidermis of the Drosophila embryo. Development 124(1):91–100

Chapter 6

Using Transgenic Zebrafish to Study Muscle Stem/Progenitor Cells

Phong D. Nguyen and Peter D. Currie

Abstract

Understanding muscle stem cell behaviors can potentially provide insights into how these cells act and respond during normal growth and diseased contexts. The zebrafish is an ideal model organism to examine these behaviors in vivo where it would normally be technically challenging in other mammalian models. This chapter will describe the procedures required to successfully conduct live imaging of zebrafish transgenics that has specifically been adapted for skeletal muscle.

Key words Zebrafish, Skeletal muscle, Muscle progenitor, Satellite cell, Live imaging, Timelapse, Transgenics, Stem cell

1 Introduction

Satellite cells were first identified to be anatomically positioned between the basal lamina and sarcolemma of muscle fibers, hence, given the name "satellite cell" [1]. Since then, there have been many insights into the molecular and cellular mechanisms in the maintenance of their population and their responses during injury and regeneration [2].

One limitation of studying this stem cell in mammals is the accessibility of these cells in a native environment. Most of the findings of satellite cell behaviors come from histological or ex vivo techniques. Since satellite cells actively interact with the surrounding tissues, it may be more beneficial to examine stem cell behaviors in vivo.

The zebrafish is an ideal vertebrate animal model to examine in vivo cell behaviors, and a muscle stem/progenitor cell population has been identified to have similar properties to a satellite cell but is not yet fully characterized [3–5]. The optical clarity and fast development of the zebrafish allows easy access to skeletal muscle during development. Additionally, transgenic fish can be generated to mark specific cell populations and track them over time, thus

Eusebio Perdiguero and DDW Cornelison (eds.), *Muscle Stem Cells: Methods and Protocols*, Methods in Molecular Biology, vol. 1556, DOI 10.1007/978-1-4939-6771-1_6, © Springer Science+Business Media LLC 2017

providing a unique opportunity to follow their live dynamics during growth, injury, and regeneration.

This chapter will provide the methodologies and techniques to establish live imaging of zebrafish muscle and stem/progenitor cells using confocal microscopes. Once established, the user can take advantage of the genetic tools in zebrafish to further modulate cells of interest in overexpression, knockdown and knockout environments. In addition, other imaging techniques can potentially be applied such as photoconversions, cell ablation, FRET/FRAP, birefringence, and light sheet microscopy to further understand the functional mechanisms of stem/progenitor cells in skeletal muscle.

2 Materials

2.1 Fish Husbandry

1. Transgenic zebrafish (*see* **Note 1**).
2. Breeder boxes.
3. Tea strainer.
4. 90 mm plastic petri dishes.
5. Incubator set to 28.5 °C.

2.2 Mounting

1. Microwave.
2. Water bath set to 42 °C.
3. Fluorescent stereo microscope.
4. Sylgard 184 Silicone Elastomer (Dow Corning).
5. Plastic Microinjection Mold (TU-1, Adaptive Science Tools).
6. Rotating centrifuge.
7. Dumont forceps INOX #04.
8. Glass Pasteur pipette.
9. Plastic Pasteur pipette.
10. 10 μl pipette tips.
11. 30 G needles attached to 1 ml syringe that has been blunted with a sharpening stone.

2.3 Confocal Imaging

1. Confocal microscope with heated stage/chamber (e.g., Zeiss LSM 710, Nikon C1 invert).
2. 20× objective.
3. 24 well plate.

2.4 Image Analysis

1. Imaging processing software (e.g., Fiji/ImageJ, Imaris, Metamorph).

2.5 Buffers and Reagents

1. System water: water in a closed zebrafish circulating system that has been filtered for biological matter, disinfected with UV, and is dechlorinated.
2. E3 water (embryo medium): add 0.292 g NaCl, 0.013 g KCl, 0.044 g CaCl, and 0.081 g $MgSO_4$ to 1 l dH_2O. Add methylene blue to the final concentration of 0.01 % as a fungicide. Store solution at room temperature.
3. 0.4 % tricaine (3-aminobenzoic acid ethyl ester): add 2.1 ml of 1 M Tris buffer (pH 9.5) and fill to 100 ml with ddH_2O. Add 0.4 g tricaine. Mix well and adjust pH to around 7. Don't autoclave. Store at 4 °C.
4. 100 mM PTU: dissolve 1.52 g in 100 ml ethanol. If precipitation occurs, briefly shake at 37 °C. Store at room temperature.
5. 1 % agarose: add 1 g of low-melt agarose to 100 ml E3 water, dissolve by heating in microwave, and mix occasionally. Once dissolved, store at room temperature where it will solidify.

3 Methods

3.1 Fish Husbandry

1. In the afternoon, fill breeder boxes containing dividers with system water.
2. Place a male and female of the desired transgenic fish line, and ensure fish are separated with the divider.
3. In the following morning, pull the dividers and allow fish to mate. Once fish have laid, collect eggs with a tea strainer, and place into a petri dish with E3 water.
4. Incubate at 28.5 °C until the desired time point. Add 50 μl of 100 mM PTU per petri dish if necessary (*see* **Note 2**).

3.2 Mounting

1. Melt 1 % agarose in E3 solution with a microwave until completely molten, and place in a 42 °C water bath to cool and maintain agarose in liquid state.
2. Take fish out of the incubator, and select the fluorophore-positive fish using a fluorescent stereo microscope.
3. Dechorionate fish if they are still within their chorions with forceps.
4. Add 0.4 % tricaine dropwise using a plastic Pasteur pipette until fish stop moving but still have beating hearts (*see* **Note 3**).
5. Using a glass Pasteur pipette (*see* **Note 4**), pick up the fish and place fish into a dry petri dish (*see* **Note 5**).
6. Remove as much liquid as possible from the petri dish without drying the fish.

7. Draw up some cooled agarose (*see* **Note 6**) with a plastic Pasteur pipette, and place onto the fish (*see* **Note 7**).
8. Using 10 μl pipette tips or 30 G needles that have been blunted, orientate the fish into the lateral position with the anterior pointed to the left, and hold in position until the agarose has solidified (*see* **Note 8**).
9. Immerse the dish with fresh E3 water, add tricaine and PTU, and transport fish to the microscope.

3.3 Confocal Imaging

1. Turn on confocal microscope (*see* **Note 9**).
2. Place the dish under the confocal and using the software, and focus onto the region of interest (*see* **Note 10**).
3. Select the lowest laser power intensity with the appropriate lasers that still allow visualization of the cells of interest to avoid photobleaching.
4. Set up the *z*-axis limits to define the *z*-stack that will be used to image the fish (*see* **Note 11**).
5. Conduct imaging (*see* **Notes 12–14**).
6. If the same fish is required to be imaged at later stages, release fish from the agarose by carefully removing agarose with blunted needles or forceps.
7. Transfer fish into a petri dish containing E3 water and PTU, or if individual fish are being tracked, place the fish in individual wells of a 24-well plate with E3 and PTU, and incubate at 28.5 °C until desired (*see* **Note 15**).

3.4 Image Analysis

1. Save images into the file type of choice (typically the default company setting).
2. Export and load the file in imaging processing software such as Fiji/ImageJ, Imaris, and Metamorph and analyze according to manufacturer's instructions (*see* **Notes 16** and **17**).

4 Notes

1. A number of transgenic fish have now been published that mark satellite cell-like populations. These include pax3a [3], pax7a [3], and myf5 [6]. Other potentially useful transgenics have the fluorophore under the control of various muscle-specific promoters. Some are pan-muscle markers such as actal [7] and unc600 [8]; others mark subsets of muscle groups such as slow muscle and adaxial cells (smyhc) [9], fast muscle (fmyhc) [10], muscle pioneers (prox1a) [11], and differentiating muscle cells (myod) [3]. New transgenic lines can also be generated if a particular one is not available. When maintaining transgenic

lines, it is important to screen for the brightest fish that correctly recapitulates the expected expression pattern as silencing of the transgene is possible over multiple generations.

2. If examining stages later than 1 day post fertilization (dpf), PTU must be added to inhibit pigmentation as melanocytes lie directly on top of the muscle and obstruct images of the underlying muscle. For further steps, if fish are moved into a fresh petri dish, PTU must be added again as pigmentation comes back in the absence of this chemical. PTU is toxic, wear gloves, and handle with care.
3. Fish become more sensitive to tricaine as they are older so the amount required varies. If the heart stops beating following addition of too much tricaine, quickly place fish into fresh E3, and wait for fish to wake up again.
4. While it only happens occasionally, fish can stick to a plastic Pasteur pipette when compared to a glass Pasteur pipette, particularly at earlier stages.
5. An alternative to mounting fish onto the petri dish is to mount them in precast silicon molds [12]. This allows fish to be mounted in the same orientation at roughly the same *z*-axis depth to avoid focusing through large *z*-axis depths, hence imaging between different fish faster. To make the mold, mix curing agent with base at a 1:10 ratio. Mix at 4 °C with a rotating centrifuge for 2 h. Once mixed, pour mixture over a plastic injection mold that is placed within an empty petri dish. Allow the mixture to cure over 2 days and gently remove the plastic mold. The silicon mold is ready to use. Bubbles often form however over the curing stage; these will naturally be eliminated. To mount fish, use a glass pipette and place a fish into the well. Remove as much liquid as possible and add some melted agarose. Use 10 μl pipette tips or 30 G needles that have been blunted to orientate the fish, and hold in position until agarose has solidified.
6. Hot agarose will kill the fish. As a general rule, it is too hot if you cannot place a drop of agarose onto your wrist. Holding the pipette that has already drawn some agarose in your palm for 10–15 s will usually quickly cool agarose to a safe temperature.
7. Try not to add too much agarose as the object lens will not be able to focus into the working distance; however, adding too little agarose will cause the fish to pop out and float away.
8. Low-melt agarose generally takes 30–60 s to solidify.
9. Many types of confocals can be adapted for zebrafish imaging. If using an upright confocal, a water-immersion dipping lens is required which is not necessary if using an inverted confocal.

The confocal microscope should include either a heated stage or chamber set to 28.5 °C. This is particularly important if conducting long-term time-lapse imaging. The appropriate lasers and filters for illuminating the fluorophore need to be present such as 405 nm for tagBFP, 488 nm for GFP, 560 nm for mCherry, and 633 nm for crimson.

10. Routine imaging of trunk muscle is usually focused at the myotomes surrounding the end of the yolk extension and start of the anal pore as this is the region that contains the most muscle without other organs impeding view (such as the gut). Different muscle populations can be anatomically identified. The external cell layer contains muscle stem/progenitor cells and is located at the lateral-most position on the myotome. Underneath this structure is a single layer of slow muscle fibers, and medial to this is fast muscle, which makes up the bulk of the myotome. Muscle stem/progenitors are also located within this fast muscle layer (*see* Fig. 1 and [13]).
11. Fish grow bigger over time. If a timelapse is being conducted, the size of the *z*-stack must be sufficiently large enough to capture the entire muscle.
12. If time-lapse imaging is desired, choose the minimum number of cycles between each *z*-stack acquisition that will still capture the event of interest. One major issue especially with long timelapses is the amount of data generated. The large file sizes as a result make it difficult to quickly analyze data. Reducing the number of cycles along with the number of *z*-stacks will aid in reducing the file size.
13. If time-lapse imaging is conducted, ensure there is sufficient E3 water present. As the confocal will be heated, there will be water evaporation over long timelapses. Excess evaporation of water will distort the image during the scan.
14. If imaging multiple fish is desired during time-lapse imaging, the multiposition function can be used as long as there is sufficient scanning time. The duration between each cycle and the time taken to complete one *z*-stack will dictate how much fish can be imaged in one session.
15. Fish kept for longer than 6 dpf for downstream experiments need to be fed. Live imaging of fish after 12 dpf is technically challenging due to their considerable size, requirement for oxygenated water to pass along their gills, and the presence of scales that causes imaging aberrations and distortions. Histological techniques will need to be conducted if later time-points are desired.
16. A data management strategy must be taken into account as file sizes will be the biggest limitation to analyzing data. If possible, try to analyze files on a high-performance computer to

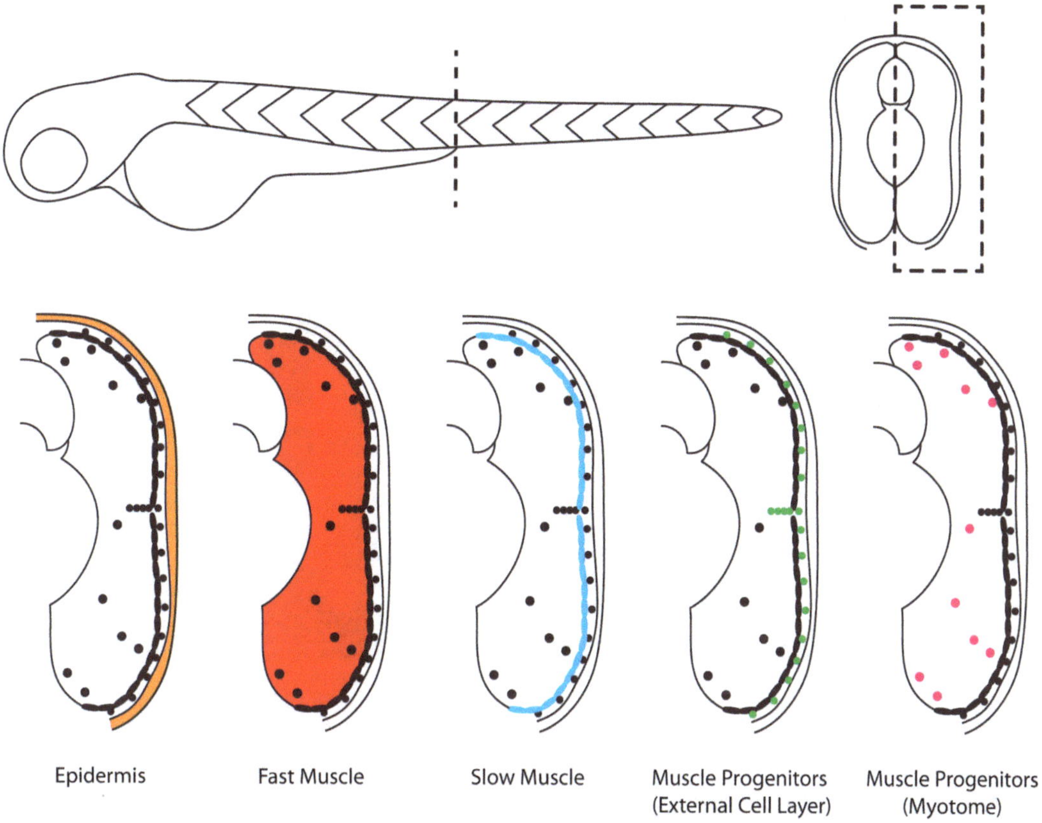

Fig. 1 Schematic of different muscle populations in zebrafish. In zebrafish, muscle populations are contained within defined anatomical positions. The epidermis (*orange*) is the most lateral structure. Lying directly beneath is a single layer of muscle progenitors also known as the external cell layer (*green*). This layer also penetrates deep into the myotome at the midpoint of the dorsal-ventral axis into a structure called the horizontal myoseptum. Medial to the external cell layer is a monolayer of slow muscle fibers (*blue*). Finally, the deepest muscle layer and the largest in fiber number and size is the fast muscle population (*red*). Within this fast muscle population contains a second pool of muscle progenitors (*pink*). Transverse sections are created at the level of the *dotted line* on the zebrafish embryo

speed up processing times. If they are too large to process, it may be necessary to split data points or modify image acquisition settings for the next imaging session to reduce file sizes.

17. A standard imaging and analysis pipeline we perform is to first decide on the transgenic line we wish to image and in what context, for example, in a native, disease, or injury model. This will dictate the number of time points required to capture the behavior of cells. At the desired time point, fish are placed under the confocal microscope (we use a Zeiss LSM 710 upright confocal with a 20× water-dipping objective). If imaging for the first time, the laser power is first set to a low setting such as 2–3 % laser power to determine the fluorescence intensity of the transgenic fish. The laser power is gradually increased

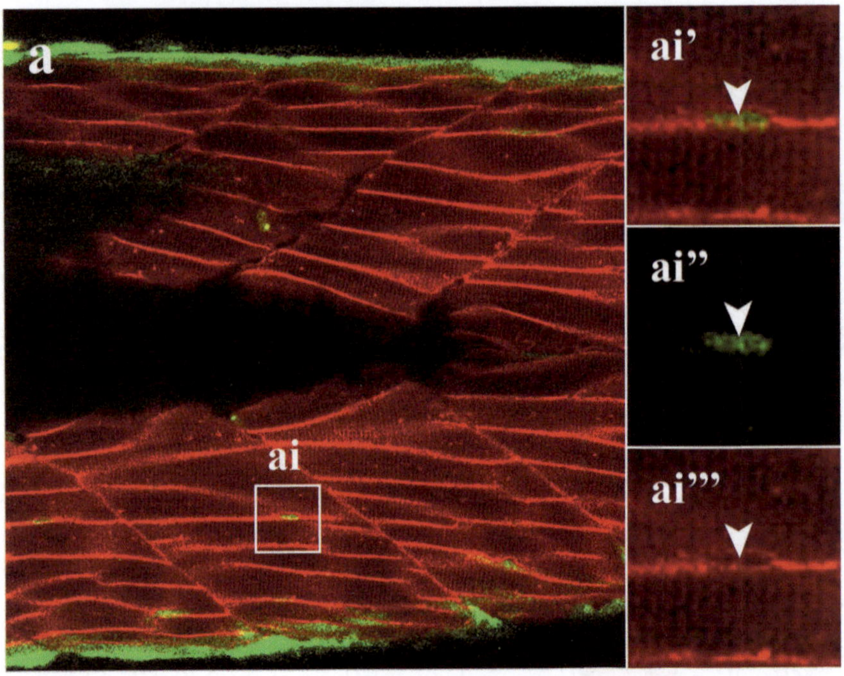

Fig. 2 Muscle progenitors within the fast muscle population. Single confocal slice of a 6-day postfertilization, double transgenic fish (TgBAC(*pax7a*: GFP); Tg(*acta1*: mCherryCAAX)) which marks muscle progenitors in *green* and muscle membranes in *red*. a. Overview image of the zebrafish myotome. ai. Boxed image within a. ai'. Image of *red* and *green channel* of boxed image in ai. ai''. Image of *green channel* of boxed image in ai. ai'''. Image of *red channel* of boxed image in ai

until a fluorescent signal is sufficiently strong while not being overexposed. Examining the saturation levels can assist this process. A *z*-stack is conducted and the file is saved (we use a .LSM extension). The file is then loaded onto an imaging processing program (we use Imaris software), and images are examined. Depending on the experimental paradigm, we typically look at cell morphology, cell number, cell size, and location. If time-lapse imaging was also conducted, we additionally analyze cell behaviors such as motility, velocity, and proliferation activity. This can all be achieved within the software. An example of a confocal image of muscle progenitors located within the fast muscle population is shown in Fig. 2 (arrowhead).

References

1. Mauro A (1961) Satellite cell of skeletal muscle fibers. J Biophys Biochem Cytol 9:493–495
2. Montarras D, L'Honoré A, Buckingham M (2013) Lying low but ready for action: the quiescent muscle satellite cell. FEBS J 280(17):4036–4050. doi:10.1111/febs.12372
3. Seger C, Hargrave M, Wang X, Chai RJ, Elworthy S, Ingham PW (2011) Analysis of Pax7 expressing myogenic cells in zebrafish muscle development, injury, and models of disease. Dev Dyn 240(11):2440–2451
4. Hollway GE, Bryson-Richardson RJ, Berger S, Cole NJ, Hall TE, Currie PD (2007) Whole-

somite rotation generates muscle progenitor cell compartments in the developing zebrafish embryo. Dev Cell 12(2):207–219

5. Siegel AL, Gurevich DB, Currie PD (2013) A myogenic precursor cell that could contribute to regeneration in zebrafish and its similarity to the satellite cell. FEBS J 280(17):4074–4088. doi:10.1111/febs.12300
6. Chen Y-H, Wang Y-H, Chang M-Y, Lin C-Y, Weng C-W, Westerfield M, Tsai H-J (2007) Multiple upstream modules regulate zebrafish myf5 expression. BMC Dev Biol 7(1):1
7. Cole NJ, Hall TE, Don EK, Berger S, Boisvert CA, Neyt C, Ericsson R, Joss J, Gurevich DB, Currie PD (2011) Development and evolution of the muscles of the pelvic fin. PLoS Biol 9(10):e1001168. doi:10.1371/journal.pbio.1001168
8. Berger J, Currie PD (2013) 503unc, a small and muscle-specific zebrafish promoter. Genesis 51(6):443–447. doi:10.1002/dvg.22385
9. Elworthy S, Hargrave M, Knight R, Mebus K, Ingham PW (2008) Expression of multiple slow myosin heavy chain genes reveals a diversity of zebrafish slow twitch muscle fibres with differing requirements for Hedgehog and Prdm1 activity. Development 135(12):2115–2126
10. Nord H, Burguiere AC, Muck J, Nord C, Ahlgren U, von Hofsten J (2014) Differential regulation of myosin heavy chains defines new muscle domains in zebrafish. Mol Biol Cell 25(8):1384–1395. doi:10.1091/mbc.E13-08-0486
11. van Impel A, Zhao Z, Hermkens DMA, Roukens MG, Fischer JC, Peterson-Maduro J, Duckers H, Ober EA, Ingham PW, Schulte-Merker S (2014) Divergence of zebrafish and mouse lymphatic cell fate specification pathways. Development 141(6):1228–1238. doi:10.1242/dev.105031
12. Masselink W, Wong JC, Liu B, Fu J, Currie PD (2014) Low-cost silicone imaging casts for zebrafish embryos and larvae. Zebrafish 11(1):26–31. doi:10.1089/zeb.2013.0897
13. Bryson-Richardson RJ, Currie PD (2008) The genetics of vertebrate myogenesis. Nat Rev Genet 9(8):632–646

Part III

Other Muscle Interstitial Stem Cells

Chapter 7

Muscle Interstitial Cells: A Brief Field Guide to Non-satellite Cell Populations in Skeletal Muscle

Francesco Saverio Tedesco*, Louise A. Moyle*, and Eusebio Perdiguero*

Abstract

Skeletal muscle regeneration is mainly enabled by a population of adult stem cells known as satellite cells. Satellite cells have been shown to be indispensable for adult skeletal muscle repair and regeneration. In the last two decades, other stem/progenitor cell populations resident in the skeletal muscle interstitium have been identified as "collaborators" of satellite cells during regeneration. They also appear to have a key role in replacing skeletal muscle with adipose, fibrous, or bone tissue in pathological conditions. Here, we review the role and known functions of these different interstitial skeletal muscle cell types and discuss their role in skeletal muscle tissue homeostasis, regeneration, and disease, including their therapeutic potential for cell transplantation protocols.

Key words Skeletal muscle, Interstitial cells, Pericytes, Mesoangioblasts, Mesenchymal progenitors, Fibro-adipogenic progenitors, *Pw1*, Interstitial cells, Stem cells, Muscle regeneration

1 Introduction

The primary role of skeletal muscle is to generate movement, maintain posture, and support soft tissues, whilst also contributing to body metabolism and temperature control. Muscle contraction and force generation are mediated by the interaction of actin and myosin proteins within the complex sarcomere unit. Aligned sarcomere units make myofibrils, bundles of which span the length of each muscle fiber (myofiber). In turn, numerous bundles of myofibers make up each muscle [1]. These multinucleated, syncytial cells are formed during the process of myogenesis [2]. However, as myofiber nuclei are postmitotic, they are unable to contribute to growth and repair [3].

It is generally accepted that satellite cells, a population of muscle stem cells that reside beneath the basal lamina of myofibers, are responsible for the regenerative capacity of adult skeletal muscle.

*The three authors contributed equally to this work.

Eusebio Perdiguero and DDW Cornelison (eds.), *Muscle Stem Cells: Methods and Protocols*, Methods in Molecular Biology, vol. 1556, DOI 10.1007/978-1-4939-6771-1_7, © Springer Science+Business Media LLC 2017

Satellite cells are a heterogeneous group of stem cells of embryonic somitic origin that normally reside in a quiescent state until activated by damage or growth signals [4–6]. The vast majority of mammalian satellite cells can be identified by expression of the paired-box transcription factor Pax7, which is satellite cell specific in skeletal muscle. Many other proteins mark the majority of satellite cells, including integrin-α7, M-cadherin, caveolin-1, CD56/NCAM, CD29/integrin-β1, and syndecans 3 and 4 (reviewed in [4, 7, 8]). However, these markers can be expressed only in subpopulations (i.e. Pax3), by other non-satellite cell populations in muscle (i.e. CD34, syndecan 3/4) or in a species-specific manner depending on activation stage (i.e. CD56). Therefore, they should be used in combination to ensure specificity (Table 1). Once activated, satellite cells undergo defined proliferation/differentiation or self-renewal processes to contribute either to tissue repair or replenishment of their stem cell pool [9–11].

Other stem/progenitor cell populations present in the adult skeletal muscle (Fig. 1) have been identified as capable to contribute to or to modulate muscle regeneration. The role of these populations in normal muscle homeostasis and function is still under investigation, although some populations, including pericytes/me soangioblasts, Pwl+ cells, and CD133+ cells, hold special interest as therapeutically useful cell types to substitute satellite cells in clinical applications, such as stem cell transplantation. Other populations, referred here as mesenchymal progenitors, have been investigated as pharmacological targets for tissue remodeling.

Here, we provide an overview and discuss the role and known functions of these non-satellite cells residing in adult skeletal muscle, focusing on studies published in the last decade. Additional information on other cell populations (e.g., muscle resident "side population") can be found in Table 1. We direct the reader to other review articles for a more comprehensive analysis of developmental origins of muscle stem cells, molecular networks, functions, and use in cell-based therapies [4, 8, 12–18]. We define satellite cells as a Pax7+ cells located underneath the basal lamina and interstitial cells as those resident between myofibers and outside their basal lamina. This will help to distinguish satellite cells from occasional Pax7+ cells in the muscle interstitium, which could either be separate interstitial stem cell populations or satellite cells trapped outside the basal lamina following myofiber remodeling (Paolo Bianco, personal communication).

2 Pericytes and Mesoangioblasts

Pericytes are a heterogeneous group of contractile cells which encircle the endothelium of microvessels, first described by Rouget in 1873 [19]. Present in all vascularized tissues, pericytes regulate

Table 1
Summary of skeletal muscle-resident stem or progenitor cellular populations in murine muscle

Cell populations in non-regenerating murine muscle					
Cell name	**Positive for**	**Negative for**	**Differentiation/function**	**Therapeutic implications**	**References**
Stem or progenitor cells					
Satellite cells (SC)	Pax7, SM/C2.6, α7-integrin, M-cadherin, CD29, CD34, Myf5, PW1, syndecan 3/4 nestin, c-Met, calveolin-1, (Pax3)	CD45, CD31, PDGFRα, c-Kit (*Sca1*)	Myogenesis in vivo and ex vivo	Genetically corrected intramuscular (IM) transplantation	[11, 29, 48, 71, 72, 94–97]. Reviewed in [4, 98]
Adult pericytes	TNAP, PDGFRβ, NG2, desmin, CD146, αSMA, (nestin: type 2)	Pax7, CD56, (nestin: type 1), (PDFRα: type 2)	Angiogenesis, myogenesis, pericyte and SC self-renewal, fibrosis/ fatty infiltration (type 1)	Genetically corrected IM and intra-arterial transplantation	[21, 24, 26]. Reviewed in [12]
PICs	PW1/Peg3, Sca1, CD34, (PDGFRα)	Pax7, (PDGFRα)	PDGFRα+ adipogenesis and PDGFRα– myogenesis. Contribute to SC pool	–	[72, 75]
FAPs/ mesenchymal progenitors	Sca1, PDGFRα, PDGFRβ, vimentin, adam 12, PW1, tcf4	CD31, CD45, α7-integrin, SM/C-2.6, αSMA, NG2, Pax7	Sca1−: adipogenesis, osteogenesis, and chondrogenesis; Sca1+, adipo/fibrogenesis	Targets for anti-fibrotic interventions with drugs inhibiting PDGFRα	[48–50, 53, 72]. Reviewed in [17]
Other stem or progenitor populations					
CD133+ cells	CD133, (CD34)	(CD34)	Myogenesis in vivo and ex vivo	Genetically corrected IM and intra-arterial transplantation	[80–82]

(continued)

Table 1
(continued)

Cell populations in non-regenerating murine muscle

Cell name	Positive for	Negative for	Differentiation/function	Therapeutic implications	References
Muscle side population (SP)	Sca1, ABCG2, (CD45), (CD31),(CD34), (*Pax7*)	Hoechst negative, (CD45), (CD31), c-Kit, (CD34)	Unclear: CD45+, hematopoiesis; CD31+, angiogenesis; negative cells myogenesis	–	[74, 99–101]. Reviewed in [4]
Myoendothelial cells	CD34, Sca1, c-Met	CD45, CD14, CD31, CD49, CD144, c-Kit, FLK, Pax3, Pax7, MyoD, Myf5, M-cadherin	Unclear: adipogenesis, angiogenesis, and myogenesis	–	[102]
Other resident cells					
Resident myeloid cells	CD45, CD11b, F4/80	CD11c, Ly-6C, CX3CR1	Recruitment of neutrophils, monocytes	–	[103–106]. Reviewed in [107, 108]
Endothelial cells	CD31, Sca1, FLK-1, CD144/VE-cadherin, CD34	CD45, CD11b, α7-integrin, Pax7	Vessel formation	–	[20, 109–111]
Fibroblasts	Tcf4, PDGFRα	CD45, CD31, CD11b, α7-integrin, Pax7	Extracellular matrix production	–	[112, 113]

Note: As these populations are heterogeneous, not all the cells express all the indicated markers, and some markers may have high/medium/low subpopulations, like Pax7 in satellite cells [9]. Markers between brackets affect only a subpopulation or depend on origin/developmental stage; markers between brackets and underlined are controversial Markers used for isolation and characterization, and known functions and therapeutic implications, are indicated

blood vessel growth, homeostasis, and permeability, in addition to other tissue-specific roles (reviewed in [14]). In skeletal muscle, blood vessels run adjacent to myofibers resulting in the close association and likely cross talk between pericytes and satellite cells [20–22]. Indeed, pericytes have been shown to regulate postnatal myogenesis and satellite cell quiescence [23]. At rest, pericytes are embedded within the vascular basement membrane, which separates them from other periendothelial mesenchymal cells [14, 22].

A major limitation in the study of muscle pericytes is the lack of a specific marker to distinguish them from other satellite and non-satellite cell populations (reviewed in [12, 14]). Therefore, they are mainly defined by their anatomical location and by the combined expression of multiple genes. Additionally, expression of accepted pericyte markers differs between species, cellular subpopulations, and developmental stage [14, 21, 24, 25]. Furthermore, not all muscle pericytes have the potential to contribute to skeletal myogenesis. The subpopulation of pericytes with skeletal myogenic capacity can be distinguished in vivo by expression of tissue nonspecific alkaline phosphatase (TNAP) in murine and human muscle [21, 26]. Skeletal muscle pericytes/perivascular cells with myogenic potential can also be identified by the expression of CD146 ([27], [28] and Paolo Bianco and Mara Riminucci, personal communication) or nestin [24], although nestin is also expressed in satellite cells [29]. The extent to which these three populations overlap is currently unknown. Notably, Sacchetti et al. reported that human CD146+ myogenic pericytes express PAX7 and behave like satellite cells in some in vitro and in vivo assays, suggesting that they could even be subsets of the same population randomly recruited to different (but close) anatomical sites (i.e. satellite cell niche or perivascular area) [28]. Interestingly, a recent report has confirmed that human pericytes isolated from skeletal and smooth muscle tissues are functionally different and that only the pericytes isolated from skeletal muscle are able to contribute to skeletal muscle regeneration [30].

Mesoangioblast is a term for vessel-associated mesodermal stem/progenitor cells expanded in vitro, initially used for cells isolated from the murine embryonic dorsal aorta [31]. Mesoangioblast markers depend on the stage of development at which they are isolated; embryonic mesoangioblasts deriving from the dorsal aorta express mostly endothelial markers such as VE-cadherin and CD34 [32, 33]. Cells similar to embryonic mesoangioblasts can be derived from adult skeletal muscle pericytes, expressing varying degrees of pericyte markers such as neuroglial 2 proteoglycan (NG2), platelet-derived growth factor receptor-beta (PDGFR-β), alpha smooth muscle actin (αSMA), desmin, and, most importantly, TNAP, while being negative for endothelial and myogenic makers [14, 21, 24–26]. Mesoangioblasts also express Pw1 (see below), which was shown to be essential for proper stem cell function [34].

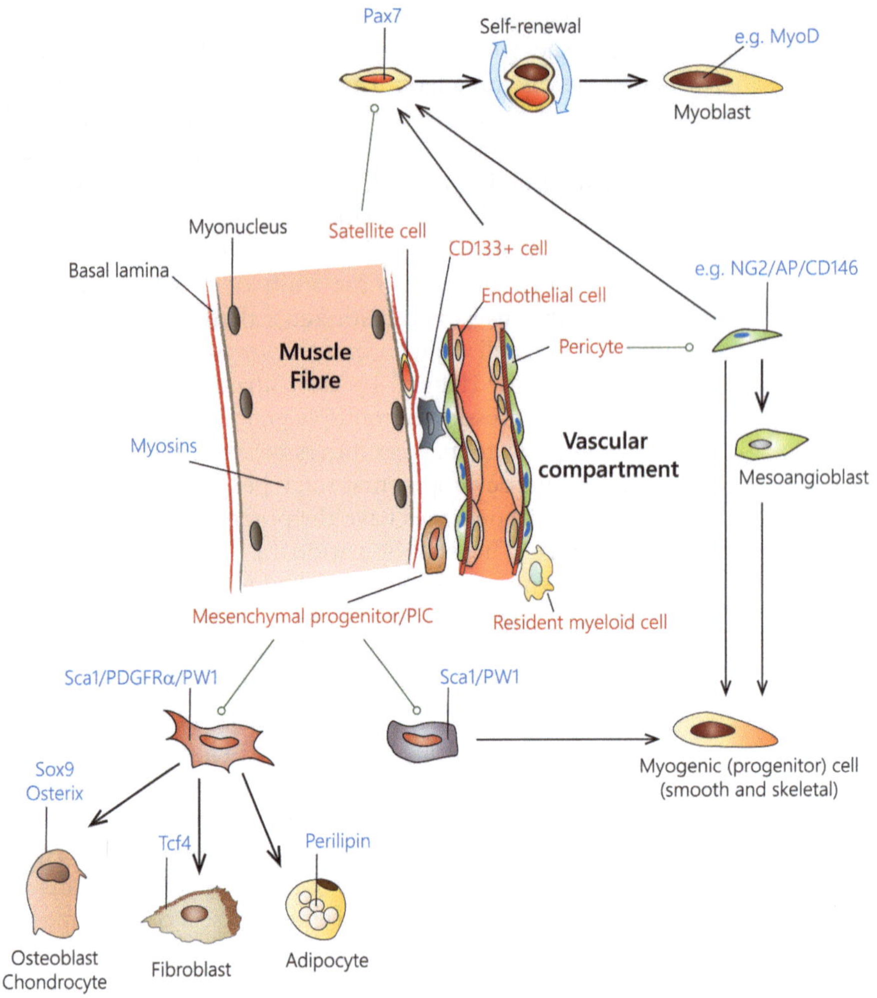

Fig. 1 Illustration of the main cellular populations in adult skeletal muscle. A skeletal muscle fiber and a blood vessel are shown. Satellite cells, interstitial populations, and vessel-associated cell populations and their main protein markers (*blue*) are indicated. Differentiation of satellite cells, pericytes, and mesenchymal progenitors/PICs is showed

To simplify relationships with other cell types resident in postnatal skeletal muscle, mesoangioblasts could be considered as the activated progeny of pericytes in the same way that myoblasts are the activated progeny of satellite cells (Fig. 1).

Although myogenic pericytes may have a lower myogenic capacity than myoblasts, their advantageous traits of expansion, migration, and extravasation upon intra-arterial delivery in dystrophic models [21, 25, 35–38] make them suitable candidates for cell therapies of muscle disorders. Additionally, activated mesoangioblasts can self-renew or migrate under the basal lamina and contribute to the satellite cell pool during skeletal muscle growth and in chronic and acute muscle regeneration [25, 26, 36, 39]. Importantly, Pax7+ cells are also found when expanded mesoangioblasts are transplanted. Notably, a first-in-human phase I/II clinical trial based upon intra-arterial delivery of pericyte-derived mesoangioblasts in five boys with Duchenne muscular dystrophy was recently completed [40]. The study showed a good safety profile (one adverse event with no clinical sequelae), and functional parameters appeared to have transiently stabilized in two out of three ambulant patients (statistical analysis was limited by the small number of patients). Although donor-derived dystrophin was detected in one patient, several aspects of the current protocol will need optimization in order to reach clinical efficacy. Examples of future improvements may include (1) enrollment of younger children, (2) increase of cell dose (e.g., by means of iPS cell-derived progenitors) [41], (3) modulation of inflammation [42, 43], and (4) enhanced engraftment or differentiation by acting on properties of donor cells [44, 45] or recipient patients [38].

Finally, several recent publications have implicated a role for pericytes in fibro-adipose infiltration of tissues (reviewed in [46]). In muscle, some pericytes may be precursors of myofibroblasts, interstitial cells which regulate fibrosis. Indeed, Birbair and colleagues have shown that type 1 non-myogenic pericytes contribute to fatty-fibrotic accumulation in aged and regenerating muscle [24, 47]. Therefore, pericyte-based therapies should focus on promoting myogenic differentiation while repressing fibro-adipogenic differentiation.

3 Fibro-adipogenic/Mesenchymal Progenitors

Other resident muscle interstitial progenitor populations may have increased propensity to differentiate toward non-myogenic cell types. One such population has been identified upon expression of PDGFR-α, CD34, and stem cell antigen-1 (Sca1). Initial reports, based on in vitro experiments, showed that these cells could be an important source of pro-differentiation signals for myoblasts during the process of muscle regeneration and that were able to differentiate into myofibroblasts and/or adipose cells [48, 49]. Consequently, they were named fibro-adipogenic progenitors (FAPs) or mesenchymal progenitors (MPs). Recent reports have demonstrated that these cells are also capable of osteogenic and chondrogenic differentiation in vivo [50]; therefore, we will refer to these cells using the more general term "mesenchymal progenitors" (Fig. 1). However, they should not be confused with "mesenchymal stem cells," whose markers and properties are still a matter of active discussion [51]. We apologize to the reader for the oversimplification of the model (and the possible occasional inappropriate nomenclature) and redirect them to excellent reviews that clarify terminology and lineage relationships of mesenchymal stem/progenitor cells in other mesodermal tissues [51, 52].

It has been shown that during acute muscle injury, these mesenchymal progenitors activate, rapidly expand, and then disappear [48, 49, 53]. Once activated, they have been shown to interact with satellite cells and with the regenerating muscle environment, promoting satellite cell differentiation and myofiber formation [48, 49]. However, it was recently shown that upon aging, mesenchymal progenitors have a deleterious effect on satellite cell function, repressing satellite cell myogenesis [54]. Moreover, epigenetic reprograming of mesenchymal progenitors by treatment with HDAC inhibitors has been shown to drive them toward a myogenic lineage and improve regeneration of dystrophic mice [55], opening a therapeutic avenue for these progenitors. Interestingly, mesenchymal progenitors have been isolated from human muscles [56] and demonstrated to contribute to adipocyte formation [57].

The expansion of resident mesenchymal progenitors has been shown to be mediated by the cytokine interleukin 4 (IL-4), produced by eosinophils during the early phases of regeneration [58]. Also, a recent report from Rossi lab has demonstrated that pro-inflammatory cytokines (i.e., tumor necrosis factor, TNF) produced by the first wave of infiltrating macrophages induce apoptosis of mesenchymal progenitors. During chronic injury, where pro-inflammatory and anti-inflammatory macrophage populations coexist [59], changes in the cytokine milieu (i.e., higher levels of transforming growth factor β1, TGFβ1) prevent the apoptosis of mesenchymal progenitors and induce their differentiation into persistent matrix-producing cells [53]. The cytokine combination to induce adipogenic differentiation remains to be determined, although certain types of injury (with different inflammatory response and therefore different cytokine environment) can highly increase differentiation to the adipogenic lineage [49]. Lemos and colleagues' findings are in line with a previous report demonstrating that a subpopulation of mesenchymal progenitors expressing the metalloproteinase ADAM12 is one of the major sources of fibrotic tissue accumulation after muscle damage [60]. This subpopulation of mesenchymal progenitors rapidly differentiates into myofibroblasts upon TGFβ1 stimulation, and it may represent a more committed fibrogenic progenitor. Interestingly, ADAM12+ mesenchymal progenitors share features with pericytes, being associated with blood vessel walls and expressing NG2, a marker of pericytes [60].

Muscle-resident fibroblasts are the cell population classically thought to be responsible for extracellular matrix remodeling and accumulation of fibrosis in pathological conditions such as muscular dystrophies. However, they also support healthy myogenesis, as ablation of transcription factor 4 (Tcf4)-positive muscle fibroblasts has been shown to impair muscle regeneration through premature differentiation of satellite cells and reduction of the satellite cell pool [61]. Interestingly, Tcf4 is expressed by both fibroblasts and mesenchymal progenitors, making it difficult to decipher the individual roles of each cell type using current experimental strategies. For example, whether some of the functions currently accounted to fibroblasts may be in fact performed by different populations of mesenchymal progenitors (and vice versa). Additionally, in chronic injury models, resident myoblasts and endothelial and hematopoietic cells have been shown to transdifferentiate into fibroblastic cells, advancing dystrophic pathology [61, 62], with a mechanism of transdifferentiation that occurs through an intermediate mesenchymal stem cell step [62]. Therefore, the definition of which cells are fibroblast or mesenchymal progenitors and their origin may be even more difficult than expected.

Besides their role during muscle regeneration and chronic pathologies, resident mesenchymal progenitors may have a role in skeletal muscle homeostasis. They secrete a number of Wnt ligands and myokines such as IL-6 [48, 49, 53]. Moreover, it has been demonstrated that interstitial mesenchymal cells are the main producers of collagen VI in resting muscle [63]. Collagen VI fibers are abundant in the endomysium of skeletal muscle and are a regular component of the satellite cell niche [64]. Mutations in collagen VI-encoding genes cause several diseases associated with muscle weakness in humans [65, 66], and collagen VI-deficient mice show myofiber degeneration, reduced strength, and deficient satellite cell self-renewal [64, 67–69]. Interestingly, mesenchymal progenitors of synovial origin secreted collagen VI when engrafted into muscle [70], indicating that this may be one of the functions of muscle-resident mesenchymal progenitors.

4 PW1+ Interstitial Cells

In 2010, Mitchell et al. isolated a Pax7- non-satellite cell muscle-resident population located in the skeletal muscle interstitial space and capable of myogenic differentiation [18, 71]. Apart from their location, these cells are characterized by the expression of the PW1/paternally expressed gene 3 (Peg3) and were named as PICs (PW1+ interstitial cells). In addition, PICs were mostly Sca1+ and CD34+. Lineage-tracing experiments demonstrated that PICs do not share the same embryonic origin of satellite cells and have increased potency, since they are capable to generate smooth and skeletal muscle cells and adipocytes [71].

Interestingly, satellite cells and mesoangioblasts also express PW1 [33, 34, 71]. Another report using a PW1 reporter mouse demonstrated that the combination of PW1, Sca1, and PDGFRα markers may be used to separate all the different stem cell populations in skeletal muscle [72]. Using this isolation strategy, the PW1+/Sca1+/PDGFRα+ subpopulation of cells results the most abundant and comprises the totality of the fibro-adipogenic mesenchymal progenitors with pro-adipogenic potential. Interestingly, this population is similar to the recently described Sca1+ brown adipocyte progenitors resident in skeletal muscle [73]. PW1−/Sca1+ cells were also functionally similar to the FAPs/MPs, although with just pro-fibrotic potential. In the referred study, this is the only cell subpopulation having a fibroblastic fate. It is therefore tempting to propose that fibro-adipogenic mesenchymal progenitors may be a heterogeneous population of muscle-resident Sca1+ cells that upon pro-fibrotic environmental cues (e.g., TGFβ) will turn into fibroblastic cells or, upon still poorly characterized signals, will acquire PW1 expression and become adipogenic. The PW1+/Sca1+/PDGFRα− subpopulation comprises a small group of cells with myogenic potential but negative for Pax7, defined by the authors as "non-satellite cell progenitors with myogenic potential", although they also hold some pro-adipogenic potential in vitro. These cells may account for the Sca1+ primary myoblast subpopulations described in some reports in the early 2000s [74–78]. Finally, the PW1+/Sca1−/PDGFRα− subpopulation included Pax7+ satellite cells and Pax7− cells which were positive for adult myogenic pericyte markers (e.g., NG2+/PDGFRβ+/Myf5−). Interestingly, PW1 is expressed in pericyte-derived mouse and human mesoangioblasts, where it regulates their myogenic ability and migration capacity [34].

5 CD133+ Cells

CD133 (prominin 1) was identified as a surface marker of both neural and hematopoietic stem and progenitor cells [79], and its expression has been used to characterize a population of human blood and muscle-derived myogenic stem cells. A small fraction of adult peripheral blood cells expressing CD133 was initially shown to display myogenic potential [80]. Muscle-resident human CD133+ cells are found both in the muscle interstitium and underneath the basal lamina of myofibers, co-expressing Pax7 [81]. When expanded in vitro, CD133+ preparations contained a heterogeneous population of cells expressing myoblast, pericyte, and mesenchymal genes [81, 82]. Additionally, expression of CD133 is unstable in culture and influenced by culture media; a thorough expression analysis has not been performed on freshly isolated cells due to their rarity [81–83]. When injected intramuscularly, human CD133+ cells effectively engraft in the muscle and contribute to myogenesis with a proportion entering the satellite cell compartment [81–83]. Transplanted human CD133+/Pax7+ cells are functional and capable of regenerating mouse muscle following injury [81]. Taken together, the variability of genes and anatomical location implies that CD133 positivity may distinguish a heterogeneous set of stem cells with high myogenic capacity. This makes them an interesting candidate for cellular therapies, and indeed they were tested in a pilot, phase I autologous clinical study for Duchenne muscular dystrophy based upon intramuscular transplantation without genetic correction [84]. However, whether the proportion of cells that extravasates and engrafts into muscles downstream of the injection site derives from the population expressing pericyte or mesenchymal markers or whether the CD133+/Pax7+ population is able to be safely injected systemically is currently unknown. Although there have been no reports on the contribution of murine CD133+ cells to skeletal muscle regeneration (probably due to technical limitations), the use of reporter mouse models for CD133 expression in other stem cell niches [85] may allow future lineage-tracing studies in murine skeletal muscle.

6 Concluding Remarks and Future Prospects

Adult muscle growth and regeneration is fueled by satellite cells. However, a growing milieu of interstitial stem or progenitor cells has been described both in resting and regenerating skeletal muscle, which are able to cross talk with satellite cells, myoblasts, myofibers, and cells of vascular and hematopoietic origin. These interstitial cells can differentiate into vascular, fibrogenic, adipogenic, osteogenic, and chondrogenic lineages in pathological conditions (e.g., [24, 56, 57, 86]), although their function and lineage relationships in healthy tissue (where non-myogenic differentiation pathways have a supportive role, or are absent, or are repressed) are still far from being understood.

Moreover, there is an urgent need to improve the characterization and distinction of the different populations of muscle interstitial progenitors, in order to determine whether particular cell types identified in different studies might actually be analogous and to find out which of them should be enhanced (or repressed) to foster efficient myogenesis. In the near future, advanced flow cytometry techniques such as spectral flow cytometry [87, 88] or flow cytometry coupled with mass spectrometry (mass cytometry or CyTOF) [89, 90], which is able to discriminate between many factors at the same time, could allow researchers to answer these questions.

A question likely to arise from this in-depth analysis is when does the differential expression of markers correspond to a subpopulation or to a separate progenitor population? Additionally, how definitive are these populations? During normal growth and regeneration, some interstitial muscle progenitors are known to have lineage plasticity. A well-characterized example of this is pericytes becoming Pax7+ stem cells residing under the basal lamina [26]. Whether these cells are identical to satellite cells and whether they can transdifferentiate back to the pericyte lineage is unknown. Moreover, lineage plasticity between many of the muscle resident cell populations has been demonstrated to increase greatly in pathogenic conditions [62, 91], implying that cellular relationships and composition of the cellular populations in uninjured, acutely, or chronically injured skeletal muscle could vary dramatically. Furthermore, it is crucial that in vivo analysis of cell populations is performed on freshly isolated cells, as changes in the physical environment and culture medium during ex vivo expansion can greatly impact surface-marker expression, as can different isolation protocols.

The majority of studies describing interstitial muscle stem cells in healthy and pathological tissue have been performed in rodents. Differences in marker expression between species are well documented in satellite cells (reviewed in [7]), and the field is now gradually improving the knowledge on the human satellite cell niche, their markers, and properties [92]. Similar characterization efforts are being performed on human mesenchymal progenitors [57]. Therefore, it is crucial to identify and characterize the comparable interstitial muscle stem cell populations in human muscle to the well-known rodent ones, in order to maximize therapeutic relevance.

Finally, a thorough characterization of the different subpopulations of muscle satellite cells and interstitial progenitors may enable the development of next-generation protocols to derive them from human pluripotent stem cells [93] for drug screening, tissue engineering, and cell therapies of skeletal muscle disorders.

Acknowledgments

We thank S. Benedetti, G. Cossu, S. Maffioletti, J. Morgan, E. Rebollo, M. Riminucci, D. Sassoon, and A. Serrano for insightful comments and discussions. E.P. acknowledges funding from MINECO, Spain (SAF2015-67369-R and "María de Maeztu" Programme for Units of Excellence in R&D MDM-2014-0370), AFM, CIBERNED (IntraCIBER 2015-2/06, InterCIBER PIE14/00061). F.S.T. and L.A.M. acknowledge funding from the European Union's Seventh Framework Programme for research, technological development and demonstration under grant agreement no. 602423 (PluriMes). F.S.T. is funded by a National Institute for Health Research (NIHR) Academic Clinical Fellowship in Paediatrics; the views expressed are those of the author and not necessarily those of the NHS, the NIHR or the Department of Health. Work in the Tedesco laboratory is also funded by the IMI joint undertaking under grant agreement n° 115582 (EU FP7 and EFPIA companies - EBiSC), Takeda New Frontier Science, the UK BBSRC and MRC, Duchenne Parent Project Onlus, Muscular Dystrophy UK, Duchenne Children's Trust, the Duchenne Research Fund and Fundació La Marató de TV3. F.S.T. dedicates this paper to the late Professor Paolo Bianco, whose ideas and vision on stem cell dynamics in mesodermal tissues have inspired part of this review and will be deeply missed.

References

1. Frontera WR, Ochala J (2015) Skeletal muscle: a brief review of structure and function. Calcif Tissue Int 96(3):183–195. doi:10.1007/s00223-014-9915-y
2. Comai G, Tajbakhsh S (2014) Molecular and cellular regulation of skeletal myogenesis. Curr Top Dev Biol 110:1–73. doi:10.1016/B978-0-12-405943-6.00001-4
3. Ciciliot S, Schiaffino S (2010) Regeneration of mammalian skeletal muscle. Basic mechanisms and clinical implications. Curr Pharm Des 16(8):906–914
4. Yin H, Price F, Rudnicki MA (2013) Satellite cells and the muscle stem cell niche. Physiol Rev 93(1):23–67. doi:10.1152/physrev.00043.2011
5. Relaix F, Zammit PS (2012) Satellite cells are essential for skeletal muscle regeneration: the cell on the edge returns centre stage. Development 139(16):2845–2856. doi:10.1242/dev.069088
6. Gros J, Manceau M, Thome V, Marcelle C (2005) A common somitic origin for embryonic muscle progenitors and satellite cells. Nature 435(7044):954–958. doi:10.1038/nature03572
7. Boldrin L, Muntoni F, Morgan JE (2010) Are human and mouse satellite cells really the same? J Histochem Cytochem 58(11):941–955. doi:10.1369/jhc.2010.956201
8. Tedesco FS, Dellavalle A, Diaz-Manera J, Messina G, Cossu G (2010) Repairing skeletal muscle: regenerative potential of skeletal muscle stem cells. J Clin Invest 120(1):11–19. doi:10.1172/JCI40373
9. Rocheteau P, Gayraud-Morel B, Siegl-Cachedenier I, Blasco MA, Tajbakhsh S (2012) A subpopulation of adult skeletal muscle stem cells retains all template DNA strands after cell division. Cell 148(1–2):112–125. doi:10.1016/j.cell.2011.11.049
10. Zammit PS, Golding JP, Nagata Y, Hudon V, Partridge TA, Beauchamp JR (2004) Muscle satellite cells adopt divergent fates: a mechanism for self-renewal? J Cell Biol 166(3):347–357. doi:10.1083/jcb.200312007jcb.200312007 [pii]
11. Seale P, Sabourin LA, Girgis-Gabardo A, Mansouri A, Gruss P, Rudnicki MA (2000) Pax7 is required for the specification of myogenic satellite cells. Cell 102(6):777–786
12. Cappellari O, Cossu G (2013) Pericytes in development and pathology of skeletal muscle. Circ Res 113(3):341–347. doi:10.1161/CIRCRESAHA.113.300203
13. Bentzinger CF, Wang YX, Rudnicki MA (2012) Building muscle: molecular regulation of myogenesis. Cold Spring Harb Perspect Biol 4(2). doi:10.1101/cshperspect.a008342
14. Armulik A, Genove G, Betsholtz C (2011) Pericytes: developmental, physiological, and pathological perspectives, problems, and promises. Dev Cell 21(2):193–215. doi:10.1016/j.devcel.2011.07.001
15. Berry SE (2015) Concise review: mesoangioblast and mesenchymal stem cell therapy for muscular dystrophy: progress, challenges, and future directions. Stem Cells Transl Med 4(1):91–98. doi:10.5966/sctm.2014-0060
16. Sambasivan R, Tajbakhsh S (2007) Skeletal muscle stem cell birth and properties. Semin Cell Dev Biol 18(6):870–882. doi:10.1016/j.semcdb.2007.09.013
17. Judson RN, Zhang RH, Rossi FM (2013) Tissue-resident mesenchymal stem/progenitor cells in skeletal muscle: collaborators or saboteurs? FEBS J 280(17):4100–4108. doi:10.1111/febs.12370
18. Pannerec A, Marazzi G, Sassoon D (2012) Stem cells in the hood: the skeletal muscle niche. Trends Mol Med 18(10):599–606. doi:10.1016/j.molmed.2012.07.004
19. Rouget C (1873) Mémoire sur le développment, la structure et les propertiés physiologiques des capillaries sanguins et lymphatiques. Arch Physiol Norm Pathol 5:60
20. Christov C, Chretien F, Abou-Khalil R, Bassez G, Vallet G, Authier FJ, Bassaglia Y,

Shinin V, Tajbakhsh S, Chazaud B, Gherardi RK (2007) Muscle satellite cells and endothelial cells: close neighbors and privileged partners. Mol Biol Cell 18(4):1397–1409. doi:10.1091/mbc.E06-08-0693 [pii] E06-08-0693

21. Dellavalle A, Sampaolesi M, Tonlorenzi R, Tagliafico E, Sacchetti B, Perani L, Innocenzi A, Galvez BG, Messina G, Morosetti R, Li S, Belicchi M, Peretti G, Chamberlain JS, Wright WE, Torrente Y, Ferrari S, Bianco P, Cossu G (2007) Pericytes of human skeletal muscle are myogenic precursors distinct from satellite cells. Nat Cell Biol 9(3):255–267. doi:10.1038/ncb1542
22. Sims DE (1986) The pericyte--a review. Tissue Cell 18(2):153–174
23. Kostallari E, Baba-Amer Y, Alonso-Martin S, Ngoh P, Relaix F, Lafuste P, Gherardi RK (2015) Pericytes in the myovascular niche promote post-natal myofiber growth and satellite cell quiescence. Development 142(7): 1242–1253. doi:10.1242/dev.115386
24. Birbrair A, Zhang T, Wang ZM, Messi ML, Mintz A, Delbono O (2013) Type-1 pericytes participate in fibrous tissue deposition in aged skeletal muscle. Am J Physiol Cell Physiol 305(11):C1098–C1113. doi:10.1152/ajpcell.00171.2013
25. Sampaolesi M, Blot S, D'Antona G, Granger N, Tonlorenzi R, Innocenzi A, Mognol P, Thibaud JL, Galvez BG, Barthelemy I, Perani L, Mantero S, Guttinger M, Pansarasa O, Rinaldi C, Cusella De Angelis MG, Torrente Y, Bordignon C, Bottinelli R, Cossu G (2006) Mesoangioblast stem cells ameliorate muscle function in dystrophic dogs. Nature 444(7119):574–579. doi:10.1038/nature05282
26. Dellavalle A, Maroli G, Covarello D, Azzoni E, Innocenzi A, Perani L, Antonini S, Sambasivan R, Brunelli S, Tajbakhsh S, Cossu G (2011) Pericytes resident in postnatal skeletal muscle differentiate into muscle fibres and generate satellite cells. Nat Commun 2:499. doi:10.1038/ncomms1508
27. Meng J, Adkin CF, Xu SW, Muntoni F, Morgan JE (2011) Contribution of human muscle-derived cells to skeletal muscle regeneration in dystrophic host mice. PloS One 6(3):e17454. doi:10.1371/journal.pone.0017454
28. Sacchetti B, Funari A, Remoli C, Giannicola G, Kogler G, Liedtke S, Cossu G, Serafini M, Sampaolesi M, Tagliafico E, Tenedini E, Saggio I, Robey PG, Riminucci M, Bianco P (2016) Mesenchymal Stem Cells. Human Committed Progenitors of Distinct Origin and Differentiation Potential Are Incorporated as Adventitial Cells in Microvessels. Stem Cell Reports. 6(6):897–913. doi:10.1016/j.stemcr.2016.05.011
29. Day K, Shefer G, Richardson JB, Enikolopov G, Yablonka-Reuveni Z (2007) Nestin-GFP reporter expression defines the quiescent state of skeletal muscle satellite cells. Dev Biol 304(1):246–259. doi:10.1016/j.ydbio.2006.12.026
30. Pierantozzi E, Vezzani B, Badin M, Curina C, Severi F, Petraglia F, Randazzo D, Rossi D, Sorrentino V (2016) Tissue-specific cultured human pericytes: perivascular cells from smooth muscle tissue have restricted mesodermal differentiation ability. Stem Cells Dev. doi:10.1089/scd.2015.0336
31. De Angelis L, Berghella L, Coletta M, Lattanzi L, Zanchi M, Cusella-De Angelis MG, Ponzetto C, Cossu G (1999) Skeletal myogenic progenitors originating from embryonic dorsal aorta coexpress endothelial and myogenic markers and contribute to postnatal muscle growth and regeneration. J Cell Biol 147(4):869–878
32. Minasi MG, Riminucci M, De Angelis L, Borello U, Berarducci B, Innocenzi A, Caprioli A, Sirabella D, Baiocchi M, De Maria R, Boratto R, Jaffredo T, Broccoli V, Bianco P, Cossu G (2002) The meso-angioblast: a multipotent, self-renewing cell that originates from the dorsal aorta and differentiates into most mesodermal tissues. Development 129(11):2773–2783
33. Tagliafico E, Brunelli S, Bergamaschi A, De Angelis L, Scardigli R, Galli D, Battini R, Bianco P, Ferrari S, Cossu G, Ferrari S (2004) TGFbeta/BMP activate the smooth muscle/bone differentiation programs in mesoangioblasts. J Cell Sci 117(Pt 19):4377–4388. doi:10.1242/jcs.01291
34. Bonfanti C, Rossi G, Tedesco FS, Giannotta M, Benedetti S, Tonlorenzi R, Antonini S, Marazzi G, Dejana E, Sassoon D, Cossu G, Messina G (2015) PW1/Peg3 expression regulates key properties that determine mesoangioblast stem cell competence. Nature Commun 6:6364. doi:10.1038/ncomms7364
35. Sampaolesi M, Torrente Y, Innocenzi A, Tonlorenzi R, D'Antona G, Pellegrino MA, Barresi R, Bresolin N, De Angelis MG, Campbell KP, Bottinelli R, Cossu G (2003) Cell therapy of alpha-sarcoglycan null dystrophic mice through intra-arterial delivery of mesoangioblasts. Science 301(5632):487–492. doi:10.1126/science.1082254
36. Tedesco FS, Hoshiya H, D'Antona G, Gerli MF, Messina G, Antonini S, Tonlorenzi R, Benedetti S, Berghella L, Torrente Y, Kazuki

Y, Bottinelli R, Oshimura M, Cossu G (2011) Stem cell-mediated transfer of a human artificial chromosome ameliorates muscular dystrophy. Sci Transl Med 3(96):96ra78. doi:10.1126/scitranslmed.3002342

37. Diaz-Manera J, Touvier T, Dellavalle A, Tonlorenzi R, Tedesco FS, Messina G, Meregalli M, Navarro C, Perani L, Bonfanti C, Illa I, Torrente Y, Cossu G (2010) Partial dysferlin reconstitution by adult murine mesoangioblasts is sufficient for full functional recovery in a murine model of dysferlinopathy. Cell Death Dis 1:e61. doi:10.1038/cddis.2010.35
38. Giannotta M, Benedetti S, Tedesco FS, Corada M, Trani M, D'Antuono R, Millet Q, Orsenigo F, Galvez BG, Cossu G, Dejana E (2014) Targeting endothelial junctional adhesion molecule-A/EPAC/ Rap-1 axis as a novel strategy to increase stem cell engraftment in dystrophic muscles. EMBO Mol Med 6(2):239–258. doi:10.1002/emmm.201302520
39. Berry SE, Liu J, Chaney EJ, Kaufman SJ (2007) Multipotential mesoangioblast stem cell therapy in the mdx/utrn−/− mouse model for Duchenne muscular dystrophy. Regen Med 2(3):275–288. doi:10.2217/17460751.2.3.275
40. Cossu G, Previtali SC, Napolitano S, Cicalese MP, Tedesco FS, Nicastro F, Noviello M, Roostalu U, Natali Sora MG, Scarlato M, De Pellegrin M, Godi C, Giuliani S, Ciotti F, Tonlorenzi R, Lorenzetti I, Rivellini C, Benedetti S, Gatti R, Marktel S, Mazzi B, Tettamanti A, Ragazzi M, Imro MA, Marano G, Ambrosi A, Fiori R, Sormani MP, Bonini C, Venturini M, Politi LS, Torrente Y, Ciceri F (2015) Intra-arterial transplantation of HLA-matched donor mesoangioblasts in Duchenne muscular dystrophy. EMBO Mol Med 7(12):1513–1528. doi:10.15252/emmm.201505636
41. Tedesco FS, Gerli MF, Perani L, Benedetti S, Ungaro F, Cassano M, Antonini S, Tagliafico E, Artusi V, Longa E, Tonlorenzi R, Ragazzi M, Calderazzi G, Hoshiya H, Cappellari O, Mora M, Schoser B, Schneiderat P, Oshimura M, Bottinelli R, Sampaolesi M, Torrente Y, Broccoli V, Cossu G (2012) Transplantation of genetically corrected human iPSC-derived progenitors in mice with limb-girdle muscular dystrophy. Sci Transl Med 4(140):140ra189. doi:10.1126/scitranslmed.3003541
42. Maffioletti SM, Noviello M, English K, Tedesco FS (2014) Stem cell transplantation for muscular dystrophy: the challenge of immune response. Biomed Res Int 2014: 964010. doi:10.1155/2014/964010
43. Noviello M, Tedesco FS, Bondanza A, Tonlorenzi R, Rosaria Carbone M, Gerli MF, Marktel S, Napolitano S, Cicalese MP, Ciceri F, Peretti G, Cossu G, Bonini C (2014) Inflammation converts human mesoangioblasts into targets of alloreactive immune responses: implications for allogeneic cell therapy of DMD. Mol Ther 22(7):1342–1352. doi:10.1038/mt.2014.62
44. Quattrocelli M, Costamagna D, Giacomazzi G, Camps J, Sampaolesi M (2014) Notch signaling regulates myogenic regenerative capacity of murine and human mesoangioblasts. Cell Death Dis 5:e1448. doi:10.1038/cddis.2014.401
45. Costamagna D, Berardi E, Ceccarelli G, Sampaolesi M (2015) Adult stem cells and skeletal muscle regeneration. Curr Gene Ther 15(4):348–363
46. Greenhalgh SN, Iredale JP, Henderson NC (2013) Origins of fibrosis: pericytes take centre stage. F1000Prime Rep 5:37. doi:10.12703/P5-37
47. Birbrair A, Zhang T, Wang ZM, Messi ML, Enikolopov GN, Mintz A, Delbono O (2013) Skeletal muscle pericyte subtypes differ in their differentiation potential. Stem Cell Res 10(1):67–84. doi:10.1016/j.scr.2012.09.003
48. Joe AW, Yi L, Natarajan A, Le Grand F, So L, Wang J, Rudnicki MA, Rossi FM (2010) Muscle injury activates resident fibro/adipogenic progenitors that facilitate myogenesis. Nat Cell Biol 12(2):153–163. doi:10.1038/ncb2015 ncb2015 [pii]
49. Uezumi A, Fukada S, Yamamoto N, Takeda S, Tsuchida K (2010) Mesenchymal progenitors distinct from satellite cells contribute to ectopic fat cell formation in skeletal muscle. Nat Cell Biol 12(2):143–152. doi:10.1038/ncb2014 ncb2014 [pii]
50. Wosczyna MN, Biswas AA, Cogswell CA, Goldhamer DJ (2012) Multipotent progenitors resident in the skeletal muscle interstitium exhibit robust BMP-dependent osteogenic activity and mediate heterotopic ossification. J Bone Miner Res 27(5):1004–1017. doi:10.1002/jbmr.1562
51. Bianco P (2014) "Mesenchymal" stem cells. Annu Rev Cell Dev Biol 30:677–704. doi:10.1146/annurev-cellbio-100913-013132
52. Bianco P, Cao X, Frenette PS, Mao JJ, Robey PG, Simmons PJ, Wang CY (2013) The meaning, the sense and the significance: translating the science of mesenchymal stem cells into medicine. Nat Med 19(1):35–42. doi:10.1038/nm.3028
53. Lemos DR, Babaeijandaghi F, Low M, Chang CK, Lee ST, Fiore D, Zhang RH, Natarajan

A, Nedospasov SA, Rossi FM (2015) Nilotinib reduces muscle fibrosis in chronic muscle injury by promoting TNF-mediated apoptosis of fibro/adipogenic progenitors. Nat Med 21(7):786–794. doi:10.1038/nm.3869

54. Mozzetta C, Consalvi S, Saccone V, Tierney M, Diamantini A, Mitchell KJ, Marazzi G, Borsellino G, Battistini L, Sassoon D, Sacco A, Puri PL (2013) Fibroadipogenic progenitors mediate the ability of HDAC inhibitors to promote regeneration in dystrophic muscles of young, but not old Mdx mice. EMBO Mol Med 5(4):626–639. doi:10.1002/emmm.201202096
55. Saccone V, Consalvi S, Giordani L, Mozzetta C, Barozzi I, Sandona M, Ryan T, Rojas-Munoz A, Madaro L, Fasanaro P, Borsellino G, De Bardi M, Frige G, Termanini A, Sun X, Rossant J, Bruneau BG, Mercola M, Minucci S, Puri PL (2014) HDAC-regulated myomiRs control BAF60 variant exchange and direct the functional phenotype of fibro-adipogenic progenitors in dystrophic muscles. Genes Dev 28(8):841–857. doi:10.1101/gad.234468.113
56. Uezumi A, Fukada S, Yamamoto N, Ikemoto-Uezumi M, Nakatani M, Morita M, Yamaguchi A, Yamada H, Nishino I, Hamada Y, Tsuchida K (2014) Identification and characterization of PDGFRalpha+ mesenchymal progenitors in human skeletal muscle. Cell Death Dis 5:e1186. doi:10.1038/cddis.2014.161
57. Arrighi N, Moratal C, Clement N, Giorgetti-Peraldi S, Peraldi P, Loubat A, Kurzenne JY, Dani C, Chopard A, Dechesne CA (2015) Characterization of adipocytes derived from fibro/adipogenic progenitors resident in human skeletal muscle. Cell Death Dis 6:e1733. doi:10.1038/cddis.2015.79
58. Heredia JE, Mukundan L, Chen FM, Mueller AA, Deo RC, Locksley RM, Rando TA, Chawla A (2013) Type 2 innate signals stimulate fibro/adipogenic progenitors to facilitate muscle regeneration. Cell 153(2):376–388. doi:10.1016/j.cell.2013.02.053
59. Villalta SA, Nguyen HX, Deng B, Gotoh T, Tidball JG (2009) Shifts in macrophage phenotypes and macrophage competition for arginine metabolism affect the severity of muscle pathology in muscular dystrophy. Hum Mol Genet 18(3):482–496. doi:10.1093/hmg/ddn376 ddn376 [pii]
60. Dulauroy S, Di Carlo SE, Langa F, Eberl G, Peduto L (2012) Lineage tracing and genetic ablation of ADAM12(+) perivascular cells identify a major source of profibrotic cells during acute tissue injury. Nat Med 18(8):1262–1270. doi:10.1038/nm.2848
61. Murphy MM, Lawson JA, Mathew SJ, Hutcheson DA, Kardon G (2011) Satellite cells, connective tissue fibroblasts and their interactions are crucial for muscle regeneration. Development 138(17):3625–3637. doi:10.1242/dev.064162
62. Pessina P, Kharraz Y, Jardi M, Fukada S, Serrano AL, Perdiguero E, Munoz-Canoves P (2015) Fibrogenic cell plasticity blunts tissue regeneration and aggravates muscular dystrophy. Stem Cell Reports 4(6):1046–1060. doi:10.1016/j.stemcr.2015.04.007
63. Zou Y, Zhang RZ, Sabatelli P, Chu ML, Bonnemann CG (2008) Muscle interstitial fibroblasts are the main source of collagen VI synthesis in skeletal muscle: implications for congenital muscular dystrophy types Ullrich and Bethlem. J Neuropathol Exp Neurol 67(2):144–154. doi:10.1097/nen.0b013e3181634ef7
64. Urciuolo A, Quarta M, Morbidoni V, Gattazzo F, Molon S, Grumati P, Montemurro F, Tedesco FS, Blaauw B, Cossu G, Vozzi G, Rando TA, Bonaldo P (2013) Collagen VI regulates satellite cell self-renewal and muscle regeneration. Nat Commun 4:1964. doi:10.1038/ncomms2964
65. Bonnemann CG (2011) The collagen VI-related myopathies: muscle meets its matrix. Nat Rev Neurol 7(7):379–390. doi:10.1038/nrneurol.2011.81
66. Cescon M, Gattazzo F, Chen P, Bonaldo P (2015) Collagen VI at a glance. J Cell Sci 128(19):3525–3531. doi:10.1242/jcs.169748
67. De Palma S, Leone R, Grumati P, Vasso M, Polishchuk R, Capitanio D, Braghetta P, Bernardi P, Bonaldo P, Gelfi C (2013) Changes in muscle cell metabolism and mechanotransduction are associated with myopathic phenotype in a mouse model of collagen VI deficiency. PloS One 8(2):e56716. doi:10.1371/journal.pone.0056716
68. Grumati P, Coletto L, Schiavinato A, Castagnaro S, Bertaggia E, Sandri M, Bonaldo P (2011) Physical exercise stimulates autophagy in normal skeletal muscles but is detrimental for collagen VI-deficient muscles. Autophagy 7(12):1415–1423
69. Grumati P, Coletto L, Sabatelli P, Cescon M, Angelin A, Bertaggia E, Blaauw B, Urciuolo A, Tiepolo T, Merlini L, Maraldi NM, Bernardi P, Sandri M, Bonaldo P (2010) Autophagy is defective in collagen VI muscular dystrophies, and its reactivation rescues myofiber degeneration. Nat Med 16(11):1313–1320. doi:10.1038/nm.2247
70. Meng J, Adkin CF, Arechavala-Gomeza V, Boldrin L, Muntoni F, Morgan JE (2010) The contribution of human synovial stem

cells to skeletal muscle regeneration. Neuromuscul Disord 20(1):6–15. doi:10.1016/j.nmd.2009.11.007

71. Mitchell KJ, Pannerec A, Cadot B, Parlakian A, Besson V, Gomes ER, Marazzi G, Sassoon DA (2010) Identification and characterization of a non-satellite cell muscle resident progenitor during postnatal development. Nat Cell Biol 12(3):257–266. doi:10.1038/ncb2025 ncb2025 [pii]
72. Pannerec A, Formicola L, Besson V, Marazzi G, Sassoon DA (2013) Defining skeletal muscle resident progenitors and their cell fate potentials. Development 140(14):2879–2891. doi:10.1242/dev.089326
73. Schulz TJ, Huang TL, Tran TT, Zhang H, Townsend KL, Shadrach JL, Cerletti M, McDougall LE, Giorgadze N, Tchkonia T, Schrier D, Falb D, Kirkland JL, Wagers AJ, Tseng YH (2011) Identification of inducible brown adipocyte progenitors residing in skeletal muscle and white fat. Proc Natl Acad Sci USA 108(1):143–148. doi:10.1073/pnas.1010929108
74. Asakura A, Seale P, Girgis-Gabardo A, Rudnicki MA (2002) Myogenic specification of side population cells in skeletal muscle. J Cell Biol 159(1):123–134
75. Mitchell PO, Mills T, O'Connor RS, Kline ER, Graubert T, Dzierzak E, Pavlath GK (2005) Sca-1 negatively regulates proliferation and differentiation of muscle cells. Dev Biol 283(1):240–252. doi:10.1016/j.ydbio.2005.04.016
76. Nicole S, Desforges B, Millet G, Lesbordes J, Cifuentes-Diaz C, Vertes D, Cao ML, De Backer F, Languille L, Roblot N, Joshi V, Gillis JM, Melki J (2003) Intact satellite cells lead to remarkable protection against Smn gene defect in differentiated skeletal muscle. J Cell Biol 161(3):571–582. doi:10.1083/jcb.200210117
77. Qu-Petersen Z, Deasy B, Jankowski R, Ikezawa M, Cummins J, Pruchnic R, Mytinger J, Cao B, Gates C, Wernig A, Huard J (2002) Identification of a novel population of muscle stem cells in mice: potential for muscle regeneration. J Cell Biol 157(5):851–864. doi:10.1083/jcb.200108150
78. Sherwood RI, Christensen JL, Conboy IM, Conboy MJ, Rando TA, Weissman IL, Wagers AJ (2004) Isolation of adult mouse myogenic progenitors: functional heterogeneity of cells within and engrafting skeletal muscle. Cell 119(4):543–554. doi:10.1016/j.cell.2004.10.021
79. Bauer N, Fonseca AV, Florek M, Freund D, Jaszai J, Bornhauser M, Fargeas CA, Corbeil D (2008) New insights into the cell biology of hematopoietic progenitors by studying prominin-1 (CD133). Cells Tissues Organs 188(1–2):127–138. doi:10.1159/000112847
80. Torrente Y, Belicchi M, Sampaolesi M, Pisati F, Meregalli M, D'Antona G, Tonlorenzi R, Porretti L, Gavina M, Mamchaoui K, Pellegrino MA, Furling D, Mouly V, Butler-Browne GS, Bottinelli R, Cossu G, Bresolin N (2004) Human circulating AC133(+) stem cells restore dystrophin expression and ameliorate function in dystrophic skeletal muscle. J Clin Invest 114(2):182–195. doi:10.1172/JCI20325
81. Meng J, Chun S, Asfahani R, Lochmuller H, Muntoni F, Morgan J (2014) Human skeletal muscle-derived CD133(+) cells form functional satellite cells after intramuscular transplantation in immunodeficient host mice. Mol Ther 22(5):1008–1017. doi:10.1038/mt.2014.26
82. Negroni E, Riederer I, Chaouch S, Belicchi M, Razini P, Di Santo J, Torrente Y, Butler-Browne GS, Mouly V (2009) In vivo myogenic potential of human CD133+ muscle-derived stem cells: a quantitative study. Mol Ther 17(10):1771–1778. doi:10.1038/mt.2009.167
83. Benchaouir R, Meregalli M, Farini A, D'Antona G, Belicchi M, Goyenvalle A, Battistelli M, Bresolin N, Bottinelli R, Garcia L, Torrente Y (2007) Restoration of human dystrophin following transplantation of exon-skipping-engineered DMD patient stem cells into dystrophic mice. Cell Stem Cell 1(6):646–657. doi:10.1016/j.stem.2007.09.016
84. Torrente Y, Belicchi M, Marchesi C, D'Antona G, Cogiamanian F, Pisati F, Gavina M, Giordano R, Tonlorenzi R, Fagiolari G, Lamperti C, Porretti L, Lopa R, Sampaolesi M, Vicentini L, Grimoldi N, Tiberio F, Songa V, Baratta P, Prelle A, Forzenigo L, Guglieri M, Pansarasa O, Rinaldi C, Mouly V, Butler-Browne GS, Comi GP, Biondetti P, Moggio M, Gaini SM, Stocchetti N, Priori A, D'Angelo MG, Turconi A, Bottinelli R, Cossu G, Rebulla P, Bresolin N (2007) Autologous transplantation of muscle-derived CD133+ stem cells in Duchenne muscle patients. Cell Transplant 16(6):563–577
85. Snippert HJ, van Es JH, van den Born M, Begthel H, Stange DE, Barker N, Clevers H (2009) Prominin-1/CD133 marks stem cells and early progenitors in mouse small intestine. Gastroenterology 136(7):2187–2194 . doi:10.1053/j.gastro.2009.03.002e2181

86. Uezumi A, Ito T, Morikawa D, Shimizu N, Yoneda T, Segawa M, Yamaguchi M, Ogawa R, Matev MM, Miyagoe-Suzuki Y, Takeda S, Tsujikawa K, Tsuchida K, Yamamoto H, Fukada S (2011) Fibrosis and adipogenesis originate from a common mesenchymal progenitor in skeletal muscle. J Cell Sci 124(Pt 21):3654–3664. doi:10.1242/jcs.086629
87. Gregori G, Patsekin V, Rajwa B, Jones J, Ragheb K, Holdman C, Robinson JP (2012) Hyperspectral cytometry at the single-cell level using a 32-channel photodetector. Cytometry A 81(1):35–44. doi:10.1002/cyto.a.21120
88. Gregori G, Rajwa B, Patsekin V, Jones J, Furuki M, Yamamoto M, Paul Robinson J (2014) Hyperspectral cytometry. Curr Top Microbiol Immunol 377:191–210. doi:10.1007/82_2013_359
89. Bandura DR, Baranov VI, Ornatsky OI, Antonov A, Kinach R, Lou X, Pavlov S, Vorobiev S, Dick JE, Tanner SD (2009) Mass cytometry: technique for real time single cell multitarget immunoassay based on inductively coupled plasma time-of-flight mass spectrometry. Anal Chem 81(16):6813–6822. doi:10.1021/ac901049w
90. Bendall SC, Simonds EF, Qiu P, Amir el AD, Krutzik PO, Finck R, Bruggner RV, Melamed R, Trejo A, Ornatsky OI, Balderas RS, Plevritis SK, Sachs K, Pe'er D, Tanner SD, Nolan GP (2011) Single-cell mass cytometry of differential immune and drug responses across a human hematopoietic continuum. Science 332(6030):687–696. doi:10.1126/science.1198704
91. Biressi S, Miyabara EH, Gopinath SD, Carlig PM, Rando TA (2014) A Wnt-TGFbeta2 axis induces a fibrogenic program in muscle stem cells from dystrophic mice. Sci Transl Med 6(267):267ra176. doi:10.1126/scitranslmed.3008411
92. Xu X, Wilschut KJ, Kouklis G, Tian H, Hesse R, Garland C, Sbitany H, Hansen S, Seth R, Knott PD, Hoffman WY, Pomerantz JH (2015) Human satellite cell transplantation and regeneration from diverse skeletal muscles. Stem Cell Reports 5(3):419–434. doi:10.1016/j.stemcr.2015.07.016
93. Loperfido M, Steele-Stallard HB, Tedesco FS, VandenDriessche T (2015) Pluripotent stem cells for gene therapy of degenerative muscle diseases. Curr Gene Ther 15(4):364–380
94. Beauchamp JR, Heslop L, Yu DS, Tajbakhsh S, Kelly RG, Wernig A, Buckingham ME, Partridge TA, Zammit PS (2000) Expression of CD34 and Myf5 defines the majority of quiescent adult skeletal muscle satellite cells. J Cell Biol 151(6):1221–1234
95. Montarras D, Morgan J, Collins C, Relaix F, Zaffran S, Cumano A, Partridge T, Buckingham M (2005) Direct isolation of satellite cells for skeletal muscle regeneration. Science 309(5743):2064–2067. doi:10.1126/science.1114758 1114758 [pii]
96. Fukada S, Uezumi A, Ikemoto M, Masuda S, Segawa M, Tanimura N, Yamamoto H, Miyagoe-Suzuki Y, Takeda S (2007) Molecular signature of quiescent satellite cells in adult skeletal muscle. Stem Cells 25(10):2448–2459. doi:10.1634/stemcells.2007-0019
97. Cheung TH, Quach NL, Charville GW, Liu L, Park L, Edalati A, Yoo B, Hoang P, Rando TA (2012) Maintenance of muscle stem-cell quiescence by microRNA-489. Nature 482(7386):524–528. doi:10.1038/nature10834
98. Boldrin F, Casonato S, Dainese E, Sala C, Dhar N, Palu G, Riccardi G, Cole ST, Manganelli R (2010) Development of a repressible mycobacterial promoter system based on two transcriptional repressors. Nucleic Acids Res 38(12):e134. doi:10.1093/nar/gkq235
99. Penton CM, Thomas-Ahner JM, Johnson EK, McAllister C, Montanaro F (2013) Muscle side population cells from dystrophic or injured muscle adopt a fibro-adipogenic fate. PloS One 8(1):e54553. doi:10.1371/journal.pone.0054553
100. Tanaka KK, Hall JK, Troy AA, Cornelison DD, Majka SM, Olwin BB (2009) Syndecan-4-expressing muscle progenitor cells in the SP engraft as satellite cells during muscle regeneration. Cell Stem Cell 4(3):217–225. doi:10.1016/j.stem.2009.01.016
101. Uezumi A, Ojima K, Fukada S, Ikemoto M, Masuda S, Miyagoe-Suzuki Y, Takeda S (2006) Functional heterogeneity of side population cells in skeletal muscle. Biochem Biophys Res Commun 341(3):864–873. doi:10.1016/j.bbrc.2006.01.037
102. Tamaki T, Akatsuka A, Ando K, Nakamura Y, Matsuzawa H, Hotta T, Roy RR, Edgerton VR (2002) Identification of myogenic-endothelial progenitor cells in the interstitial spaces of skeletal muscle. J Cell Biol 157(4):571–577. doi:10.1083/jcb.200112106
103. Honda H, Kimura H, Rostami A (1990) Demonstration and phenotypic characterization of resident macrophages in rat skeletal muscle. Immunology 70(2):272–277
104. McLennan IS (1993) Resident macrophages (ED2- and ED3-positive) do not phagocytose degenerating rat skeletal muscle fibres. Cell Tissue Res 272(1):193–196
105. Pimorady-Esfahani A, Grounds M, McMenamin PG (1997) Macrophages and

dendritic cells in normal and regenerating murine skeletal muscle. Muscle Nerve 20: 158–166

106. Brigitte M, Schilte C, Plonquet A, Baba-Amer Y, Henri A, Charlier C, Tajbakhsh S, Albert M, Gherardi RK, Chretien F (2010) Muscle resident macrophages control the immune cell reaction in a mouse model of notexin-induced myoinjury. Arthritis Rheum 62(1):268–279. doi:10.1002/art.27183

107. Chazaud B (2016) Inflammation during skeletal muscle regeneration and tissue remodeling: application to exercise-induced muscle damage management. Immunol Cell Biol 94(2):140–145. doi:10.1038/icb.2015.97

108. Kharraz Y, Guerra J, Mann CJ, Serrano AL, Munoz-Canoves P (2013) Macrophage plasticity and the role of inflammation in skeletal muscle repair. Mediat Inflamm 2013:491497. doi:10.1155/2013/491497

109. Asahara T, Murohara T, Sullivan A, Silver M, van der Zee R, Li T, Witzenbichler B, Schatteman G, Isner JM (1997) Isolation of putative progenitor endothelial cells for angiogenesis. Science 275(5302):964–967

110. Asahara T, Masuda H, Takahashi T, Kalka C, Pastore C, Silver M, Kearne M, Magner M, Isner JM (1999) Bone marrow origin of endothelial progenitor cells responsible for postnatal vasculogenesis in physiological and pathological neovascularization. Circ Res 85(3):221–228

111. Yamashita J, Itoh H, Hirashima M, Ogawa M, Nishikawa S, Yurugi T, Naito M, Nakao K, Nishikawa S (2000) Flk1-positive cells derived from embryonic stem cells serve as vascular progenitors. Nature 408(6808):92–96. doi:10.1038/35040568

112. Kardon G, Harfe BD, Tabin CJ (2003) A Tcf4-positive mesodermal population provides a prepattern for vertebrate limb muscle patterning. Dev Cell 5(6):937–944 S1534580703003605 [pii]

113. Mathew SJ, Hansen JM, Merrell AJ, Murphy MM, Lawson JA, Hutcheson DA, Hansen MS, Angus-Hill M, Kardon G (2011) Connective tissue fibroblasts and Tcf4 regulate myogenesis. Development 138(2):371–384. doi:10.1242/dev.057463 138/2/371 [pii]

Chapter 8

Isolation and Characterization of Vessel-Associated Stem/Progenitor Cells from Skeletal Muscle

Rossana Tonlorenzi, Giuliana Rossi, and Graziella Messina

Abstract

More than 10 years ago, we isolated from mouse embryonic dorsal aorta a population of vessel-associated stem/progenitor cells, originally named mesoangioblasts (MABs), capable to differentiate in all mesodermal-derived tissues, including skeletal muscle. Similar though not identical cells have been later isolated and characterized from small vessels of adult mouse and human skeletal muscles. When delivered through the arterial circulation, MABs cross the blood vessel wall and participate in skeletal muscle regeneration, leading to an amelioration of muscular dystrophies in different preclinical animal models. As such, human MABs have been used under clinical-grade conditions for a Phase I/II clinical trial for Duchenne muscular dystrophy, just concluded. Although some pericyte markers can be used to identify mouse and human MABs, no single unequivocal marker can be used to isolate MABs. As a result, MABs are mainly defined by their isolation method and functional properties. This chapter provides detailed methods for isolation, culture, and characterization of MABs in light of the recent identification of a new marker, PW1/Peg3, to screen and identify competent MABs before their use in cell therapy.

Key words Mesoangioblasts, Muscle stem cells, Pericytes, Mesodermal lineages, Cell culture, PW1/Peg3

1 Introduction

Skeletal muscle regeneration is assured by satellite cells (SCs), which are localized between the basal lamina and the sarcolemmal membrane [1]. Despite their indispensable role in the regeneration of adult skeletal muscle [2–6], in the last years, several groups identified different atypical muscle progenitors of non-somitic origin that are able to differentiate in skeletal muscle [7–18]. Among these, in the last years mesoangioblasts (MABs) demonstrated to be the most promising. MABs are blood vessel-associated stem/progenitor cells that can differentiate into mesoderm cell types, including skeletal muscle [7]. Most importantly, when systemically delivered, MABs cross the blood vessel wall and participate in skeletal muscle regeneration, leading to an amelioration of muscular dystrophies in different preclinical models [19–23]. The ability of MABs to cross the

Eusebio Perdiguero and DDW Cornelison (eds.), *Muscle Stem Cells: Methods and Protocols*, Methods in Molecular Biology, vol. 1556, DOI 10.1007/978-1-4939-6771-1_8, © Springer Science+Business Media LLC 2017

vessel wall confers an advantage as therapeutic donor stem cells as compared with SCs and myoblasts that need to be delivered directly into the muscle tissue to properly engraft [24, 25]. Consistently, human MABs, expanded under clinical-grade conditions, have been recently used for a Phase I/II clinical trial for Duchenne muscular dystrophy (EudraCT no. 2011-000176-33; [26]).

Both human and mouse MABs can be retrospectively characterized by a combination of markers not all shared by both cell populations (Table 1).

This chapter provides an updated guide for isolation, expansion, and characterization of MABs from adult human and mouse skeletal muscle (*see* Subheadings 3.1–3.3 and 3.9). Various differentiation methods are also described: spontaneous skeletal muscle differentiation (*see* Subheadings 3.5), induction of smooth muscle

Table 1
List of the positive and negative markers used to characterize human and mouse MABs

Marker	mMABs	hMABs
Sca-1	+	NA
c-kit	–	–
CD45 (leukocyte common antigen)	–	–
CD34	–	–
CD31 (PECAM-1)	–	–
VE-cadherin	–	–
Tie2	–	–
Pax3	+	+
Pax7	–	–
Myf5	–	–
MyoD	–	–
PW1/Peg3	+	+
Alkaline phosphatase	+/–	+/–
Ng2	+	+
Pdgfrb	+	+
CD44 (hyaluronan-binding protein)	+	+
CD56 (NCAM)	–	–
CD133 (prominin-1)	–	–
CD146 (MCAM)	+/–	+/–

by TGF-β (*see* Subheading 3.6), induction of osteoblasts by BMP2 (*see* Subheading 3.7), and induction of adipocytes (*see* Subheading 3.8). Additionally, in light of the recent identification of PW1/Peg3 as a key marker essential in regulating MAB competence [27], a paragraph will be dedicated to the description of the methods developed to detect PW1/Peg3 in mouse and human MABs (*see* Subheading 3.9).

Successful isolation, propagation, and characterization of MABs require basic animal handling and competences in cellular and molecular biology. Notably, for human MABs, cell cultures must be cultured under physiological O_2 conditions (3 % O_2, 5 % CO_2, 92 % N_2). Importantly, all the procedure described in this chapter should be performed under sterile conditions in either Class II biohazard flow hoods (especially for human cells) or laminar flow horizontal hoods. For mouse samples, the Institutional Animal Welfare Body (AWB), Ethics Committee, and National Local Authority must approve the protocols. Muscle biopsies must be performed under general or local anesthesia with the minimum degree of pain. Approval of Institutional Ethics Committee and patients' informed consent are necessary in case of human samples.

2 Materials

2.1 Basic Materials, Media, and Solutions

1. 5 % CO_2, 3 % O2, 92 % N2 humidified (water-saturated) incubator.
2. Biohazard flow hood.
3. Benchtop centrifuge.
4. Inverted phase-contrast microscope.
5. Stereoscopic microscope (optional).
6. Water bath.
7. +4 °C refrigerator.
8. −20° freezer.
9. −80° freezer.
10. Liquid nitrogen tank.
11. *Collagen solution (250 ml) and Petri dish coating*: transfer 250 mg of lyophilized collagen type 1 to a sterile glass bottle. Gradually add 50 ml of glacial acetic acid. Due to variable purity in different collagen preparations, the time necessary for complete dissolution of collagen may vary. Overnight incubation at room temperature is recommended. After complete collagen dissolution in acetic acid, gradually add 200 ml of ultrapure distilled water. Mix gently without shaking. Store up to 6 months at 4 °C.

To obtain an efficient solution, it is very important to wait for the collagen to be completely dissolved in acetic acid before adding ultrapure distilled water.

Membrane filtration is not recommended (a substantial portion of collagen will be lost or degraded).

For coating, place the appropriate number of Petri dishes under a biohazard flow hood. Carefully add the collagen type I solution into each Petri dish making sure the whole surface is completely covered. Use 1, 5, and 10 ml of collagen solution for 3.5-, 6-, and 9-cm petri dish, respectively, and let stand 5 min. Slowly remove most of the solution (80–90 %), leaving the surface of the dish uniformly wet and let dry completely.

12. *Collagenase/Dispase solution (50 ml)*: depending on enzyme activity U/W (specified by the manufacturer for each lot), weigh the appropriate amounts to prepare 50 ml of 1 U/ml collagenase, 0.5 U/ml Dispase II stock solution in PBS. Filter through a 0.22-μM syringe filter and store in 5-ml aliquots up to 6 months at −20 °C.

13. *D2 medium (250 ml)*: 245 ml high-glucose DMEM supplemented with 5 ml heat-inactivated horse serum (HS, Euroclone), 2 mM glutamine, 1 % penicillin/streptomycin solution (10,000 U/ml and 10 μg/ml, respectively), 1 mM sodium pyruvate.

    Store up to 4 weeks at +4 °C.

14. D20 medium (250 ml): 200 ml high-glucose DMEM supplemented with 50 ml heat-inactivated fetal bovine serum (FBS, Euroclone), 2 mM glutamine, 1 % penicillin/streptomycin solution (10,000 U/ml and 10 μg/ml, respectively), and 1 mM sodium pyruvate.

    Store up to 4 weeks at +4 °C.

15. M5 medium (250 ml): 237.5 ml Megacell DMEM supplemented with 12.5 ml heat-inactivated FBS, 2 mM glutamine, 1 % penicillin/streptomycin solution (10,000 U/ml and 10 μg/ml, respectively), 0.1 mM β-mercaptoethanol, 1 % nonessential amino acids, and 1.25 μg human recombinant bFGF (Life Technologies).

    Store up to 2 weeks at +4 °C.

16. Matrigel stock and Petri dish coating.

    Thaw a 10-ml bottle of growth factor-reduced Matrigel (Corning) overnight on ice. Prepare aliquots by using sterile microcentrifuge tubes chilled on ice and pipet tips kept at +4 °C. Store undiluted Matrigel in 100-μl aliquots up to 12 months at −20 °C.

    Concentrated Matrigel solution tends to polymerize very quickly at room temperature.

Frozen aliquots have to be thawed on ice and immediately diluted in cold sterile solutions.

Gentle and slow pipetting is recommendable for either diluted or concentrated Matrigel solutions, to avoid degradation of the components. Matrigel solutions cannot be membrane filtered.

For Petri coating, thaw Matrigel stock solution on ice and prepare the working solution by diluting the stock 1:80 in cold DMEM (without any supplement). Place the appropriate number of Petri dishes to be coated under a hood and apply Matrigel working solution carefully into each Petri dish making sure the whole surface is completely covered.

Use 1, 3, and 7 ml of Matrigel working solution for 3.5-, 5-, and 9-cm Petri dishes, respectively, and incubate 30 min–2 h at 37 °C. Just before use, remove the Matrigel working solution and gently rinse the dish surface with appropriate culture medium. Matrigel-coated Petri dishes have to be freshly prepared.

Diluted Matrigel working solution can be stored up to 24 h at +4 °C.

17. Human recombinant bFGF stock solution (1 ml): reconstitute 50 μg of human recombinant bFGF (Life Technologies) in 1 ml of 10 mM Tris–HCl, pH 7.6.

    Store in 25-μl aliquots up to 6 months at −20 °C.
18. *Freezing solution (50 ml)*: 45 ml heat-inactivated FBS supplemented with 5 ml Hybri-Max DMSO. Prepare fresh.

2.2 Isolation, Propagation, and Freezing of MABs from Human Skeletal Muscle (hMABs)

1. Human skeletal muscle fragments (at least 200 mg of tissue; *see* **Note 1**).
2. Ca^{2+}/Mg^{2+} free phosphate-buffered saline (CMF-PBS).
3. M5 medium.
4. 6- and 15-cm Petri dishes (Nunc).
5. 3.5- and 6-cm collagen-coated Petri dishes.
6. Rounded-edge disposable scalpels.
7. Curved forceps.
8. Sharp-edged straight forceps.
9. 0.05 % (w/v) trypsin/0.02 % (w/v) EDTA.
10. 25, 75-cm^2 tissue culture flasks (vented, Nunc).
11. M5 medium.
12. Freezing solution.
13. Hemacytometer.
14. 1.8-ml sterile cryovials , ice cold.
15. Cryogenic-controlled rate freezing container or insulated cardboard/polystyrene foam box.

2.3 CD56+ Cell Fraction Depletion by Magnetic Microbeads Separation

1. Trypsin/EDTA.
2. Ca^{2+}/Mg^{2+} free phosphate-buffered saline (CMF-PBS).
3. M5 medium.
4. CD56 microbeads, human (Miltenyi).
5. LD columns (Miltenyi).
6. MidiMACS MultiStand (Miltenyi).
7. MidiMACS Separator (Miltenyi).
8. 40-μm cell strainers.
9. Trypan blue.
10. Hemocytometer.
11. *Buffer for microbeads separation (500 ml)*: CMF-PBS (pH 7.2) supplemented with 0.5 % BSA and 2 mM EDTA. Store up to 4 weeks at +4 °C.

2.4 Isolation, Propagation, and Freezing of MABs from Murine Skeletal Muscle (mMABs)

1. Mouse skeletal (tibialis anterior) muscle fragments (at least 50 mg of tissue; *see* **Note 1**).
2. Ca^{2+}/Mg^{2+} free phosphate-buffered saline (CMF-PBS).
3. D20 medium.
4. 6- and 15-cm Petri dishes (Nunc).
5. 3.5- and 6-cm collagen-coated Petri dishes.
6. Matrigel-coated 48-well multiwells.
7. Rounded-edge disposable scalpels.
8. Curved forceps.
9. Sharp-edged straight forceps.
10. 0.05 % (w/v) trypsin/0.02 % (w/v) EDTA.
11. 25, 75-cm^2 tissue culture flasks (vented, Nunc).

2.5 AP Staining

1. Human or murine MAB cultures (grown in 3.5-cm Petri dishes).
2. CMF-PBS.
3. *AP buffered solution, pH 9.5 (10 ml)*: 10 ml of 100 mM Tris–HCl (pH 9.5), additioned with
   100 mM NaCl, 50 mM MgCl2, 0.1 % (v/v) Tween 20.
   Adjust pH with 0.1 N HCl or 0.1 N NaOH.
   Particular attention must be paid to the exact pH (9.5).
   Prepare fresh.
4. Alkaline phosphatase (AP) staining solution (10 ml): dissolve 135 μg of 4-nitroblue tetrazolium chloride (NTC, Roche) and 1.75 mg of 5-bromo-4-chloro-3-indolyl phosphate (BCIP, Roche) in AP buffered solution.
   Prepare fresh.

2.6 Skeletal Muscle Differentiation

1. MABs to be tested grown in a 25-cm^2 flask.
2. M5/D20 medium.
3. 3.5-cm Petri dishes, coated with growth factor-reduced Matrigel.
4. 3.5-cm elastic surface (ESS) Petri dishes (IBIDI), coated with growth factor-reduced Matrigel.
5. CMF-PBS.
6. D2 medium.

2.7 Induction into Smooth Muscle by TGF-β1 Treatment

1. MABs to be tested grown in a 25-cm^2 flask.
2. M5 medium/D20 medium.
3. D2 medium.
4. *Human recombinant TGF-β stock solution (1 ml)*: reconstitute 5 μg human recombinant TGF-β (Life Technologies) in 1 ml of 10 mM citric acid, additioned with 0.1 % BSA.

   Store in 25-μl aliquots up to 6 months at −20 °C.

2.8 Induction of Osteoblast Differentiation by BMP2 Treatment

1. MABs to be tested grown in a 25-cm^2 flask.
2. M5 medium/D20 medium.
3. D2 medium.
4. *Human recombinant BMP2 stock solution (1 ml)*: reconstitute 100 μg human recombinant BMP2 (Life Technologies) in 1 ml of 20 mM Acetic Acid.

   Store in 25-μl aliquots up to 6 months at −20 °C.

2.9 Induction of Adipocyte Differentiation

1. MABs to be tested grown in a 25-cm^2 flask.
2. M5 medium/D20 medium.
3. Adipogenic medium (LONZA cat. no. 3004).
4. *Oil Red O solution*: dissolve 350 mg of Oil Red O powder in 100 ml of 2-propanol in a glass bottle. Let stand overnight at room temperature, protected from light. Do not mix. Filter on 3-MM chromatography paper into a new glass bottle. Add 75 ml of distilled water. Let stand overnight at +4 °C, protected from light. Do not mix. Filter two times through 3-MM chromatography paper into a new glass bottle.

   Store up to 6 months at room temperature, protected from light. Oil Red O is a very strong staining agent and should be handled carefully according to the manufacturer's product data sheet instructions.

2.10 PW1/Peg3 as a Screening Molecule to Identify the Best Mesoangioblast Populations

For general information, refer to [27, 28].

2.10.1 Real-Time PCR for Murine and Human PW1/ Peg3

Maintain all components on ice while preparing PCR mix; store solutions containing Sybr Green protected from light. Clean bench and pipettes before starting with the procedure and eventually use DNAZap for cleaning.

1. cDNA retrotranscribed from 1 μg of RNA, using random primers (iScript reverse transcription supermix, Biorad). cDNA should be diluted in nuclease-free water 1:10 before use.
2. Nuclease-free water.
3. PW1/Peg3-specific primers:

   Mouse Forward: GAGAATCCTCCATTTATATC

   Mouse Reverse: TCATGAATCTTCTGGTGCTC

   Human Forward: GATCCAAGAGAAGTGCCTACC

   Human Reverse: GGAAGATTCATCTTCACAAATCCC

   Reference gene-specific primers; we use GAPDH to normalize expression:

   Forward: TTCACCACCATGGAGAAGGC

   Reverse: GGCATGGACTGTGGTCATGA
4. Real-Time PCR Thermal Cycler (CFX-96 Connect, Biorad).
5. Optically clear PCR tubes or plates.
6. Sybr Green Master Mix (iTaq Universal SYBR Green Supermix (Biorad) for murine PW1 and Sso Advanced Universal SYBR Green Supermix (Biorad) for human Peg3).

2.10.2 Immuno-fluorescence for Murine PW1

1. mMABs Petri dish.
2. 4 % paraformaldehyde solution in PBS.
3. +4 °C refrigerator.
4. −20 °C freezer.
5. Precooled (at −20 °C) methanol.
6. *Blocking solution*: PBS additioned with 5 % goat serum, 2 % Bovine Serum Albumin (BSA).
7. *0.1 % BSA solution: PBS* additioned with 0.1 % BSA.
8. *5 % BSA solution: PBS* additioned with 5 % BSA.
9. Rabbit anti-mouse PW1 primary antibody (*see* **Note 2**).
10. Fluorophore-conjugated goat anti-rabbit secondary antibody.
11. DAPI solution.

12. Fluorescence mounting medium (Dako fluorescence mounting medium).
13. Fluorescence microscope.

2.10.3 Western Blot for Murine PW1

1. Murine MABs.
2. *RIPA lysis buffer*: PBS additioned with 1 % NP40, 0.5 % sodium deoxycholate, 0.1 % SDS, 1 mM PMSF, 3.5 mM Na3VO4, proteases, and phosphatases inhibitors.
3. Ice.
4. Refrigerated centrifuge.
5. 1.5-ml tubes.
6. *2× loading buffer*: 100 mM Tris–HCl pH 6.8 additioned with 200 mM DTT, 4 % SDS, 0.2 % bromophenol blue, and 20 % glycerol (MilliQ water to volume).
7. 6 % acrylamide gel.
8. *Running buffer pH 8.3*: 125 mM Tris hydroxymethilaminomethane additioned with 1.25 M glycine and 0.5 % SDS (MilliQ water to volume).
9. Western blot electrophoretic apparatus.
10. iBlot Dry Blotting System (Life Technologies).
11. *5 % skimmed milk solution*: PBS additioned with 5 % skimmed milk and 0.05 % Tween 20.
12. Rabbit anti-mouse PW1 primary antibody (the same antibody used for the immunofluorescence).
13. HRP-conjugated goat anti-rabbit secondary antibody (Biorad).
14. *T-PBS*: PBS additioned with 0.05 % Tween 20.
15. ECL Western blot detection reagent.
16. X-ray developer, developing liquids, or digital chemiluminescence imaging system (ChemiDoc, Biorad).

3 Methods

The preliminary step for the isolation of both human and murine mesoangioblasts (hMABs and mMABs) is a short-time primary culture of skeletal muscle fragments.

Aim of this outgrowth phase is to increase, by selective culture conditions and procedures, the proportion of MAB populations originating from tissue explants.

Notably, the presence of CD56 expression reveals myogenic progenitors that may be derived from contaminating satellite cells or, alternatively, from in vitro differentiation of hMABs into satellite cells (lineage-tracing studies in *Tg:TN-AP-CreERT2* mouse model indicate that this event may occur spontaneously during postnatal unperturbed development of skeletal muscle) [29]. CD56+ cell fraction depletion can be performed either by magnetic microbeads separation (as described in the Subheading 3.2) or by FACS sorting procedures. Notably, magnetic microbead separation is a fast purification step that does not affect cell viability, can be repeated more than once (to increase cell purity), and is adaptable to GMP protocols.

Efficiency in derivation of MABs may vary depending on age and intrinsic characteristics of skeletal muscle fragments. The accurate selection of tissue portions may however help to obtain optimal yield for each sample.

Sterile and tissue culture-grade reagents are recommended for all steps described.

All reagents are provided by Sigma, unless otherwise specified.

3.1 Isolation, Propagation, and Freezing of MABs from Human Skeletal Muscle (hMAB)

1. Rinse each skeletal muscle fragment in CMF-PBS to remove residual blood.
2. Set a 10-ml Petri dish containing 3–4 ml of M5 medium to start dissection.
3. Cut each fragment in 2 mm pieces using a round-edge scalpel, with the help of straight forceps (*see* **Note 3**).
4. Pretreat 6-cm diameter collagen-coated Petri dish by pipetting 2 ml of M5 medium into each dish, making sure that the surface is completely covered. Remove most of the medium, leaving the dish thoroughly wet.
5. Transfer the selected fragments into each dish (5–10 fragment/dish) with the help of curved forceps.
6. Add 2 ml of pre-warmed M5 medium pipetting it along the edge of the dish to prevent detachment and floating of fragments.

   Incubate overnight in a 37 °C, 5 % CO_2, 3 % O2, and 92 % N2 humidified incubator (*see* **Notes 4** and **5**).
7. Around 24 h after initiation of cultures, carefully add two additional ml of pre-warmed M5 medium to each dish.
8. After 5–7 days, examine the cultures for preliminary growth of adherent cells.
9. Add 1.5 ml of pre-warmed M5 medium to each dish.
10. After additional 2–3 days, examine the cultures for hMABs: MABs initially look round and small, and then they tend to attach to the substrate as more flat/spindle-shaped cells. With passages, the floating fraction tends to reduce, and the adherent fraction increases in percentage (Fig. 1A).

11. Carefully transfer culture medium and floating cells to a new, uncoated 6-cm Petri dish.

    In case of poor recovery of floating cells, transfer them to a smaller Petri dish (3.5 cm) to increase cell density and favor proliferation.
12. Add pre-warmed fresh medium if necessary (to reach a total volume of 5 ml).
13. After 24 h, examine the cultures. Around 50–70 % of the cells should adhere to the plastic surface, but a floating fraction should always be clearly distinguishable (*see* **Notes 6** and 7).
14. When the adherent fraction of the cell population reaches 70–80 % confluence, proceed to trypsinization and transfer to 25-cm^2 flask.
15. At 70–80 % confluence of the adherent cell population, remove culture medium and set aside in 15-ml centrifuge tubes.
16. Rinse the growing surface with 2 ml CMF-PBS.
17. Add 2 ml of trypsin/EDTA and incubate 3–5 min at room temperature. Check under a microscope for complete detachment of cells.
18. Use the medium set aside to collect all cells. In this way, both floating and adherent populations are recovered.
19. Centrifuge 10 min at 188 × *g*, room temperature.
20. Gently suspend the pellet in 6-ml M5 medium and transfer to a single 75-cm^2 tissue culture flask or dispense 2-ml aliquots of cell suspension into each of the three 25-cm^2 flasks (1:3 split).
21. Add M5 medium to reach a final volume of 5 ml for 25-cm^2 tissue culture flasks (12 ml for 75-cm^2 tissue culture flasks) (*see* **Notes 8–10**).
22. To freeze hMABs, detach cells with trypsin/EDTA according to corresponding steps described for cell propagation (*see* **Note 11**).
23. Spin 5 min at 227 × *g*, discard supernatant and suspend the cell pellet in 5 ml of M5 medium.
24. Count cells with a hemocytometer.
25. Centrifuge 5 min at 227 × *g*, room temperature.
26. Discard supernatant and gently suspend cells in appropriate volume of cold freezing solution (optimal range of cell concentration: 1–2 × 10^6 cells/ml).
27. Set up the appropriate number of 1.8-ml cryovials and dispense 1 ml of cell suspension into each.

    Each cryovial should be clearly labeled with date, cell line code, and passage number.
28. Transfer vials into a freezing container and place overnight at −80 °C.

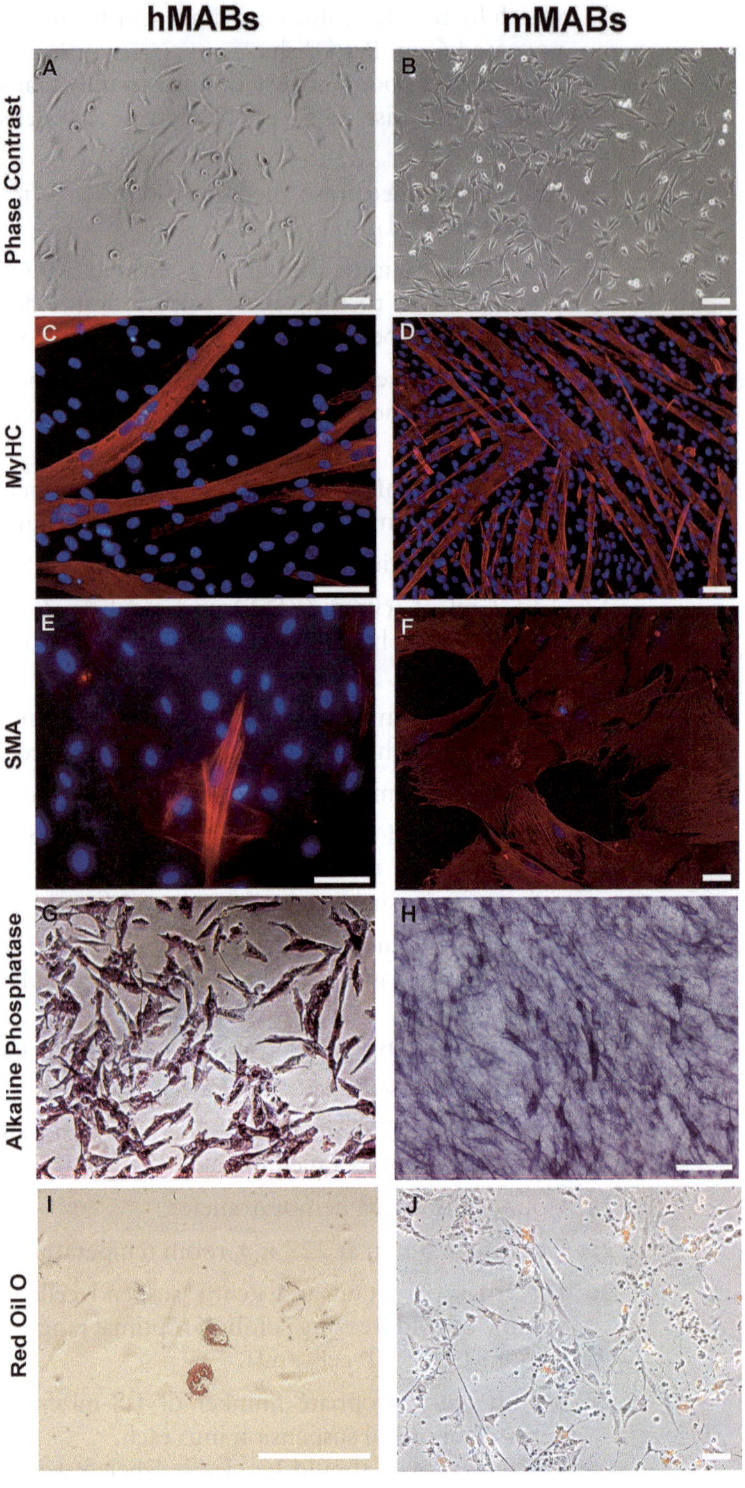

Fig. 1 Differentiation of mouse and human MABs into different mesodermal-derived cell types. (**A, B**) Phase-contrast images of growing human (**A**) and mouse (**B**) MABs; (**C, D**) immunofluorescence for all sarcomeric myosins (MyHC, in *red*) on differentiated human (**C**) and mouse (**D**) MABs; (**E, F**) smooth muscle actin (SMA, in *red*) immunofluorescence analysis of human (**E**) and mouse (**F**) MABs after TGF-β1 stimulation; (**G, H**) alkaline phosphate staining after BMP2 treatment of human (**G**) and mouse (**H**) MABs; (**I, J**) Oil Red O staining of human (**I**) and mouse (**J**) MABs after culturing in adipogenic medium. Scale bars 100 μm

29. On the following day, transfer vials to −135 °C or to a liquid nitrogen container (*see* **Note 12**).

    The morphology of hMABs after few passages in culture at the inverted phase contrast microscope is shown in Fig. 1A.

3.2 CD56+ Cell Fraction Depletion by Magnetic Microbead Separation

1. Trypsinize hMABs as described in Subheading 3.1.
2. Centrifuge at 188 × *g* for 10 min.
3. Resuspend the pellet in 10 ml of depletion buffer and pass the suspension through a cell strainer to remove any clump (*see* **Note 13**).
4. Count viable cells by trypan blue exclusion.
5. Centrifuge at 188 ×*g* for 10 min.
6. Discard completely the supernatant (remove the last drops using carefully a P200 pipettor without touching the cell pellet).
7. Resuspend the cell pellet in 80 μl of depletion buffer.
8. Add 20 μl of CD56 microbeads.
9. Mix by pipetting carefully and incubate at 4°–8 °C (fridge or cold room) for 15 min. Mix rapidly by flicking during incubation (once). Higher temperature and/or longer incubations lead to nonspecific labeling.
10. Add 2 ml of depletion buffer to wash the cells. Centrifuge at 188 ×*g* for 10 min.
11. Rinse the column with 500 μl of depletion buffer (*see* **Note 14**).
12. Resuspend the cells in 500 μl of depletion buffer and apply the suspension onto the column.
13. Wash the column with 500 μl three times, once the reservoir is empty (*see* **Note 15**).
14. Collect the total effluent, containing the CD56− depleted cell fraction.
15. Centrifuge the total effluent at 188 ×*g* for 10 min.
16. Resuspend the cell pellet in 5 ml of M5 medium and proceed to counting of viable cells by trypan blue exclusion.
17. Plate the cells in M5 medium (6000 cells/cm^2) and put immediately in the incubator at 37 °C.

    A scheme of the CD56 depletion is shown in Fig. 2.

3.3 Isolation, Propagation, and Freezing of MABs from Murine Skeletal Muscle (mMABs)

At variance with the protocol set for hMABs, a cloning step by limiting dilution has been introduced for mMABs isolation.

Primary polyclonal population, derived from murine skeletal muscle culture under physiological O_2 tension, can be cloned without the support of any feeder layer, on Matrigel-coated plasticware.

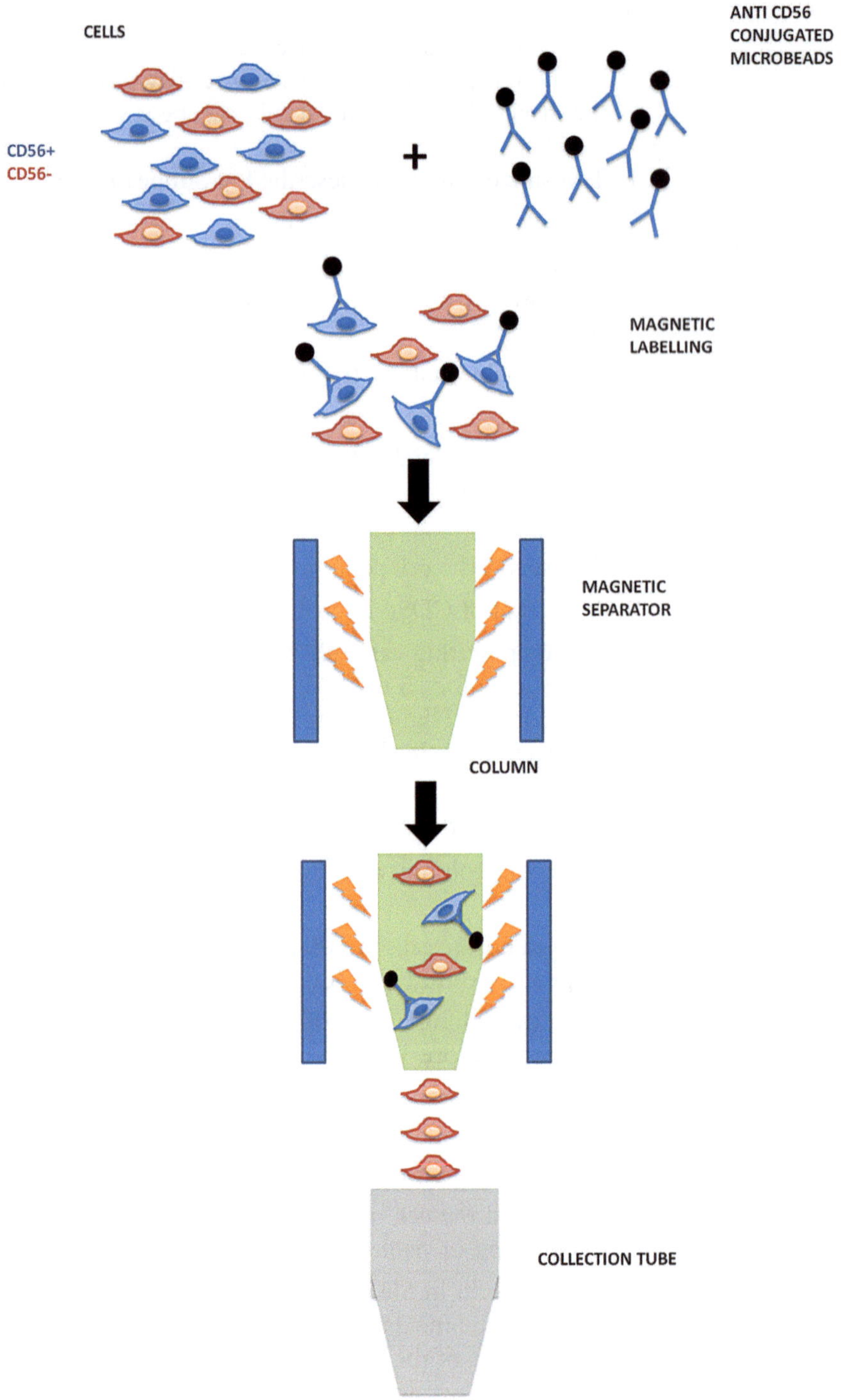

Fig. 2 Depletion of CD56+ cells using magnetic beads. Scheme of the procedure used to separate mesoangioblasts from CD56+ satellite cells

Isolated populations are subsequently screened for PW1/Peg3 to select the ones with highest efficiency in differentiation (*see* Subheading 3.9). On top of physiological O_2 tension, the use of Matrigel, which mimics an extracellular matrix-like microenvironment, represents an additional strategy to improve culture conditions during cloning steps.

Primary culture of murine skeletal muscle is initiated following the same procedures described for human skeletal muscle, (**steps 1–8**), using D20 medium instead of M5.

After 5–7 days, proceed to dissociate the culture as follows:

1. Remove D20 culture medium and carefully rinse twice with 1 ml of CMF-PBS at room temperature, avoiding to touch either the tissue fragments or the surrounding cells.
2. Remove PBS and add 2 ml of collagenase/Dispase solution (*see* recipe in Subheading 2.1) to each dish.
3. Carefully remove cells first and then the tissue fragments by gentle pipetting and mild scraping using a pipettor.
4. Transfer the cells and tissue suspension into 15-ml centrifuge tube.
5. Repeat **steps 2** and **3** three additional times.
6. Add an additional 2 ml of collagenase/Dispase solution directly to centrifuge tube (the digestion of tissue fragments and cells is performed in a final volume of 10 ml for each dish).
7. Incubate 15 min in a 37 °C water bath.
   Flick and invert the tube three times during incubation, monitoring the dissociation of tissue.
8. Stop the reaction by adding 3 ml FBS to the tube. Centrifuge at 188 × *g* for 15 min at room temperature.
9. Discard supernatant and suspend the pellet in 300 μl of D20 medium.
10. Pipet up and down several times using a 1000-μl pipettor with filtered tips to disaggregate muscle fragments as much as possible. Let the larger debris sediment for few seconds and transfer the upper mere homogeneous cell suspension in a new 15-ml centrifuge tube.
11. Count viable cells by trypan blue exclusion and proceed to cloning (*see* **Note 16**).
12. Dilute cells in D20 medium to obtain 150 ml of each of the following concentrations:
    1 cell/ml.
    10 cells/ml.
    20 cells/ml.
    30 cells/ml.

13. For each cell concentration, plate 1 ml/well in three 48-well plates.
14. Prepare a humidified chamber by placing the 48-well plates into a clean plastic box, along with two open 6-cm Petri dishes filled with sterile water. Cover the box with aluminum foil.
15. Place the cultures in a 37 °C, 5 % CO_2, 3 % O2, and 92 % humidified incubator for at least 1 week.
16. After 1 week, carefully inspect the cultures with a microscope to distinguish the first clone. If clones appear in dishes plated with 1 cell/well, discard dishes plated at higher density.
17. Add 200 μl D20 medium to each well.
18. Passage the clones when the cells have covered at least 50 % of the well surface.
19. At the time of first passage, carefully aspirate the medium and rinse each well with 1 ml of CMF-PBS at room temperature.
20. Add 200 μl of 0.025 % trypsin/EDTA to each well. Incubate 5–10 min at 37 °C, monitoring under microscope for complete detachment of cells.
21. Inactivate trypsin by adding 800 μl D20 medium down the growing surface of each well. Carefully collect all cells.
22. Transfer cells and medium to a 15-ml centrifuge tube and centrifuge 5 min at 188 × *g*, room temperature.
23. Discard supernatant, suspend the pellet in 1 ml fresh D20 medium, and plate in uncoated well of a 24-well plate (*see* **Note 17**).
24. When 70–80 % confluent, proceed to trypsinization of established mMABs and transfer to uncoated 25-cm^2 flask.
25. To propagate mMABs, aspirate and discard the medium and rinse with 2 ml of CMF-PBS for 25-cm^2 flask (5 ml for 75-cm^2 flasks).
26. Add 1 ml trypsin/EDTA to 25-cm^2 flask (2 ml for 75-cm^2 flasks), and incubate 3–5 min at 37 °C. Check under a microscope for complete detachment of cells.
27. Collect the cells with 4 ml of D20 medium (8 ml for 75-cm^2 flasks).
28. Centrifuge cells 5 min at 227 × *g*, room temperature. Discard supernatant.
29. Suspend the pellet thoroughly in 6–10 ml of D20 medium and dispense 2-ml aliquots of cell suspension into each of three or five flasks (1:3 or 1:5 split, depending upon proliferation rate). Add D20 medium to reach a final volume of 5 ml for 25-cm^2 flasks (12 ml for 75-cm^2 flasks). Drag the flasks with a cross movement on the incubator shelf, to ensure homogeneous distribution of cells. Incubate at 37 °C.

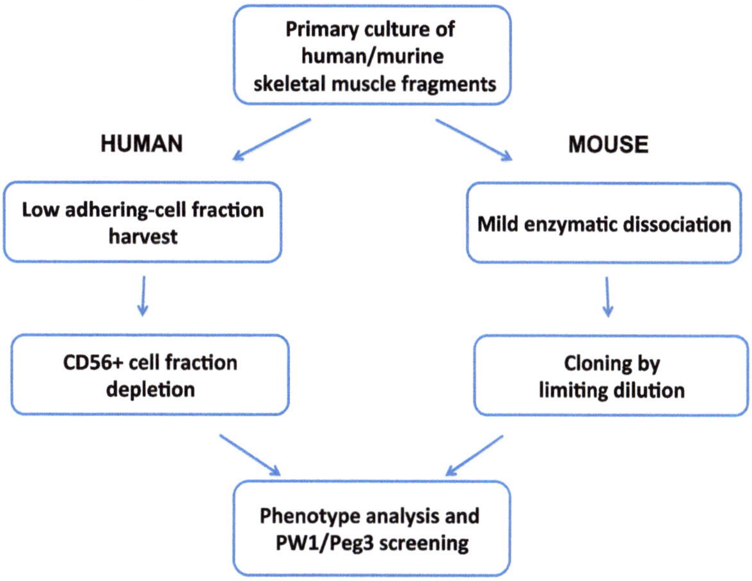

Fig. 3 Schematic framework recapitulating the different steps used for isolation and characterization of human and mouse MABs

Murine MABs are frozen according to the procedure described for human MABs, using D20 instead of M5 medium.

The morphology of mMABs after few passages in culture at the inverted phase contrast microscope is shown in Fig. 1B.

A general scheme on the procedures used for isolation and characterization of human and mouse MABs is shown in Fig. 3.

3.4 AP Staining

The protocols described in this unit lead to the isolation of MAB populations that, at early passage, express AP at a percentage usually ≥50 %.

In case of very low expression percentage (≤10 %), a FACS step can be introduced [30]. Notably, AP expression physiologically presents high variability and may decrease with passages during in vitro propagation.

1. Remove medium from Petri dishes, rinse the growing surface with 1 ml CMF-PBS.
2. Fix with 1 ml of 4 % PFA 5 min at room temperature.
3. Remove 4 % PFA and rinse with1 ml CMF-PBS.

   Proceed immediately to alkaline phosphatase staining.

   Fixed cultures may be stored up to 48 h at 4 °C. If stored, add 0.5 ml CMF-PBS to each dish and seal with Parafilm to avoid drying out and/or contamination.
4. Remove CMF-PBS and add 1 ml of alkaline phosphatase staining solution to each dish.
5. Incubate 2 h at room temperature in the dark.

6. Examine cultures under inverted phase-contrast microscope for a brown cytoplasmic stain, whose intensity is roughly proportional to the level of enzymatic activity.

3.5 Skeletal Muscle Differentiation

Selected MAB populations can be tested for their capability to spontaneously differentiate into skeletal muscle on Matrigel-coated plastics.

1. Plate 0.5–1 × 10^5 m MABS or 1–2 × 10^5 h MABs on Matrigel-coated dishes in 2 ml D20 or M5 medium, respectively.Slight adjustment in cell number/dish may be necessary, due to variability in cell proliferation rate and differentiation efficiency.
2. Incubate overnight at 37 °C, 5 % CO_2.
3. Remove medium and rinse each dish with 1 ml CMF-PBS.
4. Add 2 ml D2 differentiation medium to each dish.
5. Incubate at least 1 week in at 37 °C, 5 % CO_2.
   At that time, first myotubes should be evident. A time period of 7–8 days is usually sufficient for mMAB differentiation, while 10–14 days may be necessary for hMABs.
6. Remove medium from Petri dishes and carefully rinse the growing surface with 1 ml CMF-PBS.
7. Fix with 1 ml of 4 % PFA, 5 min at room temperature.
   Remove 4 % PFA and rinse with 1 ml CMF-PBS. Proceed immediately to immunofluorescence for the expression of sarcomeric MyHC by MF20 antibody (Hybridoma Bank), or store up to 48 h at 4 °C.
8. Calculate percentage of myogenic differentiation as the number of MAB cell nuclei (detected by DAPI) inside myosin positive cells or myotubes divided by the total MAB nuclei multiplied by 100. Examples of human and mouse MAB skeletal muscle differentiation are shown in Fig. 1C, D.

3.6 Induction into Smooth Muscle by TGF-β1 Treatment

1. Plate 5 × 10^4 cells/3.5-cm Petri dish in 2 ml of medium (D20 for murine, M5 medium for human cells). For each cell line to be tested, plate at least two dishes (one test and one control dish).
2. Incubate overnight in a 37 °C, 5 % CO_2 incubator.
3. Remove medium and rinse each dish with 1 ml CMF-PBS.
4. Add 1.5 ml D2 medium to each dish.
5. Add 1.5 μl TGF-β1 stock solution to test dishes (5 ng/ml final concentration). No addition has to be made to control dishes.
6. Each other day add 1.5 μl TGF-β1 stock solution to test dishes.
7. Check cultures for smooth muscle differentiation, which should be complete after 7–8 days.

8. Remove medium from Petri dishes and carefully rinse the growing surface with 1 ml CMF-PBS.
9. Fix with 1 ml of 4 % PFA 5 min at room temperature.
10. Remove 4 % PFA and rinse with 1 ml CMF-PBS. Proceed immediately to immunofluorescence, or store up to 48 h at 4 °C.
11. Calculate percentage of smooth muscle differentiation as the number of MABs expressing a smooth muscle phenotype (detected by an antibody directed against smooth alpha actin) (Sigma cat. no. A2547) or calponin (Sigma cat.no. C2687) divided by total number of mesoangioblast nuclei multiplied by 100.

    Examples of human and mouse MAB smooth muscle differentiation are shown in Fig. 1E, F.

3.7 Induction of Osteoblast Differentiation by BMP2 Treatment

BMP2 treatment results in a strong increase of original alkaline phosphatase activity of MABs, which can be easily detected by AP staining (*see* Subheading 3.4).

Follow **steps 1–4** described for Subheading 3.6.

5. Add 15 μl BMP2 stock solution to each test dish (100 ng/ml final concentration).

    No addition has to be made to control dishes.
6. Each other day add 15 μl fresh BMP2 stock solution to test dishes.
7. Assess the cultures for differentiation, which should be complete after 7–8 days.
8. Remove medium from Petri dishes, and carefully rinse the growing surface with 1 ml CMF-PBS.
9. Fix with 1 ml of 4 % PFA 5 min at room temperature.
10. Remove 4 % PFA and rinse with 1 ml CMF-PBS.

    Proceed immediately to alkaline phosphatase staining (*see* Subheading 3.4) (*see* **Note 18**).

    Examples of human and mouse MAB osteoblast differentiation tested as AP staining are shown in Fig. 1G, H.

3.8 Induction of Adipocyte Differentiation

Follow **steps 1–3** described for Subheading 3.6.

4. Add 1.5 ml of adipogenic induction medium to each test dish.

    Add 1.5 ml D2 medium to each control dish.
5. Check cultures for differentiation after 6 to 7–10 days (*see* **Note 19**).

    Examples of human and mouse MAB adipocyte differentiation tested as Oil Red O staining are shown in Fig. 1I, J.

3.9 Effect of Different O_2 Tension and Substrate Stiffness on hMAB Growth and Differentiation

Several processes crucial to cell fate determination, such as adhesion, migration, spreading, and differentiation, have been demonstrated to be deeply influenced not only by biochemical but also by physical and mechanical characteristics of surrounding microenvironment.

Among physical parameters, the impact of low oxygen on stemness, proliferation, and differentiation potential has been widely described for several cell type models [31–35]. As far as hMABs are concerned, physiological O_2 tension has been proved to be essential for their isolation, proliferation, and preservation of chromosomal stability, which are crucial parameters for efficient and safe high-scale propagation in cell therapy. Figure 4 shows differences in growth rate of hMABs grown at different conditions of O_2 tension (Fig. 4A–C).

Regarding mechanical parameters, substrate stiffness has been indicated as a primary modulator for stem cell commitment, being capable of directing the differentiation of pluripotent and multipotent stem cells [36–38]. It has been indeed shown how the use of gels with stiffness typical of normal muscle for myoblasts differentiation leads to the formation of myotubes exhibiting highly defined sarcomeric organization [39, 40].

We tested the effect of commercially available elastic substrates on hMAB skeletal muscle differentiation. Specifically, we used a special kind of Petri dish (IBIDI), whose bottom presents a 40-μm-thick polydimethylsiloxane layer. The layer confers stiffness very close to that of mammalian skeletal muscle (28 KPa, according to Young's Modulus) and, differently from most plastic for cell culture (approximately 300,000 times stiffer), may provide in vitro culture conditions closer to physiology. Figure 4 shows the remarkable effect of this kind of substrate on hMAB orientation and differentiation (Fig. 4D–G).

3.10 PW1/Peg3 as a Screening Molecule to Identify the Best Mesoangioblast Populations

As mentioned in the Abstract, we have recently identified a new marker of MABs, PW1/Peg3, whose expression is shared by all the MABs isolated so far, regardless of the species and the developmental stages [27]. PW1/Peg3 expression strongly correlates with the two main MAB properties, their ability to differentiate in skeletal muscle and to cross the blood vessel wall (Fig. 5). Therefore, we are currently using PW1/Peg3 to screen and identify competent MABs before their use in cell therapy.

3.10.1 Real-Time PCR for Murine and Human PW1/ Peg3

1. Thaw all reagents on ice.
2. Prepare the PCR mix; for a final reaction volume of 20 μl, prepare 15 μl of premix for each sample. For each well/tube, calculate to add the following components: 10 μl iTaq Universal or Sso Advanced SYBR Green Supermix (2×); 300 nM of each primer; water to 15 μl. Always consider an additional well to be used as no template control. Prepare the mix in excess.

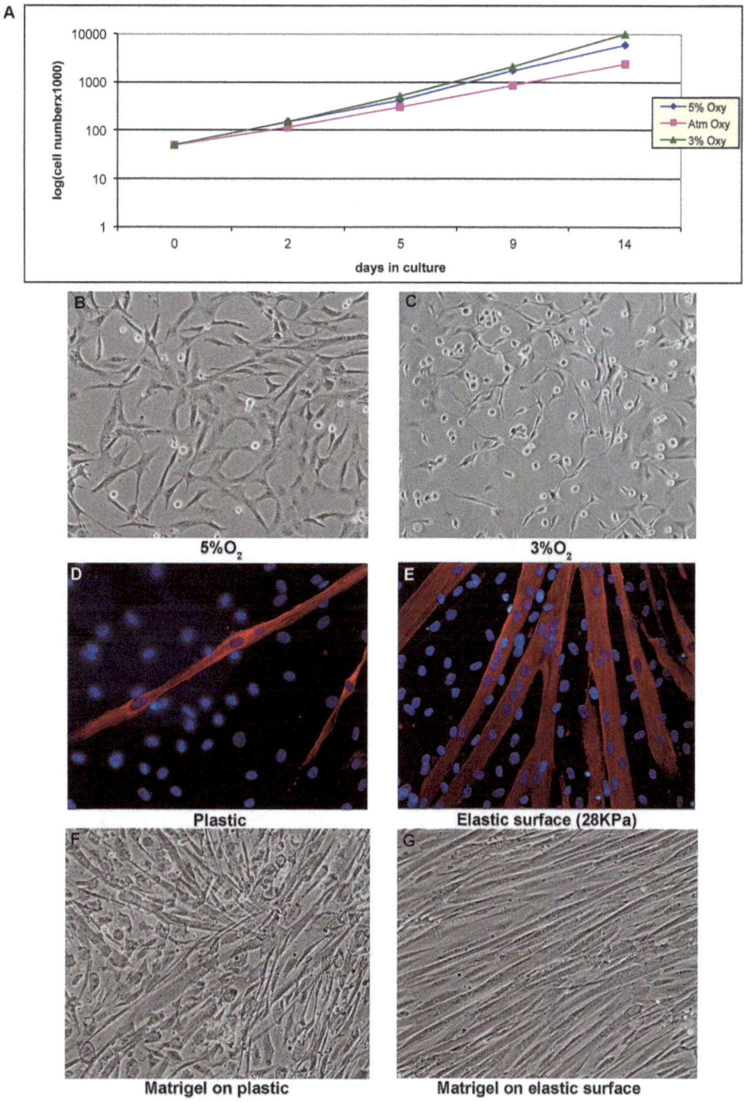

Fig. 4 Effects of stiffness and oxygen levels on MAB cell cultures. (**A**) Example of a growth curve of hMABs in different oxygen conditions (5 %, 3 %, and atmospheric oxygen levels); (**B, C**) appearance of hMABs when cultured at 5 % (**B**) or 3 % (**C**) oxygen; (**D–G**) differentiation properties of hMABs when cultured on different substrates. (**D, E**) Immunofluorescence analysis for all sarcomeric myosins (MyHC, in *red*) of hMABs differentiated on regular plastic (**D**) or on an elastic surface (stiffness 28 KPa) (**E**). (**F, G**) Phase-contrast images of hMABs differentiated on regular plastic (**F**) or on a Matrigel substrate on elastic surface (**G**)

3. Aliquot 15 μl of premix in each well. High precision while pipetting is necessary (*see* **Note 20**).
4. Load each sample in triplicate. For each well, add 5 μl of diluted sample cDNA. Mix by pipetting up and down. High precision while pipetting is necessary (*see* **Note 21**).

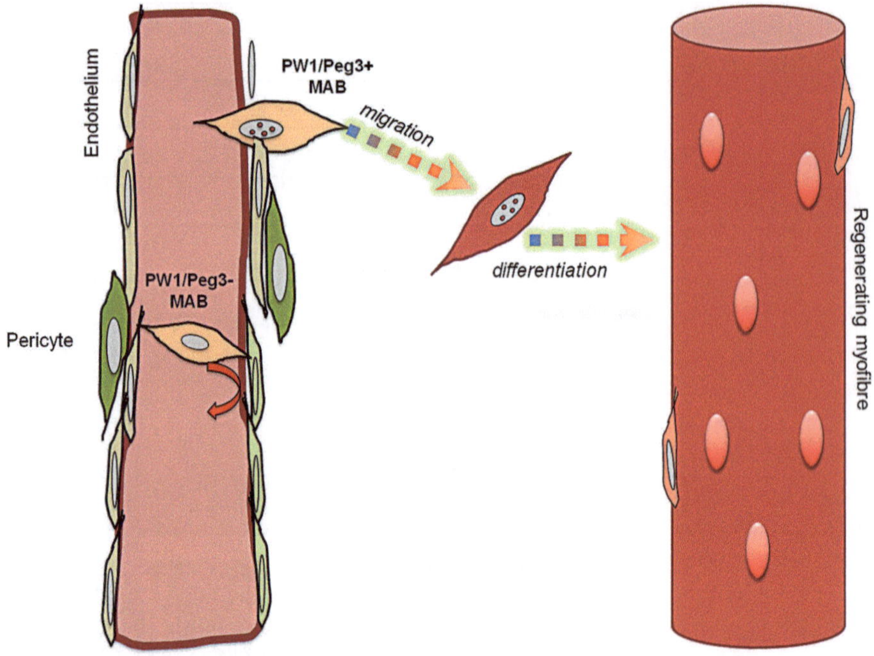

Fig. 5 Scheme exemplifying the importance of PW1/Peg3 activity for both human and mouse MAB main properties: skeletal muscle differentiation and vessel transmigration

5. Close tubes/plate and load in the Real-Time PCR Thermal Cycler.
6. Apply the following thermal protocols:
   (a) For murine PW1: 98 °C, 30 s; (95 °C, 3 s; 54 °C 30 s) for 40 cycles; melt curve (65–95 °C, 0.5 °C increments at 2 s/step).
   (b) For human Peg3: 98 °C, 30 s; (98 °C, 3 s; 60 °C 10 s; 72 °C 15 s) for 40 cycles; melt curve (65–95 °C, 0.5 °C increments at 2 s/step).
7. Normalize each sample to the housekeeping gene (GAPDH) and calculate expression levels using the comparative threshold cycle [41] method (*see* **Note 22**).

3.10.2 Immunofluorescence for Murine PW1

1. Rapidly wash cells with PBS to eliminate cell culture medium excess.
2. Fix cells with 4 % paraformaldehyde in PBS, for 10 min at 4 °C.
3. Eliminate paraformaldehyde and wash three times with PBS (5 min each at RT).
4. Permeabilize cells with precooled methanol at −20 °C for 6 min.
5. Block with blocking solution for 3 h at RT.

6. Incubate cells overnight at 4 °C with rabbit anti-mouse PW1 primary antibody (1:3000) diluted in blocking solution (*see* **Note 2**).
7. The day after, wash cells three times (15 min each at RT) with 0.1 % BSA in PBS.
8. Wash again in 5 % BSA in PBS for 15 min at RT.
9. Incubate cells with secondary antibody (we use Jackson Laboratories Antibodies, 1:500), together with DAPI (1:500, Sigma) diluted in blocking solution for 2 h at RT (*see* **Note 23**).
10. Rinse cells with 0.1 % BSA in PBS for three times (15 min each at RT).
11. Rinse cells with 5 % BSA in PBS for 15 min at RT.
12. Mount cells with fluorescence mounting medium and cover with a coverslip.
13. Watch under fluorescence microscope (*see* **Note 24**).

3.10.3 Western Blot for Murine PW1

1. Proceed with protein extraction. Lyse samples in RIPA lysis buffer, scraping cells from the Petri dish and transferring the sample to a 1.5-ml tube. Let samples on ice for 30 min, vortexing every 10 min. Centrifuge for 10 min at 12000 × *g*, 4 °C. Transfer protein-enriched supernatant to a new tube.
2. For every sample, load 30 μg of protein extracts along with 2× loading buffer (pre-boil samples for 5 min at 95 °C) on a 6 % acrylamide gel.
3. Perform the run in running buffer, at constant voltage (100 V).
4. Transfer the gel on nitrocellulose membrane using the iBlot Dry Blotting System (Life Technologies), for 10 min.
5. Block the nitrocellulose membrane with 5 % skimmed milk, 0.05 % Tween 20 in PBS for 1 h at RT on a rotating platform.
6. Incubate overnight at 4 °C on a rotating platform, with rabbit anti-mouse PW1 primary antibody (1:10,000) diluted in 5 % skimmed milk, 0.05 % Tween 20 in PBS.
7. The day after, wash four times (10 min each) with T-PBS at RT while shaking.
8. Incubate with HRP-conjugated goat anti-rabbit secondary antibody (1:10,000) diluted in 5 % skimmed milk in T-PBS for 45 min at RT while shaking.
9. Repeat the washes as described in **step 7**.
10. Incubate with ECL chemiluminescence reagents.
11. Expose the membrane to an X-ray film and develop or acquire signal with the digital chemiluminescence imaging system (*see* **Note 25**).

4 Notes

1. Both human and murine skeletal muscle fragments can be stored in D20 medium, at 4 °C, up to 24 h before being processed.
2. A reliable antibody that recognizes human Peg3 is not available commercially. The one used for mouse PW1/Peg3 has been described in the following studies [27, 28]. This group has recently generated a monoclonal antibody for the human form of Peg3 as well. For further information, please contact Dr. Sassoon at david.a.sassoon@gmail.com.
3. The use of a stereomicroscope may help for best selection in case of damaged tissue (especially for samples obtained from donors undergoing post-traumatic surgery). Try to identify portions of interstitial tissue containing small vessels. Remove as much adipose tissue as possible (since its presence may delay hMAB proliferation). Proceed with sharp and neat cuts (avoid any shearing/dissociation of the tissue, which may lower the yield of hMAB isolation).
4. Due to small volume of medium used at early stages of culture initiation, it is critical to prevent its evaporation (since this will cause tissue suffering and cell death), setting a humidified chamber. This can be obtained by placing the dish containing the dissected fragments into a 15-cm Petri dish, along with an open Petri dish filled with sterile distilled water.
5. Early phase of hMAB derivation requires a constant O_2/CO_2 tension and humidity level. Therefore use a dedicated incubator or at least a rarely opened incubator.
6. Gentle pipetting may help to detach the weakly adhering cells around the explants. Plate detached cells in uncoated Petri dish (same size of dish used for primary culture assembling). Incubate 1–3 days. At this point, the floating population of hMABs should be easily distinguishable. Transfer medium and floating cells to a new dish or 25-cm^2 tissue culture flask. Discard the primary mixed population of adherent cells.
7. hMABs cultured in vitro in M5 medium grow as floating cells when in duplicative phase. Consequently the floating fraction represents a stable and renewable source of uncommitted hMABs. When transferred to untreated plasticware, hMABs can adhere, but a floating/low-adhering fraction of dividing cells should always be present in healthy cultures, particularly in early stage populations. The recovery of an early floating/low-adhering fraction is a crucial step for proper hMAB isolation. The coating of plasticware is not necessary after primary culture step.

8. hMABs are very sensitive to trypsin. If cells are healthy, their detachment should be very quick and complete at room temp. The day after trypsinization, the floating population may be reduced. Normally, this fraction should start to expand again after 48 h.
9. When MABs have been expanded to 25-cm^2 tissue culture flasks (approximately 500,000 cells/flask), proceed to:
   - Karyotype analysis.
   - Phenotype analysis and eventual CD56+ cell fraction depletion.
   - PW1/Peg3 screening.
   - Freezing of early passage cells, for backup and further propagation.
10. hMABs can be expanded up to 20 passages under 3 % O_2 tension. At pre-senescence a strong reduction in the floating population of cells is observed, in addition to the presence of large, flat, or elongated, vacuolated cells. Late passage hMABs cell lines can be easily screened for the presence of senescent cells by assays based on histochemical stain for β-galactosidase activity at pH 6 (senescence histochemical staining kit, SIGMA cat. No. CS0030).
11. At moment of freezing hMABs should be in active proliferation (a floating fraction should be evident), and adherent fraction of cells should be at 70–80 % of confluence.
12. Upon thawing, which has to be performed quickly in a 37 °C water bath, transfer the vial content into a 15-ml centrifuge tube containing 5 ml of pre-warmed M5 culture medium; centrifuge 5 min at 227 × *g*, room temperature; discard supernatant to remove DMSO; and resuspend cells in M5 medium and plate.
13. In order to obtain an efficient depletion, it is very important to obtain a single cell suspension. Pre-wet the cell strainer with 2 ml of depletion buffer before use, to prevent the attachment of cells to the filter membrane.
14. During the centrifugation of the sample, insert the LD column in the magnetic separator, and put the separator on its stand and a suitable collection tube under the column. All the depletion procedure has to be performed under a flow hood.
15. To increase purity, the procedure can be repeated with a new column, without any effect on cells viability.
16. In parallel to cloning, a small aliquot (30 μl) of total cell suspension should be plated in a single well of 48 Matrigel-coated multiwell to check for cell survival rate. It is recommendable to expand and freeze an aliquot of polyclonal mix for backup and further analysis.

17. From this step on, no Matrigel coating will be necessary, but particular attention will have to be paid to the density of cells. Until the third/fourth passage, cells must be grown at high density and must be split when confluent into progressively larger wells (from 48- to 24- to 12- to 6-well plates). This phase is the most critical for mesoangioblast derivation; in fact, many clones may differentiate or go to senescence and/or stasis; if culture conditions are inadequate, all clones may be lost at this stage. The successful, continuously proliferating clones usually represent a small percentage of all subcultured clones (from 5 to 10 %). A clone can be considered "established" if cells proliferate at a regular rate (approximately 12 h doubling time) and maintain a typical morphology. Once established, all clones (or at least a significant number) need to be propagated and characterized (phenotype analysis, AP staining, PW1/Peg3 screening).
18. For a more rigorous test of osteoblast differentiation, in vitro formation of von Kossa positive, calcified nodules (Chaplin and Grace, 1975) should be characterized.
19. Differentiation into adipocytes is morphologically easy to detect. It is characterized by the presence of gradually enlarging, translucent vacuoles in the cytoplasm of a percentage of cells (up to 60–70 %). The presence of lipid content in these vacuoles must be confirmed by appropriate staining (Oil Red O staining).
20. To ensure maximal precision while pipetting the PCR mix and avoid bubble formation, we usually apply the repetitive pipetting technique. This technique consists in overloading the tip while aspirating the mix, pressing the operating button to the second stop, and dispensing the mix by pressing the operating button to the first stop. When using this pipetting technique, a single tip can be used to aliquot the PCR mix in all the wells/tubes used.
21. Change tip for every well/tube loaded with samples. While dispensing cDNA, mix several times by pipetting up and down. To avoid bubble formation, mix and dispense samples pressing the operating button just to the first stop.
22. It is always necessary to assess primers efficiency in the temperature and condition used before proceeding with the analysis.
23. To avoid the waste of antibody, delimit the immunofluorescence area with a liquid-blocker pen (we use PAP pen, Sigma). Use the pen to draw a hydrophobic circle around the area of interest. This will limit the amount of antibody to be used (typically 200 μl for 35-mm petri dishes).

24. The immunostaining for PW1 in murine MABs is clearly nuclear, with nucleolus exclusion.
25. Typically, incubation with PW1 antibody results in a predominant, aspecific band around 100 kDa and a correct band around 250 kDa.

Acknowledgment

This work was supported by from the European Community, ERC StG2011 (RegeneratioNfix 280611) and the Italian Ministry of University and Research (MIUR-Futuro in Ricerca 2010).

References

1. Mauro A (1961) Satellite cell of skeletal muscle fibers. J Biophys Biochem Cytol 9:493–495
2. Sambasivan R, Yao R, Kissenpfennig A, Van Wittenberghe L, Paldi A, Gayraud-Morel B, Guenou H, Malissen B, Tajbakhsh S, Galy A (2011) Pax7-expressing satellite cells are indispensable for adult skeletal muscle regeneration. Development 138(17):3647–3656. doi:10.1242/dev.067587
3. McCarthy JJ, Mula J, Miyazaki M, Erfani R, Garrison K, Farooqui AB, Srikuea R, Lawson BA, Grimes B, Keller C, Van Zant G, Campbell KS, Esser KA, Dupont-Versteegden EE, Peterson CA (2011) Effective fiber hypertrophy in satellite cell-depleted skeletal muscle. Development 138(17):3657–3666. doi:10.1242/dev.068858
4. Lepper C, Partridge TA, Fan CM (2011) An absolute requirement for Pax7-positive satellite cells in acute injury-induced skeletal muscle regeneration. Development 138(17):3639–3646. doi:10.1242/dev.067595
5. Murphy MM, Lawson JA, Mathew SJ, Hutcheson DA, Kardon G (2011) Satellite cells, connective tissue fibroblasts and their interactions are crucial for muscle regeneration. Development 138(17):3625–3637. doi:10.1242/dev.064162
6. Relaix F, Zammit PS (2012) Satellite cells are essential for skeletal muscle regeneration: the cell on the edge returns centre stage. Development 139(16):2845–2856. doi:10.1242/dev.069088
7. Minasi MG, Riminucci M, De Angelis L, Borello U, Berarducci B, Innocenzi A, Caprioli A, Sirabella D, Baiocchi M, De Maria R, Boratto R, Jaffredo T, Broccoli V, Bianco P, Cossu G (2002) The meso-angioblast: a multipotent, self-renewing cell that originates from the dorsal aorta and differentiates into most mesodermal tissues. Development 129(11):2773–2783
8. Dellavalle A, Sampaolesi M, Tonlorenzi R, Tagliafico E, Sacchetti B, Perani L, Innocenzi A, Galvez BG, Messina G, Morosetti R, Li S, Belicchi M, Peretti G, Chamberlain JS, Wright WE, Torrente Y, Ferrari S, Bianco P, Cossu G (2007) Pericytes of human skeletal muscle are myogenic precursors distinct from satellite cells. Nat Cell Biol 9(3):255–267. doi:10.1038/ncb1542
9. Mitchell KJ, Pannerec A, Cadot B, Parlakian A, Besson V, Gomes ER, Marazzi G, Sassoon DA (2010) Identification and characterization of a non-satellite cell muscle resident progenitor during postnatal development. Nat Cell Biol 12(3):257–266. doi:10.1038/ncb2025
10. Ferrari G, Cusella-De Angelis G, Coletta M, Paolucci E, Stornaiuolo A, Cossu G, Mavilio F (1998) Muscle regeneration by bone marrow-derived myogenic progenitors. Science 279(5356):1528–1530
11. LaBarge MA, Blau HM (2002) Biological progression from adult bone marrow to mononucleate muscle stem cell to multinucleate muscle fiber in response to injury. Cell 111(4):589–601
12. McKinney-Freeman SL, Jackson KA, Camargo FD, Ferrari G, Mavilio F, Goodell MA (2002) Muscle-derived hematopoietic stem cells are hematopoietic in origin. Proc Natl Acad Sci USA 99(3):1341–1346. doi:10.1073/pnas.032438799
13. Benchaouir R, Meregalli M, Farini A, D'Antona G, Belicchi M, Goyenvalle A, Battistelli M, Bresolin N, Bottinelli R, Garcia L, Torrente Y (2007) Restoration of human dystrophin following transplantation of exon-

skipping-engineered DMD patient stem cells into dystrophic mice. Cell Stem Cell 1(6):646–657. doi:10.1016/j.stem.2007.09.016

14. Torrente Y, Belicchi M, Sampaolesi M, Pisati F, Meregalli M, D'Antona G, Tonlorenzi R, Porretti L, Gavina M, Mamchaoui K, Pellegrino MA, Furling D, Mouly V, Butler-Browne GS, Bottinelli R, Cossu G, Bresolin N (2004) Human circulating AC133(+) stem cells restore dystrophin expression and ameliorate function in dystrophic skeletal muscle. J Clin Invest 114(2):182–195
15. Dezawa M, Ishikawa H, Itokazu Y, Yoshihara T, Hoshino M, Takeda S, Ide C, Nabeshima Y (2005) Bone marrow stromal cells generate muscle cells and repair muscle degeneration. Science 309(5732):314–317
16. Pesce M, Orlandi A, Iachininoto MG, Straino S, Torella AR, Rizzuti V, Pompilio G, Bonanno G, Scambia G, Capogrossi MC (2003) Myoendothelial differentiation of human umbilical cord blood-derived stem cells in ischemic limb tissues. Circ Res 93(5):e51–e62
17. Zheng B, Cao B, Crisan M, Sun B, Li G, Logar A, Yap S, Pollett JB, Drowley L, Cassino T, Gharaibeh B, Deasy BM, Huard J, Peault B (2007) Prospective identification of myogenic endothelial cells in human skeletal muscle. Nat Biotechnol 25(9):1025–1034
18. Meliga E, Strem BM, Duckers HJ, Serruys PW (2007) Adipose-derived cells. Cell Transplant 16(9):963–970
19. Sampaolesi M, Torrente Y, Innocenzi A, Tonlorenzi R, D'Antona G, Pellegrino MA, Barresi R, Bresolin N, De Angelis MG, Campbell KP, Bottinelli R, Cossu G (2003) Cell therapy of alpha-sarcoglycan null dystrophic mice through intra-arterial delivery of mesoangioblasts. Science 301(5632):487–492
20. Sampaolesi M, Blot S, D'Antona G, Granger N, Tonlorenzi R, Innocenzi A, Mognol P, Thibaud JL, Galvez BG, Barthelemy I, Perani L, Mantero S, Guttinger M, Pansarasa O, Rinaldi C, Cusella De Angelis MG, Torrente Y, Bordignon C, Bottinelli R, Cossu G (2006) Mesoangioblast stem cells ameliorate muscle function in dystrophic dogs. Nature 444(7119):574–579. doi:10.1038/nature05282
21. Gargioli C, Coletta M, De Grandis F, Cannata SM, Cossu G (2008) PlGF-MMP-9-expressing cells restore microcirculation and efficacy of cell therapy in aged dystrophic muscle. Nat Med 14(9):973–978
22. Diaz-Manera J, Touvier T, Dellavalle A, Tonlorenzi R, Tedesco FS, Messina G, Meregalli M, Navarro C, Perani L, Bonfanti C, Illa I, Torrente Y, Cossu G (2010) Partial dysferlin reconstitution by adult murine mesoangioblasts is sufficient for full functional recovery in a murine model of dysferlinopathy. Cell Death Dis 1:e61. doi:10.1038/cddis.2010.35
23. Tedesco FS, Hoshiya H, D'Antona G, Gerli MF, Messina G, Antonini S, Tonlorenzi R, Benedetti S, Berghella L, Torrente Y, Kazuki Y, Bottinelli R, Oshimura M, Cossu G (2011) Stem cell-mediated transfer of a human artificial chromosome ameliorates muscular dystrophy. Sci Transl Med 3(96):96ra78. doi:10.1126/scitranslmed.3002342
24. Tajbakhsh S (2009) Skeletal muscle stem cells in developmental versus regenerative myogenesis. J Int Med 266(4):372–389
25. Tedesco FS, Dellavalle A, Diaz-Manera J, Messina G, Cossu G (2010) Repairing skeletal muscle: regenerative potential of skeletal muscle stem cells. J Clin Invest 120(1):11–19. doi:10.1172/JCI40373
26. Cossu G et al. (2015) EMBO Mol Med. 7(12):1513–1528. doi:10.15252/emmm.201505636
27. Bonfanti C, Rossi G, Tedesco FS, Giannotta M, Benedetti S, Tonlorenzi R, Antonini S, Marazzi G, Dejana E, Sassoon D, Cossu G, Messina G (2015) PW1/Peg3 expression regulates key properties that determine mesoangioblast stem cell competence. Nat Commun 6:6364. doi:10.1038/ncomms7364
28. Coletti D, Yang E, Marazzi G, Sassoon D (2002) TNFalpha inhibits skeletal myogenesis through a PW1-dependent pathway by recruitment of caspase pathways. EMBO J 21(4):631–642
29. Dellavalle A, Maroli G, Covarello D, Azzoni E, Innocenzi A, Perani L, Antonini S, Sambasivan R, Brunelli S, Tajbakhsh S, Cossu G (2011) Pericytes resident in postnatal skeletal muscle differentiate into muscle fibres and generate satellite cells. Nat Commun 2:499. doi:10.1038/ncomms1508
30. Quattrocelli M, Palazzolo G, Perini I, Crippa S, Cassano M, Sampaolesi M (2012) Mouse and human mesoangioblasts: isolation and characterization from adult skeletal muscles. Methods Mol Biol 798:65–76. doi:10.1007/978-1-61779-343-1_4
31. Xie Y, Zhang J, Lin Y, Gaeta X, Meng X, Wisidagama DR, Cinkornpumin J, Koehler CM, Malone CS, Teitell MA, Lowry WE (2014) Defining the role of oxygen tension in human neural progenitor fate. Stem Cell Rep 3(5):743–757. doi:10.1016/j.stemcr.2014.09.021
32. Leijten J, Georgi N, Moreira Teixeira L, van Blitterswijk CA, Post JN, Karperien M (2014) Metabolic programming of mesenchymal

stromal cells by oxygen tension directs chondrogenic cell fate. Proc Natl Acad Sci USA 111(38):13954–13959. doi:10.1073/pnas.1410977111

33. Correia C, Serra M, Espinha N, Sousa M, Brito C, Burkert K, Zheng Y, Hescheler J, Carrondo MJ, Saric T, Alves PM (2014) Combining hypoxia and bioreactor hydrodynamics boosts induced pluripotent stem cell differentiation towards cardiomyocytes. Stem Cell Rev 10(6):786–801. doi:10.1007/s12015-014-9533-0
34. Choi JR, Pingguan-Murphy B, Wan Abas WA, Noor Azmi MA, Omar SZ, Chua KH, Wan Safwani WK (2014) Impact of low oxygen tension on stemness, proliferation and differentiation potential of human adipose-derived stem cells. Biochem Biophys Res Commun 448(2):218–224. doi:10.1016/j.bbrc.2014.04.096
35. Estrada JC, Albo C, Benguria A, Dopazo A, Lopez-Romero P, Carrera-Quintanar L, Roche E, Clemente EP, Enriquez JA, Bernad A, Samper E (2012) Culture of human mesenchymal stem cells at low oxygen tension improves growth and genetic stability by activating glycolysis. Cell Death Diff 19(5):743–755. doi:10.1038/cdd.2011.172
36. Discher DE, Janmey P, Wang YL (2005) Tissue cells feel and respond to the stiffness of their substrate. Science 310(5751):1139–1143. doi:10.1126/science.1116995
37. Cameron AR, Frith JE, Cooper-White JJ (2011) The influence of substrate creep on mesenchymal stem cell behaviour and phenotype. Biomaterials 32(26):5979–5993. doi:10.1016/j.biomaterials.2011.04.003
38. Heras-Bautista CO, Katsen-Globa A, Schloerer NE, Dieluweit S, Abd El Aziz OM, Peinkofer G, Attia WA, Khalil M, Brockmeier K, Hescheler J, Pfannkuche K (2014) The influence of physiological matrix conditions on permanent culture of induced pluripotent stem cell-derived cardiomyocytes. Biomaterials 35(26):7374–7385. doi:10.1016/j.biomaterials.2014.05.027
39. Serena E, Zatti S, Reghelin E, Pasut A, Cimetta E, Elvassore N (2010) Soft substrates drive optimal differentiation of human healthy and dystrophic myotubes. Integr Biol (Camb) 2(4):193–201. doi:10.1039/b921401a
40. Engler AJ, Griffin MA, Sen S, Bonnemann CG, Sweeney HL, Discher DE (2004) Myotubes differentiate optimally on substrates with tissue-like stiffness: pathological implications for soft or stiff microenvironments. J Cell Biol 166(6):877–887. doi:10.1083/jcb.200405004
41. Yousef S, Benden C, Boyer D, Elidemir O, Frischer T, Goldfarb S, Lopez-Mitnik G, Mallory G, Visner G, Westall G, Schecter MG (2012) Lung transplantation in children following bone marrow transplantation: a multi-center experience. Pediatr Transplant doi:10.1111/petr.12023

Chapter 9

Fibro/Adipogenic Progenitors (FAPs): Isolation by FACS and Culture

Marcela Low, Christine Eisner, and Fabio Rossi

Abstract

Fibro/adipogenic progenitors (FAPs) are tissue-resident mesenchymal stromal cells (MSCs). Current literature supports a role for these cells in the homeostasis and repair of multiple tissues suggesting that FAPs may have extensive therapeutic potential in the treatment of numerous diseases. In this context, it is crucial to establish efficient and reproducible procedures to purify FAP populations from various tissues. Here, we describe a protocol for the isolation and cell culture of FAPs from murine skeletal muscle using fluorescence-activated cell sorting (FACS), which is particularly useful for experiments where high cell purity is an essential requirement. Identification, isolation, and cell culture of FAPs represent powerful tools that will help us to understand the role of these cells in different conditions and facilitate the development of safe and effective new treatments for diseases.

Key words Fibro/adipogenic progenitors, FACS, Cell culture, Skeletal muscle, Mesenchymal stromal cell

1 Introduction

Almost all postnatal tissues contain MSCs, which are normally quiescent but quickly respond to damage [1]. These cells, isolated from skeletal muscle by various research groups including ours, are capable of differentiating along the fibrogenic and adipogenic lineages in vitro and in vivo and have therefore been described as fibro/adipogenic progenitors (FAPs) [2, 3]. Furthermore, FAPs also can generate osteogenic lineage cells in a BMP-dependent manner [4]. While strong evidence exists supporting a role for FAPs in normal regeneration and tissue degeneration, the signals regulating their growth, survival, and differentiation remain largely unknown and under intense investigation. Due to their multipotency, FAPs have emerged as a potential therapeutic target for the treatment of reparative disorders, such as adipocyte accumulation (fatty degeneration), ectopic bone formation, and fibrous tissue deposition. Traditionally, methods used to isolate muscle and non-muscle tissue-resident MSCs have resulted in heterogeneous populations of cells, making it

Eusebio Perdiguero and DDW Cornelison (eds.), *Muscle Stem Cells: Methods and Protocols*, Methods in Molecular Biology, vol. 1556, DOI 10.1007/978-1-4939-6771-1_9, © Springer Science+Business Media LLC 2017

difficult to untangle the phenotypic and functional relationships between different populations (*see* **Note 1**). In order to investigate the properties and therapeutic potential of these cells, it is crucial to establish efficient and reproducible protocols for isolation and culture of a homogeneous population of FAPs. Utilizing antibodies to cell-specific surface antigens and FACS, we are able to identify and isolate the various tissue-resident cell types including FAPs, which represent 2–3 % of the cells in skeletal muscle.

FACS is a standard technique for the purification of subpopulations of cells. It allows separation of a heterogeneous sample into distinct groups of cells based on size, granularity, and fluorescence emission when using fluorophore-conjugated antibodies. The ability to differentiate between cells expressing different levels of specific antigens using FACS permits the isolation of highly pure and specific populations. Tissue-specific differences in antigen expression, target cell abundance, and extracellular matrix composition further complicate isolation and purification of FAPs by FACS and require protocol modifications. For this reason, tissue-specific protocols for the isolation of FAPs need to be established.

In this chapter we provide a detailed protocol for the isolation and culture of FAPs from skeletal muscle. FAPs were originally described based on the expression of the surface markers CD34, Sca1, and PDGFRα [2, 3]. In the following protocol, skeletal muscle FAPs are identified as a population of CD45−/CD31−/α7 integrin−/Sca1+ cells. It is important to remember that in other tissues, additional markers may be required for their purification. Cells expressing fibroblast markers (ER-TR7/FSP1/α-SMA), adipogenic markers (perilipin/Oil red), and osteogenic markers (alkaline phosphatase/Osterix) arise from individual multipotent progenitors contained within this population [2, 3].

FACS, coupled with cell culture, is a potent tool to evaluate FAP growth capabilities, differentiation properties, and response to environmental stimuli. Cell-based experiments such as drug screening and expansion for tissue transplant are crucial to establish a role for FAPs in regenerative medicine. In this work we offer guidelines and suggestions on how to reproducibly isolate and culture murine FAPs.

2 Materials

2.1 Reagents

Mice (C57BL/6 J or equivalent strains). Note: All protocols involving the use of rodents must adhere to all relevant institutional and governmental regulations.

Ice.

Ethanol.

Conical centrifuge tubes 15 ml (BD Falcon).

Conical centrifuge tubes 50 ml (BD Falcon).

Conical microcentrifuge tubes 1.5 ml (Eppendorf or equivalent).

Round bottom polypropylene tubes 5 ml (BD Falcon).

Round bottom polystyrene tubes 5 ml with cell strainer cap (BD Falcon).

Plastic Pasteur Pipettes.

Sterile surgical tools.

Sterile cell strainer 40 μm (Fisherbrand).

Fetal bovine serum (FBS) (HyClone).

Collagenase II (Sigma) 250 CDU/ml.

Collagenase D (Roche Biochemicals) 1.5 U/ml/Dispase II (Roche Biochemicals) 2.4 U/ml.

Phosphate-buffered saline (PBS).

FACS buffer (PBS, 2 % FBS, 2 mM EDTA pH 7.9).

Hoechst 33342 (Sigma).

Propidium iodide (PI) (Sigma).

FITC anti-mouse CD31 (eBioscience).

FITC anti-mouse CD45 (AbLab).

PECy7 anti-mouse Sca1 (eBioscience).

APC anti-mouse α7-integrin (AbLab).

Growth medium: Dulbecco's modified eagle medium (DMEM) (Invitrogen), 10 % FBS, penicillin 10,000 U/ml, streptomycin 10,000 μg/ml (Gibco), 2.5 ng/ml basic fibroblast growth factor (βFGF) (Invitrogen).

Cell culture dishes (Falcon).

2.2 Equipment

2.2.1 Cell Sorters

FACS Aria IIu, equipped with three lasers (488, 405, and 633 nm) (Becton Dickinson).

Influx, equipped with five lasers (488, 561, 405, 640, and 355 nm) (Becton Dickinson).

2.2.2 Centrifuges

Eppendorf benchtop centrifuge, model: 5417R.

Eppendorf benchtop centrifuge, model: 5810R.

2.2.3 Cell Culture

Biosafety cabinet (class II type A/B3).

Incubator set to 37 °C, 5 % CO_2.

Aspiration pump (vacuum pump).

3 Methods

The protocol is divided into two main sections: FAP isolation and cell culture. The first section is subdivided into tissue collection, enzymatic digestion, antibody staining, and cell sorting. Figure 1 provides an overview of the experimental procedure.

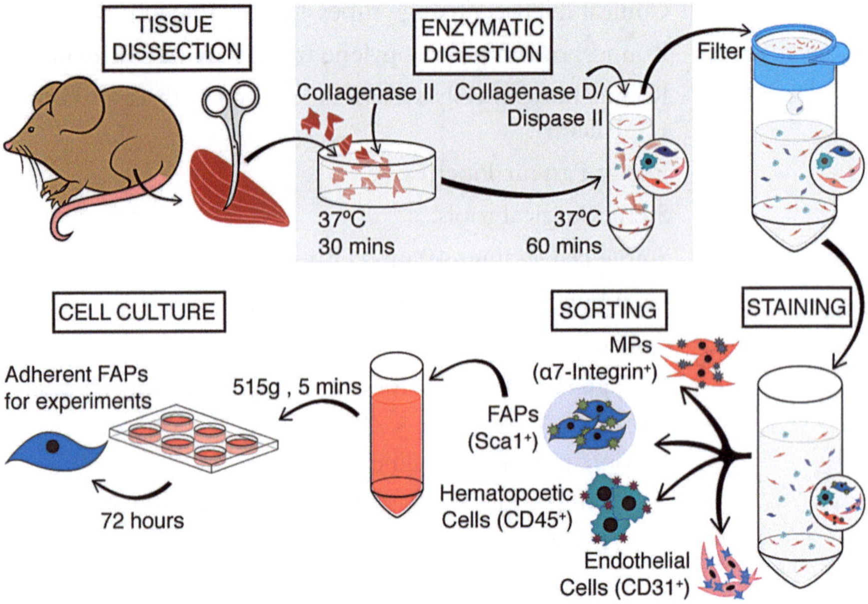

Fig. 1 Overview of FAP isolation from skeletal muscle. Skeletal muscle is isolated from adult mice and followed by sequential enzymatic digestions. Staining is then performed using fluorochrome-conjugated antibodies to label different populations, allowing isolation and culture of the desired FAP population

All procedures are performed at room temperature unless otherwise specified.

3.1 FAP Isolation (4–5 h)

3.1.1 Tissue Collection

1. Euthanize 7–12-week-old mice by CO_2 inhalation or according to approved institutional guidelines. A pair of tibialis anterior (TA) muscles and/or a calf muscle from the same or an extra mouse is used to set up the unstained, single-color controls and fluorescence-minus-one (FMO) control samples. Place mice on a paper towel, and spray coat liberally with 70 % (v/v) ethanol (*see* **Note 2**).
2. Using forceps, lift the skin of the abdominal region with forceps and make a small pelvic incision using sharp scissors.
3. Grasping the skin on either side of the incision firmly, pull in opposite directions (cranial and caudal), to completely expose the hind limb muscles underneath.
4. Dissect the muscle tissue away from the tibia and femur using blunt dissection and place in a six-well plate, using one well for every muscle group (two TA, two quadriceps, and one calf muscle per dish or well). Arrangement of the separate muscle groups is shown in Fig. 2a. If a large number of mice need to be processed, combine all muscle groups from each mouse in a 60 mm petri dish instead of separating them to allow easier processing. Keep the muscles on ice until enzymatic digestion.

3.1.2 Enzymatic Digestion

Throughout this section, use sterile reagents and tools and work under sterile conditions.

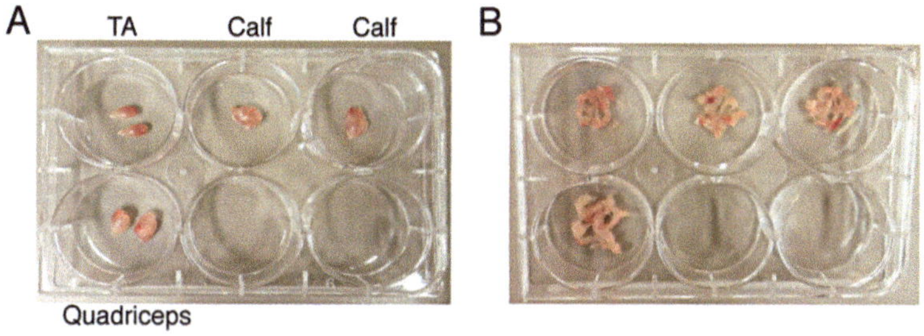

Fig. 2 Organization of the group of muscles after dissection. Muscles, once dissected, should be kept on ice in individual groups. Panel (**a**) shows the separate muscle groups collected from a single mouse and kept in a six-well plate on ice. Panel (**b**) shows the same muscles after being gently torn into smaller muscle fibers prior to the first enzymatic digestion

1. Using two sets of forceps (one serrated, one toothed), gently tear the tissue along the muscle fibers until the muscle is in small pieces (about 1 mm^3) (Fig. 2b). All non-muscle tissue, including fat and tendons, should be removed at this time. Gentle handling at this stage is critical (*see* **Note 3**).
2. Prepare the Collagenase II for the first digestion by adding 10 μl of 250 mM $CaCl_2$ per ml of Collagenase II. Add 0.5 ml of the activated Collagenase to each dish or well, containing two TAs, one calf, or two quadriceps. For pooled muscles in 60mm petri dishes, use 2 ml of Collagenase II (*see* **Note 4**).
3. Cover the well plate and incubate at 37 °C for 30 min (*see* **Note 5**).
4. After incubation, use a 1 ml syringe plunger to thoroughly mash the tissue pieces. Use one plunger for each sample to avoid contamination (*see* **Note 6**).
5. Add ~3 ml sterile ice-cold PBS to each dish or well and transfer the tissue sludge into a 15 ml conical tube, rinsing the dish as necessary to recover all the tissue. For pooled samples, use a 50 ml conical tube and increase the volume of PBS used in the wash steps accordingly. Keep the tubes on ice.
6. Top the tubes with PBS to 10–15 ml and centrifuge at 130 × *g* for 5 min at 4 °C.
7. While the muscle is spinning, prepare the enzyme solution for the second digestion by adding 10 μl of 250 mM $CaCl_2$ per 1 ml of Collagenase D/Dispase II.
8. Discard the supernatant and add 0.5 ml of the activated Collagenase D/Dispase II solution to each tube. Add 1 ml of enzyme to pooled samples
9. Incubate the tubes with rotation at 37 °C for 1 h. Vortexing the tubes every 15 min is highly recommended to help keep the sample homogenous and aid in proper enzymatic digestion.

10. Add ~5–10 ml cold FACS buffer and triturate using Pasteur pipettes. For pooled samples, increase the volume of PBS to 20–30 ml/tube
11. Put the tubes on ice and filter the suspension through a 40 μm cell strainer placed on top of a 50 ml conical tube.
12. Top each of the tubes with FACS buffer to 40–50 ml then centrifuge at 515 × *g* for 5 min at 4 °C.
13. Discard the supernatant. If the pellet is large, an additional wash step is recommended before proceeding to staining. The cells are now ready for antibody staining.

3.1.3 Antibody Cell Staining

1. To generate compensation and gating controls, resuspend the control sample in 1 ml of FACS buffer and aliquot 100 μl per tube into nine pre-labeled 1.5 ml microcentrifuge tubes. Controls should include an unstained sample, a single-color staining control each (five total) for Hoechst, PI, FITC, PeCy7, and APC and three "fluorescence minus one" (FMO) controls for FITC, PeCy7, and APC.
2. Add antibodies to the appropriate tubes, mix well, and incubate cells on ice in the dark for 30 min.
3. For the main sample, prepare the antibody cocktail at predetermined concentrations; CD31-FITC 1:500, CD45-FITC 1:1000, Sca1-PeCy7 1:5000, and α7 integrin-APC 1:1500 (*see* **Note 7**).
4. At this stage, cells from the individual muscle groups of each mouse can be consolidated if desirable (i.e., combine TAs, calves, and quadriceps of each mouse or genotype in one tube). Resuspend the main samples in 100–200 μl of the antibody cocktail/2TA (volume dependent on cell number to obtain optimal antibody concentration), e.g., five mice (TA, 1 ml; TA + calves, 2–3 ml). Staining of the main sample can be performed in the 50 ml conical tube.
5. Mix well and incubate on ice in the dark for 30 min.
6. Wash: Add 500 μl of FACS buffer to each of the nine control tubes; add 20–30 ml of FACS buffer to the main sample.
7. Centrifuge the Eppendorf tubes at 850 × *g* for 5 min 4 °C. Centrifuge the 50 ml sample tube at 515 × *g* for 5 min 4 °C.
8. Discard supernatant and resuspend cells in FACS buffer (0.5 ml for each of the controls, 1–3 ml for the main sample, depending on the size of the pellet). Stain with a viability dye such as PI (1:1000) and/or Hoechst (1:500) (*see* **Note 8**).
9. Filter cells through a 40 μm cell strainer cap into 5 ml polypropylene round bottom collection tubes to remove cell clumps and debris (*see* **Note 9**). For maximum cell yield, it is best to proceed immediately to cell sorting.

3.1.4 Sorting

1. Set up the cell sorter according to the manufacturer's instructions, using the unstained sample to ensure the voltage settings are correct. Single-color and FMO controls should be used to

ensure that the population of interest is visible in all channels and to set the compensation values (*see* **Note 10**). Gating strategies can vary, but generally, we select first for live cells and for lineage negative cells (CD45/CD31-) and finally for α7 integrin-/Sca1+ cells to identify FAPs (*see* **Note 11**). Figure 3 shows an example of the gating strategy used to identify FAPs using the markers outlined above.

2. Sort cells into 2 ml of collection buffer (FAP growth medium) in pre-cooled 5 ml polystyrene round bottom tubes (*see* **Note 12**). Whole muscle (two TAs, two calves, two quadriceps) from a single mouse usually yields approximately 70,000–10,000 FAPs (*see* **Note 13**). The viability of the sorted cells is generally well above 95 % when reanalyzed by FACS (*see* **Note 14**).

3.2 Cell Culture

All the steps described in this section should be performed in sterile conditions:

1. Immediately after sorting, centrifuge the cells at 515 × *g* for 5–10 min at 4 °C.
2. Working within a biosafety cabinet, open the tube and aspirate the supernatant carefully. Leaving some collection buffer (~100–200 μl) is recommended to avoid aspirating pelleted cells, as small cell pellets may not be clearly visible.
3. Resuspend the cell pellet in 1–3 ml of FAP growth media. The volume of growth media used will depend on the size of the cell pellet and expected cell number.
4. Plate cells at a density of 10,000 cells per cm^2 in non-treated culture dishes (*see* **Note 15**).
5. Cells should be maintained in an incubator kept at standard growth conditions of 37 °C and 5 % CO_2.

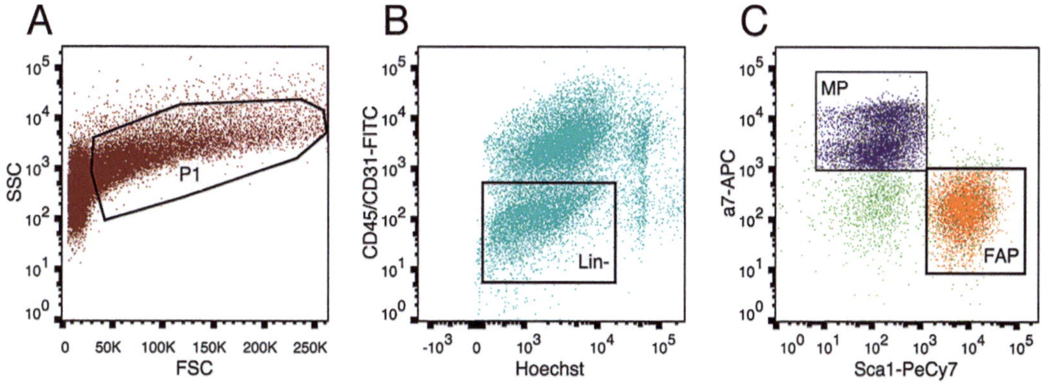

Fig. 3 Gating strategy for isolate FAPs from skeletal muscle by FACS. The sequential gating strategy used to isolate FAPs from a heterogeneous skeletal muscle preparation. (**a**) General gating using FSC (linear scale) and SSC (logarithmic scale) to discard debris and identify the primary cell population (P1). (**b**) Hoechst vs CD31/CD45 selects for lineage, (CD31−/CD45−), live (Hoechst+) cells. (**c**) Sca1 vs α7-integrin gating allows distinction between α7-integrin+ muscle progenitors and the Sca1+ FAPs

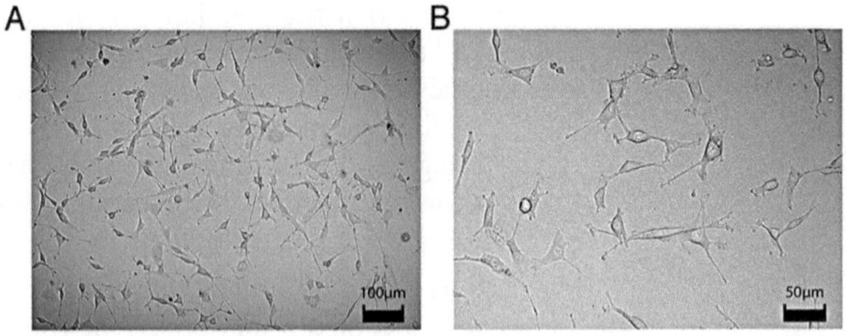

Fig. 4 Bright field images of FAPs in culture. FAPs isolated from adult mice at 72 h in culture. The cells were cultured in growth medium to promote cell growth and survival. Note the elongated cell processes and small cell body characteristic of FAPs in culture, (**a**) 100 μm and (**b**) 50 μm

6. After 72 h in culture, most FAPs should have adhered to the plate and adopted a spindle shape with short projections (Fig. 4). A small amount of floating debris and/or dead cells is considered normal and should be removed by changing the growth media.
7. Cells can continue to be cultured with FAP growth media under standard conditions until the desired confluence is reached. Media should be changed every 48 h to maintain growth of the FAPs (*see* **Note 16**). Alternatively, for differentiation assays, media can be changed to differentiation media as previously described [2–4].

3.3 Conclusions

Proper identification of FAPs is a crucial step for understanding the biology and physiological functions of this cell population and permits multiple downstream applications including cell culture, transplantation, gene expression, and single cell analysis. Establishing a universal protocol for the isolation and culture of FAPs is essential to standardize experiments and aid in the interpretation of results obtained by various research groups.

The protocol described here provides guidelines for obtaining a highly pure population of FAPs using FACS. Additionally, the high viability of the cells makes this protocol suitable for cell culture and further downstream applications. It should be noted however that this method does not yield absolute purity. In most cases it is prudent to assume that up to 1 % of the cells may be contaminants that could selectively expand in specific culture conditions.

4 Notes

1. Traditionally the isolation of these cells relied on their plastic-adherent properties, because of their low frequency in tissue and the lack of specific surface markers. However there are several limi-

tations to isolating mouse MSCs by plastic adherence. Cells isolated in this way can contain contaminating hematopoietic cells, and to avoid contamination several passage steps, repeated medium changes and prolonged culture times (several weeks) are required, which often modify the properties of MSCs. These in vitro-induced changes have made it difficult to uncover the in vivo identity, physiological function, and biological properties of mouse MSCs. The development of more efficient isolation methods such as those based on combinations of surface markers and FACS, yielding a near-homogeneous cell population of multipotent MSCs, is able to overcome these limitations. Due to its sensitivity and quantitative precision, FACS is particularly advantageous over other available methods when a cell population needs to be purified based on a weakly expressed surface marker or when two or more populations need to be purified which have different levels of expression of the same surface marker.

2. 70,000–100,000 cells can be isolated from the whole hind limb muscle of 7–12-week-old mice. Certain factors can affect the cell yield including the age of the animal, genetic mutations, and muscle damage. To obtain the maximum number of FAPs possible, we suggest damaging the TA muscle using cardiotoxin or notexin and collecting tissue on day 3 after damage when FAPs peak in the damaged tissue [5].
3. While it is important to mechanically liberate cells from the starting tissue, care must be taken to avoid excessive tearing as this will dramatically reduce cell viability and final yield.
4. Each muscle group should be individually enzymatically digested and washed (ideally using a six-well plate). Pooling the muscle groups together necessitates additional wash steps and results in increased debris and lower viable cell yield. However, when processing several mice, pooling muscle groups is the preferred option. Pooling the mucle groups reduces processing time and therefore results in increased viable cell yield.
5. Depending on the source and batch of enzyme, enzymatic activity may vary and affect digestion times. We suggest each lab test their batch of enzyme to optimize the digestion. Conditions that affect muscle stiffness/fibrosis such as age or genetic mutations may also affect the amount of time required for digestion and will require individual optimization. Muscle is considered properly digested by Collagenase II when the fibers easily dissociate. The first enzymatic digestion only partly digests the muscle; pieces of tissue that are not fully dissociated are normal.
6. If you cannot immediately proceed to mechanical dissociation using the syringe plunger, place the muscle digest on ice to slow enzymatic activity and avoid over-digestion.
7. Selecting the appropriate antibody combination and determining the optimal antibody concentration prior to starting

multicolor flow cytometry is a critical step. Fluorophore combinations should be chosen to minimize spectral overlap and to enhance signal from weak or low abundance antigens. Each antibody should be titrated prior to use to determine the optimal concentration to use. Using the wrong concentration of antibody can result in nonspecific staining and/or poor resolution of the desired populations.

8. The exclusion of dead cells is a key part of the protocol, as cellular debris can interfere with subsequent culture and growth of FAPs. Typical dyes used to distinguish between dead and live cells are Hoechst (identifying nucleated cells, which helps excluding debris) and propidium iodide (PI) (identifying dead cells). Double labeling with Hoechst and PI is ideal if it does not interfere with other fluorochromes.
9. Clumps in the sample induced by reagents and procedures used throughout the protocol can clog the nozzle in the cell sorter and affect cell yield. Thus, filtering the stained sample should be the last step before cells are sorted and is critical to obtain a good yield.
10. The process of setting up a flow cytometer, including setting cross channel compensation values, varies depending on the model and needs to be performed by appropriately trained personnel. For sterile sorts, sterilize the instrument according to the manufacturer's instructions.
11. The gating strategy for the protocol described here is as follows: (a) FSC vs SSC gate to discriminate between cells and debris, (b) FITC vs Hoechst to identify live cells and linage negative (Lin−) population (CD31- and CD45-), and (c) APC vs PeCy7 to discriminate between muscle progenitors (α7-integrin+) and FAPs (Sca1+).
12. The collection medium can vary depending on the downstream application. For cell culture, growth medium works well. However, for genomic DNA extraction or RNA extraction, PBS+10 % FBS is preferable.
13. Cell viability and yield can be compromised during the sorting process and can later affect the quality of the cell culture. Recovery of the cells can be increased collecting sorted cells in serum-rich media and inverting collection tubes during the process. Keep the cells on ice and seed them as soon as possible after the sorting help to improve the cell viability.
14. Performing a post-sort analysis on the sorted cells helps to determine their purity and viability. The final cell yield when sorting is highly influenced by the time taken to process the sample. A proper balance between the number of mice processed, and the time required for processing/sorting, is crucial in obtaining highly pure and viable cell populations.

15. To enhance cell viability and efficiency of expansion, cells should be plated at the recommended density or higher. Plating FAPs at a lower density can negatively affect cell viability and expansion in culture. FAPs may be plated on specially coated plates for specialized assays or applications as required.
16. In general, passaging FAPs is not recommended, as it is associated with loss of multipotency. However, if absolutely required, FAPs can be passaged 1–2 times at confluence of ≤70 % and still maintain the ability to differentiate along multiple lineages.

Acknowledgments

We thank Claudia Hopkins for her collaboration and figure construction. This work was funded by CIHR MOP-97856. M.L. was supported by a fellowship from the Chilean government, and C.E. was supported by a Four-Year Doctoral Fellowship (4YF) by the University of British Columbia.

Conflict of Interest: The authors declare no conflict of interest.

References

1. da Silva Meirelles L, Chagastelles PC, Nardi NB (2006) Mesenchymal stem cells reside in virtually all post-natal organs and tissues. J Cell Sci 119:2204–2213
2. Joe AWB, Yi L, Natarajan A et al (2010) Muscle injury activates resident fibro/adipogenic progenitors that facilitate myogenesis. Nat Cell Biol 12:153–163
3. Uezumi A, Ito T, Morikawa D et al (2011) Fibrosis and adipogenesis originate from a common mesenchymal progenitor in skeletal muscle. J Cell Sci 124:3654–3664
4. Wosczyna MN, Biswas AA, Cogswell CA et al (2012) Multipotent progenitors resident in the skeletal muscle interstitium exhibit robust BMP-dependent osteogenic activity and mediate heterotopic ossification. J Bone Miner Res 27:1004–1017
5. Lemos DR, Babaeijandaghi F, Low M et al (2015) Nilotinib reduces muscle fibrosis in chronic muscle injury by promoting TNF-mediated apoptosis of fibro/adipogenic progenitors. Nat Med 21:786–794

Chapter 10

Single Cell Gene Expression Profiling of Skeletal Muscle-Derived Cells

Sole Gatto, Pier Lorenzo Puri, and Barbora Malecova

Abstract

Single cell gene expression profiling is a fundamental tool for studying the heterogeneity of a cell population by addressing the phenotypic and functional characteristics of each cell. Technological advances that have coupled microfluidic technologies with high-throughput quantitative RT-PCR analyses have enabled detailed analyses of single cells in various biological contexts. In this chapter, we describe the procedure for isolating the skeletal muscle interstitial cells termed Fibro-Adipogenic Progenitors (FAPs) and their gene expression profiling at the single cell level. Moreover, we accompany our bench protocol with bioinformatics analysis designed to process raw data as well as to visualize single cell gene expression data. Single cell gene expression profiling is therefore a useful tool in the investigation of FAPs heterogeneity and their contribution to muscle homeostasis.

Key words Fibro-Adipogenic Progenitors (FAPs), Muscle regeneration, Single cell gene expression, Fluorescence Activated Cell Sorting (FACS), Flow cytometry, Data analysis, Fluidigm® technology, Principal component analysis, Hierarchical clustering heatmap, Violin plots, Bar plots

1 Introduction

Evaluation of gene expression in cells has been crucial information that molecular biologists have pursued in the last 50 years. Major technical advances in the area of next-generation sequencing have allowed for high-throughput, unbiased, genome-wide analysis of gene expression. While these technologies helped to overcome the obstacles of analyzing more than one gene at the time, they are unable to address the more fundamental and technically challenging question that is how different are cells from each other. What we could observe until recently was gene expression from bulk population, that is the average expression profile from a pool of cells. Technological advances that have coupled microfluidic technologies with high-throughput quantitative RT-qPCR (Real-Time

Electronic supplementary material: The online version of this chapter (doi:10.1007/978-1-4939-6771-1_10) contains supplementary material, which is available to authorized users.

Eusebio Perdiguero and DDW Cornelison (eds.), *Muscle Stem Cells: Methods and Protocols*, Methods in Molecular Biology, vol. 1556, DOI 10.1007/978-1-4939-6771-1_10, © Springer Science+Business Media LLC 2017

quantitative Polymerase Chain Reaction) analyses (and next-generation sequencing, not covered in this chapter) have revolutionized the way we can describe cell populations. Using current state-of-the art technologies, specifically Fluidigm®, we can learn single cell-specific gene expression profiles. This knowledge will allow us to untangle the most heterogeneous cell populations into cell-specific gene expression profiles and maybe have a more detailed and plastic definition of the function of those cells.

Fibro-adipogenic progenitors (FAPs) are a population of skeletal muscle interstitial cells whose contribution during muscle regeneration is greatly influenced by the microenvironment. In a healthy muscle, FAPs support skeletal muscle regeneration by releasing pro-myogenic trophic factors [1–4]. However, in pathological conditions, FAPs are the major drivers of fibrotic scarring and intramuscular fatty infiltration [1, 5–9].

On account of the functional heterogeneity and plasticity of FAPs in response to surrounding signals [10–13], the assessment of FAPs behavior by studying the bulk population may mask their distinct functions. Therefore, single cell gene expression profiling is a powerful tool to resolve FAPs functional heterogeneity, allowing the study of each single cell as a distinct biological entity.

In this chapter, we describe in detail single cell RT-qPCR analysis of FAPs. Briefly, FAPs are first isolated from murine skeletal hind-limb muscles by fluorescence-activated cell sorting (FACS) based on expression of cell surface markers CD34 and Sca1, and absence of CD45, Ter119, CD31, and Itga7 (alpha7 Integrin) (modified protocol from[14, 15]). Afterward, Fluidigm® automated microfluidic system is used to rapidly and reliably isolate single cells, extract RNA, synthesize cDNA, and perform multiple qPCRs on each isolated single cell. The generated dataset contains the expression levels of 96 genes for the 96 cells. Primary data analysis and visualization are then performed combining Fluidigm® RT-PCR analysis software and custom analysis in R.

2 Materials

In this section, we list the specific materials, reagents, laboratory equipment, instruments, and software required for conducting the single cell gene expression analysis starting from mouse muscle dissection and single cell isolation with FACS. Single cell capture and cDNA synthesis is performed with Fluidigm® C1 Single-Cell Auto Prep system and single cell RT-qPCR is run on Fluidigm® BioMark system. Standard lab equipment, such as 1.5 ml tubes, conical centrifuge 15 and 50 ml tubes, 0.2–1000 μl pipettes, tips, pipettor, serological pipettes are needed, but not specifically listed here.

2.1 Muscle Isolation and Homogenization into Mononuclear Cell Suspension

1. FACS buffer: HBSS (Hank's Balanced Salt Solution) containing 0.2 % (w/v) BSA and 10 Units/ml Penicillin + 10 μg/ml Streptomycin. Filter sterile using 0.22 μm bench-top filter units. (*see* **Note 1**).
2. Digestion Solution: Under sterile conditions prepare 4 ml Digestion buffer for one mouse (muscles from two hindlimbs), containing 2 mg/ml Collagenase A, 2.4 U/ml Dispase I and 10 μg/ml DNase I (Roche) in FACS buffer: in the tissue culture cabinet mix 3.2 ml FACS buffer with 0.4 ml 20 mg/ml Collagenase A, 0.4 ml 24 U/ml Dispase I and 4 μl 10 mg/ml DNAse I. (*see* **Note2**).
3. Parafilm.
4. Tweezers.
5. Scissors.
6. 70 % Ethanol.
7. 100 mm petri dish.
8. 35 mm petri dish.
9. 100 μm cell strainer.
10. 40 μm cell strainer.
11. 0.22 μm bench top filter units for large volumes over 50 ml.
12. 0.22 μm filters with syringes for small volumes up to 50 ml.
13. Centrifuge with a cooling system for 15–50 ml conical tubes.
14. Water bath set to 37 °C.
15. Cell culture sterile cabinet (hood).

2.2 Fluorescence Activated Cell Sorting (FACS)-Assisted Purification of FAPs

1. FACS buffer (*see* Subheading 2.1).
1. Propidium Iodide (Sigma) (*see* **Note3**).
2. Goat serum, sterile.
3. Anti-CD31-PacificBlue conjugated antibody (RM5228, Life Technologies).
4. Anti-CD45-eFluor450 (Pacific Blue replacement) conjugated antibody (clone 30-F11, eBioscience).
5. Anti-Ter119-eFluor450 (Pacific Blue replacement) conjugated antibody (clone TER-119, eBioscience).
6. Anti-Sca1-FITC (Fluorescein isothiocyanate) conjugated antibody (clone E13–161.7, BD Pharmingen).
7. Anti-CD34 conjugated with Alexa Fluor® 647 conjugated antibody (clone RAM34, BD Pharmingen).
8. Anti-α7integrin-PE (Itga7, R-Phycoerythrin) conjugated antibody (clone R2F2, AbLab).
9. 5 ml Polystyrene round-bottom tubes (FACS tubes).

10. 5 ml round-bottom tubes with cell strainer cap.
11. Hemocytometer.
12. Centrifuge with a cooling system for 15–50 ml conical tubes.
13. Centrifuge with a cooling system for 1.5 ml tubes.
14. Cell culture cabinet (hood).
15. Flow Cytometry analyzer (e.g., FACSAria—BD Biosciences).

2.3 C1-Assisted Single Cell Capture, RNA Isolation and cDNA Preparation

1. Pre-PCR sterile cabinet.
2. Post-PCR sterile cabinet.
3. C1 System (Fluidigm, cat. #100-7000).
4. C1 System User Guide (Fluidigm, protocol #100–4977).
5. Confocal microscope (*see* **Note 4**).
6. PCR plates and Thermal Seals.
7. Eight channels multichannel pipette 0.5–10 μl.
8. C1 Single-Cell Auto Prep Array for Pre-Amp (10–17 μm) (cat. #100-5479)—Integrated Fluidic Circuit (IFC).
9. DELTAgene assays (100 μM F+R primers) (Fluidigm®) (*see* **Note 5**).
10. C1 Single-Cell Auto Prep Module 1 and Module 2 kits (Fluidigm, cat. #100–5319).
11. Single Cell-to-CT Kit (Life Technologies, cat. #4458237).
12. Live/Dead Viability/Cytotoxicity Kit, for mammalian cells (Life Technologies, cat. #L-3224) (optional).

2.4 BioMark Real-Time PCR on Single Cell C1 Harvested cDNA

1. Pre-PCR sterile cabinet.
2. Post-PCR sterile cabinet.
3. BioMark HD system (*see* Fluidigm® Real-Time PCR User Guide (#68000088)).
4. Fluidigm® IFC Controler (*see* Fluidigm® Real-Time PCR User Guide (#68000088)).
5. multichannel pipette 0.5–10 μl (eight channels).
6. Dynamic Arrays for Gene Expression (we have used 96 × 96 Dynamic Array, Fluidigm, cat. #BMK-M-96.96, for 96 genes × 96 assays) (*see* **Note 6**).
7. GE 96.96 Dynamic Array DNA Binding Dye Sample and Assay Loading Reagent Kit—ten chips (Fluidigm, cat. #100-3415).
8. 96.96 Control Line Fluid (Fluidigm, cat. #89000021).
9. 2×Sso Fast EvaGreen SuperMix with Low ROX (Bio-Rad Laboratories, cat. #172-5211).
10. DELTAgene assays (100 μM Forward + Reverse primers) (Fluidigm) (*see* **Note 5**).

11. 1× DNA Suspension Buffer: 10 mM TRIS–HCl pH 8.0, 0.1 mM EDTA pH 8.0, 0.2 μm sterile-filtered.
12. Refrigerated benchtop centrifuge with plate adaptors (*see* **Note7**).

2.5 Software

1. FACSDiva software (only on a Windows environment—http://www.bdbiosciences.com/us/instruments/research/software/flow-cytometry-acquisition/bd-facsdiva-software/m/111112?cc=US).
2. FlowJo software (available for Windows and Mac http://www.flowjo.com/download-flowjo/).
3. Fluidigm® BioMark HD Data Collection Software (only on a Windows environment—Fluidigm, User Guide: 100–2451).
4. Fluidigm® BioMark RT PCR analysis software (only on a Windows environment—http://207.218.201.42/home/fluidigm/downloads/BioMark.zip).
5. R (runs on Linux, Mac and Windows—http://cran.r-project.org).

3 Methods

In this section, we describe in detail the experimental procedures for FAPs single cell-qPCR analysis from FAPs isolation (modified protocol from [14]) to data analysis.

3.1 Muscle Isolation and Homogenization into Mononuclear Cell Suspension

Isolation of primary cells requires working in a clean and designated laboratory space, preferably in the tissue culture sterile cabinet. Sterilize the tools—scissors and forceps—in 70 % ethanol, and allow them to dry on a clean tissue (*see* **Note 8**).

1. Anesthetize mice with isoflurane and sacrifice by cervical dislocation. Spray skin of the mouse with 70 % ethanol. Cut the skin around the abdomen and pull the skin down exposing the hind-limb muscles. Cut the legs off the body, avoid cutting through the quadriceps, and place in the 100 mm petri dish with ice cold FACS buffer (Fig. 1a). Remove unwanted fat and muscle tissue, exposing Tibialis Anterior (TA), Quadriceps and Gastrocnemius (Fig. 1b). Excise hind-limb muscles Tibialis Anterior, Quadriceps, and Gastrocnemius from the bones and place in a clean 35 mm petri dish (Fig. 1c).
2. Mechanically mince muscles to an almost homogeneous consistency using forceps and scissors for about 5–10 min. At this stage the remaining muscle pieces should be about 1 mm^3 in size, or smaller (Fig.1d, e).

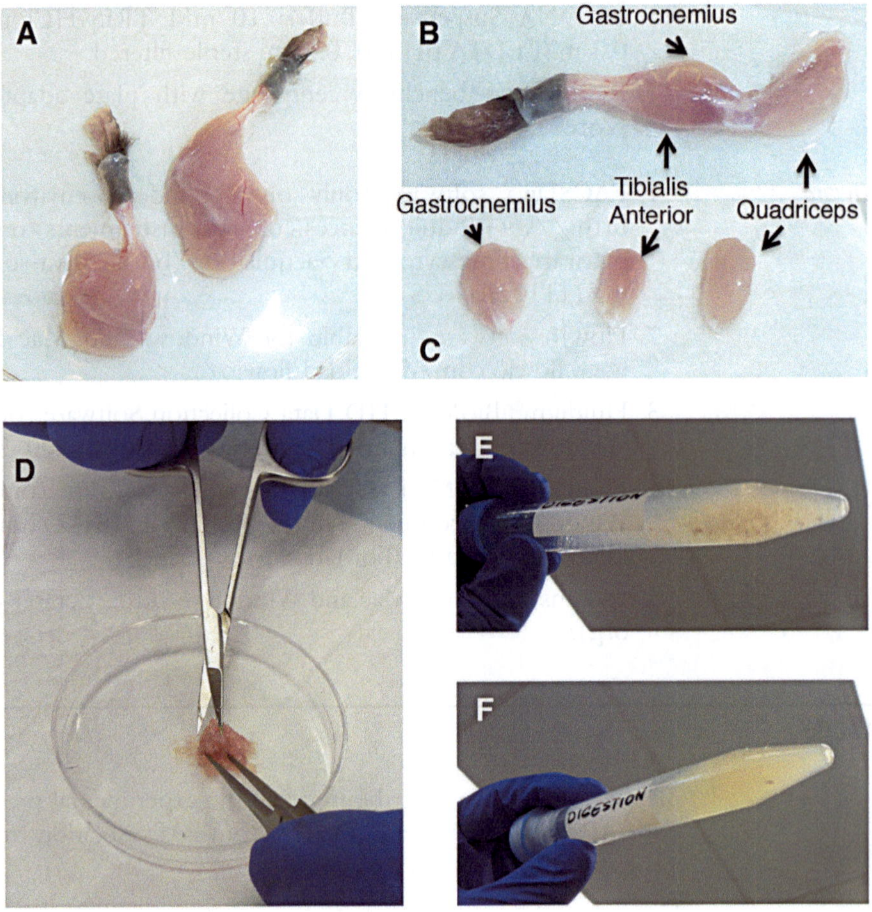

Fig. 1 Muscle isolation. (**a**) Hind-limbs isolated from mouse. (**b**) Exposed Gastrocnemius, Tibialis Anterior, and Quadriceps. (**c**) Isolated and excised muscles Gastrocnemius, Tibialis Anterior, and Quadriceps. (**d**) Mechanical mincing of muscles. (**e**) Minced muscles in the Digestion Solution at the beginning of the enzymatic digestion. (**f**) Digested muscles at the end of enzymatic digestion

3. Transfer muscle slurry with forceps into a 15 ml conical tube containing 4 ml Digestion solution (Fig. 1e) (*see* Subheading 2.1 for recipe). Close the tube tightly, seal the cap with parafilm, and incubate for 40–90 min at 37 °C with gentle horizontal rocking (about 1 rpm/s) (*see* **Note 9**). During digestion the extracellular matrix of muscle tissue is disrupted and single mononucleated cells are released. Digestion solution contains Collagenase A (protease that degrades the triple-helical native collagen fibrils) and Dispase I (protease that cleaves fibronectin and type IV collagen). It is critical to monitor the level of tissue digestion regularly. As soon as the solution becomes homogeneous and small pieces of muscles appear to have been efficiently digested (compare Fig. 1e, f), stop the reaction by placing the tube on ice, and proceed to the next step immediately. From now on, always keep cells on ice.

4. In a sterile cell culture cabinet, add 10 ml ice cold FACS buffer (*see* Subheading 2.1 for recipe) to the digested muscle homogenate and pipette up-and-down five times. Filter the cell suspension through 100 μm cell strainer placed on the top of 50 ml conical tube, collect flow-through containing cells in the 50 ml tube. Wash the 15 ml tube, where muscles were digested, with 10 ml cold FACS buffer and pass through the filter on the top of 50 ml tube. Repeat one more time. The final volume of cells in FACS buffer in 50 ml tube is about 35 ml (*see* **Note 10**).
5. Centrifuge cells at 300 × *g* for 5 min at 4 °C. Discard supernatant by careful decantation.
6. Repeat **step 4** using 40 μm cell strainer.
7. After centrifugation decant the supernatant carefully. Remaining liquid with cells will be about 300–500 μl. Adjust to 555 μl total volume with FACS buffer (55 μl of cells will be used for flow cytometry controls—*see* Subheading 3.2, final sample volume will be 500 μl), resuspend cells pipetting gently up-and-down and transfer cell suspension to a sterile 1.5 ml tube (*see* **Note 11**).
8. Block unspecific cellular surface protein affinity by adding 13.9 μl of goat serum per 555 μl cell suspension (final concentration 2.5 %) and incubate 5 min on ice.

3.2 Fluorescence Activated Cell Sorting (FACS)-Assisted Purification of FAPs

1. When establishing a Flow Cytometry protocol, it is important to set up single color controls (SCC) and Fluorescence Minus One (FMO) controls for each fluorophore to confirm the specificity of the antibodies and for the proper interpretation of flow cytometry data. Prepare eleven 15 ml FACS control tubes (*see* Table 1).
2. Stain remaining cells in 1.5 ml tube (from now on called "sample" to distinguish it from the controls) in 500 μl suspension with primary antibodies. Add 10 μl anti-CD31, 10 μl anti-CD45, 10 μl anti-Ter119, 10 μl anti-Sca1, 20 μl anti-CD34, and 1 μl anti-Itga7. Do not add Propidium iodide yet (*see* **Note 12**).
3. Incubate controls on ice for 30 min, and the sample at 4 °C on a rotator for 30 min. Keep in the dark.
4. Add 4 ml FACS buffer to each control tube to wash cells. Transfer sample to a 15 ml tube and add 14 ml FACS buffer. Gently mix by inversion. Centrifuge cells at 300 × *g* for 5 min at 4 °C. Decant the supernatants carefully leaving small amount of residual buffer. Adjust sample to 500 μl total volume with FACS buffer and pipette gently up-and-down (do not adjust controls). Resuspend the cell pellets in the controls by gentle vortexing for 5 s. Keep all tubes on ice. The controls are now ready to be analyzed by FACS.

Table 1
Pipetting scheme for tubes containing single color controls (SCC) and FMO controls

	Tubes	Cells + FACS buffer	Antibodies/propidium iodide
	1. no staining control	5 μl of cells, 95 μl of FACS buffer	
SCC	2. Propidium iodide (*see* **Note3**)	5 μl of cells, 95 μl of FACS buffer	1 μl Propidium iodide
	3. Pacific Blue	5 μl of cells, 95 μl of FACS buffer	2 μl of each anti-CD45, anti-Ter119 and anti-CD31 antibody
	4. FITC	5 μl of cells, 95 μl of FACS buffer	2 μl of anti-Sca1 antibody.
	5. Alexa Fluor® 647	5 μl of cells, 95 μl of FACS buffer	4 μl of anti-CD34 antibody.
	6. PE	5 μl of cells, 495 μl of FACS buffer	1 μl of anti-Itga7 antibody.
FMO	7. Pacific Blue	5 μl of cells, 495 μl of FACS buffer	10 μl anti-Sca1, 20 μl anti-CD34 and 1 μl anti-Itga7 antibodies, 1 μl Propidium iodide
	8. FITC	5 μl of cells, 495 μl of FACS buffer	10 μl anti-CD31, 10 μl anti-CD45, 10 μl anti-Ter119, 20 μl anti-CD34 and 1 μl anti-Itga7 antibodies, 1 μl Propidium iodide
	9. Alexa Fluor® 647	5 μl of cells, 495 μl of FACS buffer	10 μl anti-CD31, 10 μl anti-CD45, 10 μl anti-Ter119, 10 μl anti-Sca1, 20 μl and 1 μl anti-Itga7 antibodies, 1 μl Propidium iodide
	10. PE	5 μl of cells, 495 μl of FACS buffer	10 μl anti-CD31, 10 μl anti-CD45, 10 μl anti-Ter119, 10 μl anti-Sca1, 20 μl anti-CD34 antibodies, 1 μl Propidium iodide
	11. Propidium iodide (*see* **Note3**)	5 μl of cells, 495 μl of FACS buffer	10 μl anti-CD31, 10 μl anti-CD45, 10 μl anti-Ter119, 10 μl anti-Sca1, 20 μl anti-CD34 and 1 μl anti-Itga7 antibodies.

5. Add 1 μl of Propidium iodide to the sample. Dead cells in the sample will stain with Propidium iodide and will be eliminated by flow cytometry analysis. Filter the sample through the cell strainer cap containing round-bottom tube to eliminate cell aggregation. There is no need to filter the SSC and FMO controls. The sample is now ready to be analyzed by FACS (*see* **Note 3**).
6. We typically use the FACSAria instrument for sorting, and we analyze the data using the FACSDiva software (Fig. 2).
7. Analyze no color control and SCC controls to set the gates for FACS populations.
8. Analyze FMO controls to ensure that the combination of remaining fluorophores in each FMO control does not lead to

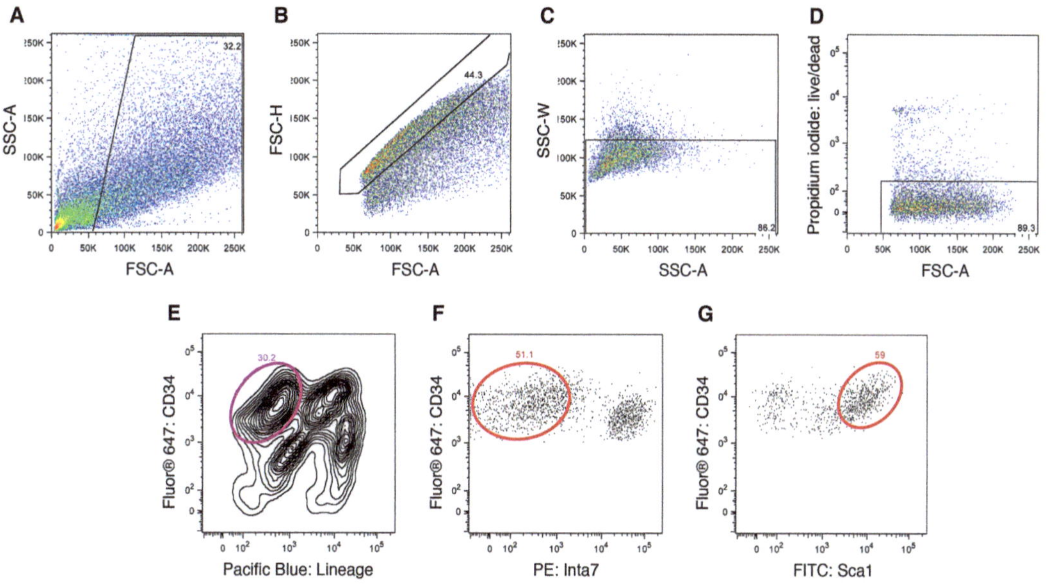

Fig. 2 Representative flow cytometry data to illustrate the gating strategy for FACS-based purification of FAPs: Mononucleated (SSC and FCS plots) and live (Propidium iodide negative) cells were further selected as negative for Lineage markers CD45, Ter119, CD31 (Pacific Blue and Fluor450 (Lin-)) and negative for alpha7integrin (PE) (a7int-), while positive for CD34 (CD34+, Alexa Fluor® 647) and Sca1 (Sca1+, FITC) markers. Thus, FAPs cell population is gated as Lin-/a7int-/CD34+/Sca1+. *SSC* = side scatter, *FSC* = forward scatter

an emission with the wavelength of the absent fluorophore tested (compensation).

9. Analyze the sample. Here, cell granularity (side scatter, SSC), dimension (forward scatter, FSC), and Propidium iodide staining is used to enclose in the gate events corresponding to live cells (Fig. 2a–d). Collect FAPs as events corresponding to cells that are negative for CD45, Ter119, CD31 and a7-int, and positive for CD34 and Sca1 (Fig. 2e–g), in 5 ml FACS tubes containing 300 μl FACS buffer.
10. After the FACS-based isolation of FAPs, centrifuge FAPs at 300 × *g* for 5 min at 4 °C. Resuspend cells at the concentration of about 5 × 10^5/ml, based on the flow cytometry events count for FAPs. FAPs concentration should be at the range 2.5–5 × 10^5/ml for optimal loading onto a C1 IFC in the next step. Double check the cell concentration by counting isolated FAPs cell using hemocytometer.

3.3 C1-assisted Single Cell Capture, RNA Isolation and cDNA Preparation

For isolation of single cell RNA and its reverse transcription to cDNA, we use C1 Single-Cell Auto Prep System developed by Fluidigm. The C1 system uses microfluidic technology to isolate up to 96 single cells on an Integrated Fluidic Circuit (IFC) solid surface. Cells are then sequentially lysed, RNA is isolated, and gene target-specific pre-amplified cDNA is synthesized. This system is highly efficient and involves minimal manipulation of the isolated single cells. However, the fact that the C1 instrumentation can only run one chip at the time, and that the single cell cDNA preparation process lasts for about 8 h, limits throughput to one C1 run per day. This limitation is particularly important since each C1 IFC chip has to be loaded with live cells and therefore only one biological sample per day can be processed. We balance this drawback by combining samples from different C1 runs on the same 96.96 Dynamic Array when performing RT-PCR on the BioMark system. We consider this step very important to control for biological sample variations due to processing in different days and for technical variations among individual BioMark runs.

In our experimental setting, after isolating FAPs with FACS sorting, we followed the detailed protocol "Using C1 to Capture Cells from Cell Culture and Perform Preamplification Using Delta Gene Assays" for cell isolation and cDNA synthesis (protocol#100–4904 from Fluidigm). FAPs are on average about 12–13 μm in diameter; therefore, we determined that 10–17 μm C1 IFC size is optimal for capturing live FAPs.

Briefly, after priming the C1 IFC chip, cells are loaded and stained for Live/Dead cytotoxicity and viability assay. Fluidigm® guidelines recommend cell concentration to be 2.5×10^5/ml. We have observed high capture efficiencies when starting with an initial suspension of FAPs at concentration 5×10^5/ml in FACS buffer. Live/Dead cytotoxicity and viability kit used with C1 Auto Prep System is based on fluorescent detection of calcein (labeling live cells) and Ethidium homodimer (labeling dead cells). In our experimental setting FAPs are labeled with FITC (anti-Sca1 FITC-conjugated antibody) and Alexa Fluor® 647 (anti-CD34 Alexa Fluor® 647-conjugated antibody). FITC and calcein have similar excitation and emission spectra profile; therefore, it is not possible to distinguish between FITC-labeled and calcein-labeled cells. Ethidium homodimer enters dead cells with damaged membranes and produces a bright red fluorescence upon interaction with DNA with excitation at 517 nm and emission peak at 617 nm. Alexa Fluor 647 is excited at 650 nm and emits at 668 nm. Since the excitation spectra for Alexa Fluor® 647 and Ethidium homodimer are different, it is possible to use these settings to monitor dead Alexa Fluor® 647-labeled FAPs (CD34) by confocal microscopy using Ethidium homodimer-based detection. In conclusion, in our setting, we cannot discriminate for the "live" signal, but we can reliably detect cells marked by Ethidium homodimer as "dead." Nevertheless, in our experience, we do not exclude a priori the cells marked as "dead." These cells usually pass all our quality controls and show a gene expression profile comparable to live cells. This could probably be due to the resilient nature of the FAPs and, even if the membrane is damaged enough to allow access to the dye, it does not interfere with the RNA quality (*see* **Note 4**).

Inspection of captured cells on the C1 IFC under the confocal microscope is a crucial quality control step. We carefully distinguish the situation in which a single cell is captured from the one in which there is no cell or more than one cell captured. These situations will be used to label cells for the purpose of the data analysis (*see* **Note 13**). After this step, Lysis mix, Reverse Transcription mix, and PreAmp mix (containing the DELTAgene Assays—primers) are loaded onto the chip (*see* **Note 14**).

After the C1 run is finished, cDNA has to be harvested in a 96-well plate containing 25 μl/well of C1 DNA dilution reagent (*see* **Note 15**).

At this point we store the cDNA plate at −20 °C.

3.4 BioMark Real-Time PCR on Single Cell C1 Harvested cDNA

For real-time PCR analysis of cDNA prepared from single FAPs cells we have used Fast Gene Expression Analysis Using EvaGreen® on the BioMark HD System analyzing 96 cells with 96 gene-specific assays. The user guide for Fluidigm® Real-Time PCR is detailed and regularly updated (#68000088). Here, we will describe our observations and tips. Briefly, the 96.96 BioMark Dynamic Array exploits a microfluidic technology to perform RT-PCR on cDNA from up to 96 single cells with 96 different gene assays. This technique allows the evaluation of the transcription level of 96 genes in the same cells for 96 cells at the same time.

The Fluidigm protocol suggests an optional exonuclease I treatment of C1 harvested cDNA to eliminate unincorporated primers. We skip this step, since we have not observed an interference of unincorporated primers from the cDNA pre-amplification step with the gene-specific qPCR reactions on BioMark. When setting up the first run we prepared control tubes according to BioMark user guide, from bulk cell population of FAPs as well as from muscle Satellite cells, to control for the amplification quality and melting curves of genes typically expressed in FAPs and Satellite cells or in both. Based on the control tubes we select the correct amplicon melting temperature (Tm) range for each gene analyzed in BioMark, and save these settings for all BioMark runs (*see* also Subheading 3.5.1 Quality control with BioMark RT-PCR analysis software). We combine in one BioMark run cDNA from two or three C1 experiments, to control for the technical variations between BioMark plates (that are usually minimal) and for technical variations among C1 runs. Moreover, we typically run a mixture of cDNA prepared from several single cells, as our standard positive control for each BioMark PCR run.

3.5 Data Analysis

The aim of this section is to focus on describing the single cell gene expression analysis performed on FAPs in our experience, rather than offering an exhaustive overview of all the available methods for data analysis or visualization.

The general workflow describing how to get from the raw qPCR run data to a matrix containing gene expression values for each cell from all the experiments performed is illustrated in Fig. 3. In the first part of the analysis, we perform quality control of the run using the Fluidigm® RT-PCR analysis software and export the file with the Ct (cycle threshold) for each assay (green box). Then, for the primary analysis, we upload our data in R, transform the data into a matrix, filter the failed assays and the outlier cells, calculate expression levels, and pool all data. The ultimate goal of the single cell primary data analysis is to have gene expression values that can be compared across cells and samples. Single-cell data is typically represented as expression level higher than Limit of Detection (LoD) on a log scale (Log2Ex).

```
Log2Ex=LoD Ct- Assay Ct; Log2Ex=0 if Assay
Ct > LoD Ct.
```

Here, Log2Ex represents transcript level above background expressed in log base 2 [16]. The LoD is the highest Ct accepted for the gene to be considered expressed.

3.5.1 Quality Control with BioMark RT-PCR Analysis Software

Once the BioMark qPCR run is completed, the first round of quality control can be performed using the BioMark RT-PCR analysis software (*see* Fluidigm® Real-Time PCR User Guide (#68000088)).

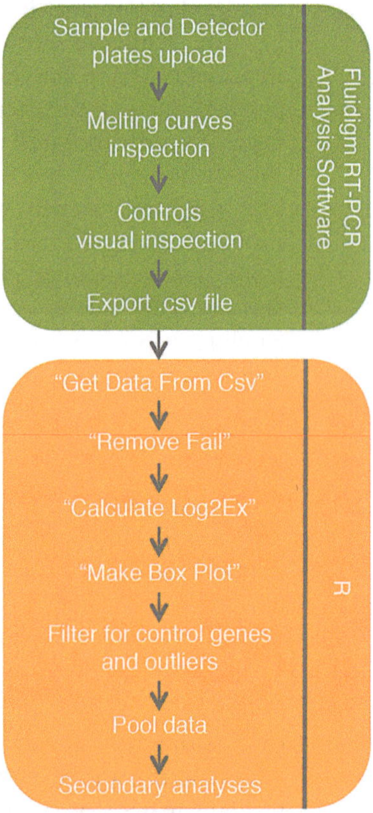

Fig. 3 Analysis workflow of single cell gene expression data from FAPs

The sample plate and detector (primer) plate have to be loaded in the software to label the data. The plates contain the 96 cells identifiers and the 96 genes identifiers in the order in which they have been loaded on the BioMark chip. It is important to name the single cells in a unique way to be able to assign them to the right biological samples and eliminate the control cells, without too much manipulation and with standardization of the procedure (*see* **Note 13**).

A visual inspection of the heatmap in the BioMark RT-PCR analysis software will help to evaluate the overall quality of the run (Fig. 4). Moreover, it is easy to monitor how the positive and negative controls performed by looking at their expression pattern (* and ** in Fig. 4 mark, respectively, negative and positive controls). When running single-cell qPCR in FAPs with DELTAgene assays for the first time, we suggest adding also cDNA prepared from bulk cell populations from both FAPs and satellite cells (or any other cell population, the "tube control"). This will allow all primer sets to be tested, including those amplifying genes known to be expressed or not in FAPs (i.e., positive: Ly6a (Sca1); negative: Pecam1 (CD31) and Itga7). As in a standard qPCR, melting curves have to be visually inspected for each gene and the melting temperature threshold information (Tm range) has to be manually set by the user based on the test run. When saving the detector plate to be used for the following runs the Tm range is saved along with it. This will ensure that the information is saved for future runs, keeping the range standard among several BioMark qPCR experiments. This is particularly important as each set of biological

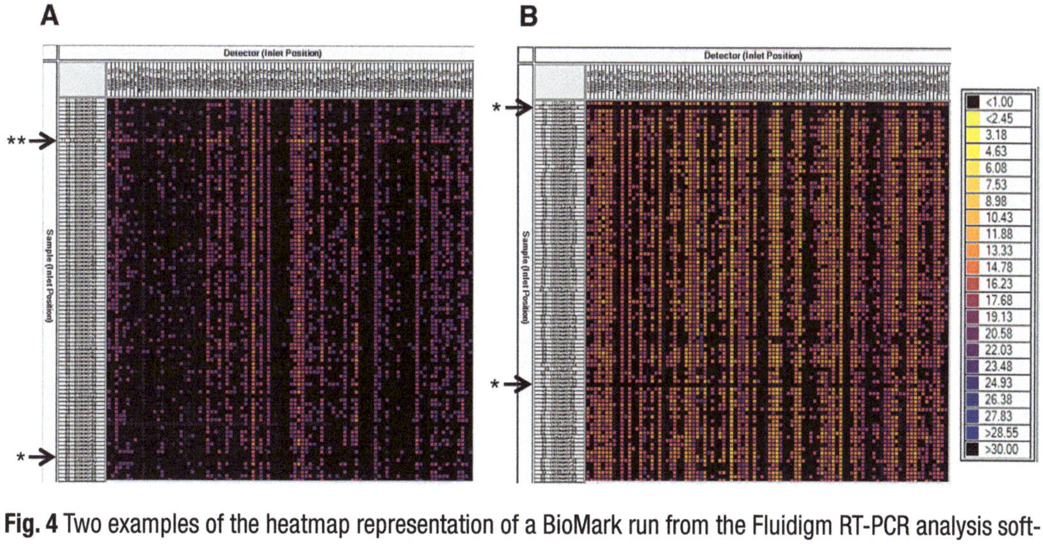

Fig. 4 Two examples of the heatmap representation of a BioMark run from the Fluidigm RT-PCR analysis software. The two plates have been run with the same samples at different times. (**a**) The overall high Ct of all the cells with all primers in this run shows that the quality of this run was very low. (**b**) Example of a good run. * marks negative controls, ** positive controls

samples will be analyzed on more than one BioMark array (to combine several C1 runs on one BioMark run), but the data from all the BioMark runs will have to be ultimately analyzed all together. It is important to note that at this stage many cell-primer couples with valid Ct values (threshold cycle lower than 999, that generally marks failed runs or no expression) could be marked as "Failed" in the results table if their melting curve does not fit the defined Tm range. This is important for the subsequent steps of the analysis, since the values from failed PCRs will be manually set to 999 and the gene will be considered as not expressed in the given cell.

Data can be saved and exported from the Fluidigm® Real-Time PCR Analysis Software as a csv file (comma separated value) containing the Ct for each gene in each cell in a table format (Fig. 5 and sample data—Plate 1.csv, Plate 2.csv, Plate 3.csv).

3.5.2 Primary Data Analysis with R

In order to perform the primary data analysis Fluidigm® offers the SiNGuLAR™ Analysis Toolset (the latest version of the complete manual for the software can be found here: https://www.fluidigm.com/binaries/content/documents/fluidigm/resources/singular-analysis-toolset-v3.0-100-5066/singular-analysis-toolset-v3.0-100-5066/fluidigm:file). This software is built on R [17] and enables simple primary data analyses. However, data analysis can be done in R in a customized way, with the advantage of designing the analysis to be more comprehensive and tailored to specific experimental needs. In the following example data analysis workflow, we will describe the steps to run in R to obtain a comprehensive data matrix of Log2Ex values starting from the csv tables obtained from the Real-Time PCR Analysis Software (in this example, "Plate1.csv"). The example data is from FAPs single cell gene expression data in three different mice (Samples A, B, C) run on three different BioMark Arrays (Plates 1–3). For convenience, all

Chip Run Info	\\VBOXSVR\filename.bml	1361967569	96.96 (136x)	GE 96x96 Fast PCR+Melt v2	ROX	EvaGreen	12/2/14 17:07	00:32:33	BIOMARKHD019
Application Version	4.1.2								
Application Build	20140110.17								
Export Type	Table Results								
Quality Threshold	0.65								
Baseline Correction Method	Linear (Derivative)								
Ct Threshold Method	Auto (Global)								

Experiment Information Chamber ID	Experiment Information Sample Name	Experiment Information Sample Type	Experiment Information Sample rConc	Experiment Information EvaGreen Name	Experiment Information EvaGreen Type	EvaGreen Ct Value	EvaGreen Ct Quality	EvaGreen Ct Call	EvaGreen Ct Threshold	EvaGreen Tm In Range	EvaGreen Tm Out Range	EvaGreen Tm Peak Ratio	User Defined Comments
S96-A01	1-SampleA-94	Unknown	1	Gene1	Test	999	0	Fail	0.021223128	999	999	0	
S96-A02	1-SampleA-94	Unknown	1	Gene2	Test	19.29886456	1	Pass	0.021223128	85.61136593	999	1	
S96-A03	1-SampleA-94	Unknown	1	Gene3	Test	19.08848912	1	Pass	0.021223128	87.78600062	999	1	
S96-A04	1-SampleA-94	Unknown	1	Gene4	Test	15.32110287	1	Pass	0.021223128	82.65244626	999	1	
S96-A05	1-SampleA-94	Unknown	1	Gene5	Test	25.27131744	0	Fail	0.021223128	999	81.90686012	0	
S96-A06	1-SampleA-94	Unknown	1	Gene6	Test	15.70925041	1	Pass	0.021223128	87.10718034	999	1	
S96-A07	1-SampleA-94	Unknown	1	Gene7	Test	15.45652804	0.959303646	Pass	0.021223128	85.29189759	999	1	
S96-A08	1-SampleA-94	Unknown	1	Gene8	Test	20.57301675	0	Fail	0.021223128	999	87.97056939	0	
S96-A09	1-SampleA-94	Unknown	1	Gene9	Test	999	0	Fail	0.021223128	999	999	0	
S96-A10	1-SampleA-94	Unknown	1	Gene10	Test	17.28415337	1	Pass	0.021223128	86.08959674	999	1	
S96-A11	1-SampleA-94	Unknown	1	Gene11	Test	999	0	Fail	0.021223128	999	999	0	
S96-A12	1-SampleA-94	Unknown	1	Gene12	Test	16.47496898	0.971891319	Pass	0.021223128	84.12639923	999	1	
S96-A13	1-SampleA-94	Unknown	1	Gene13	Test	999	0	Fail	0.021223128	999	999	0	
S96-A14	1-SampleA-94	Unknown	1	Gene14	Test	999	0	Fail	0.021223128	999	999	0	
S96-A15	1-SampleA-94	Unknown	1	Gene15	Test	17.38020717	1	Pass	0.021223128	84.6149384	999	1	
S96-A16	1-SampleA-94	Unknown	1	Gene16	Test	18.25501215	1	Pass	0.021223128	78.8536953	999	1	
S96-A17	1-SampleA-94	Unknown	1	Gene17	Test	999	0	Fail	0.021223128	999	999	0	
S96-A18	1-SampleA-94	Unknown	1	Gene18	Test	13.62746968	1	Pass	0.021223128	88.53102652	999	1	
S96-A19	1-SampleA-94	Unknown	1	Gene19	Test	999	0	Fail	0.021223128	999	999	0	
S96-A20	1-SampleA-94	Unknown	1	Gene20	Test	16.14158067	1	Pass	0.021223128	87.01334339	999	1	
S96-A21	1-SampleA-94	Unknown	1	Gene21	Test	21.20130813	0	Fail	0.021223128	999	86.13928683	0	
S96-A22	1-SampleA-94	Unknown	1	Gene22	Test	999	0	Fail	0.021223128	999	999	0	
S96-A23	1-SampleA-94	Unknown	1	Gene23	Test	27.06320764	0	Fail	0.021223128	999	999	0	
S96-A24	1-SampleA-94	Unknown	1	Gene24	Test	14.28614769	1	Pass	0.021223128	79.11674982	999	1	
S96-A25	1-SampleA-94	Unknown	1	Gene25	Test	19.00273924	1	Pass	0.021223128	83.333339	999	1	
S96-A26	1-SampleA-94	Unknown	1	Gene26	Test	18.66071207	1	Pass	0.021223128	83.21257955	999	1	
S96-A27	1-SampleA-94	Unknown	1	Gene27	Test	14.13518271	1	Pass	0.021223128	83.22358633	999	1	
S96-A28	1-SampleA-94	Unknown	1	Gene28	Test	15.34604366	1	Pass	0.021223128	85.4455134	999	1	
S96-A29	1-SampleA-94	Unknown	1	Gene29	Test	19.24856916	0.980681153	Pass	0.021223128	84.91191072	999	1	
S96-A30	1-SampleA-94	Unknown	1	Gene30	Test	12.92873957	1	Pass	0.021223128	83.22398175	999	1	
S96-A31	1-SampleA-94	Unknown	1	Gene31	Test	999	0	Fail	0.021223128	999	999	0	
S96-A32	1-SampleA-94	Unknown	1	Gene32	Test	14.50925028	1	Pass	0.021223128	79.69022732	999	1	

Fig. 5 Header of the result table from the BioMark Real-Time PCR Analysis Software (csv file)

the functions listed below can be collected in a single file (i.e., “SingleCellAnalysis. R”) and loaded in R as a source file:

```
> source("/Users/SCGE/SingleCellAnalysis.R")
```

Doing so will allow the user to directly use the functions without having to explicitly run them every time. (Refer to “source” function in R documentation).

First, set the working directory in R workspace:

```
> setwd("/Users/SCGE")
```

Now the csv file can be uploaded in R. First, load the function GetDataFromCsv, which allows uploading x as a data.frame in R workspace:

```
> GetDataFromCsv=function(x)
{
col=c("WellID","SampleName","SampleType","Samp
leConc","GeneName","GeneType","CtValue","CtQua
l","CtCall","CtTreshold","TmInRange","TmOutRan
ge","TmPeakRatio","Comments")
return(read.csv(x, skip=12, header=FALSE, col.
names=col))
}
```

In “GetDataFromCsv” function “x” is the csv table from the Fluidigm® RT PCR Analysis Software (See supplementary Table “Plate1.csv”). To upload the file in the R workspace as an object with a data.frame structure, with the file in the working directory, run the function:

```
> MyPlate1=GetDataFromCsv("Plate1.csv")
```

Once the dataset is loaded, the data marked as failed by Fluidigm® RT PCR Analysis Software (column “CtCall”) having numeric Ct values different from 999 (column “CtValue”) needs to be changed to 999. This process will generate a “genes by cells” matrix containing all the Ct values, including those from reactions that failed the quality control of the PCR product and are set to 999 (*see* Tm range above). We believe that in these cases where the quality control failed, there is a high probability that the given gene is not expressed in that cell and the produced melting curve is an artifact.

```
> RemoveFail=function(x,y,z)
{
l = nrow(x)
tableFail=x
for(i in 1:l)
      {
            if(x[i,9]=="Fail")
            {
            tableFail[i,7]=999
}
```

```
}
if (z=="ON")
      {
write.table(tableFail,file=paste(y,"_fail-
Removed.csv",sep=""), sep=",", dec=".",
quote=FALSE, row.names=F)
}
return(tableFail)
}
```

In "RemoveFail" function "x" is the data frame object uploaded in R with GetDataFromCsv function, "y" is the name of the file that will be saved with the suffix: "_failRemoved.csv" and "z" can be "ON" or "OFF," depending on if the user wants to save (ON) or not (OFF) the final csv file. Running this function returns an object with data.frame structure:

```
> MyPlate1NoFail=RemoveFail(MyPlate1,"MyPla
te1_NoFail","ON")
```

The next step includes the transformation of the data frame into a matrix and the calculation of Log2Ex values. LoD (Limit of Detection) is used to convert Ct values to Log_2 values (Log_2Ex), which represent expression levels above background. For BioMark System the recommended LoD value is 24, as suggested by Livak and colleagues [16]. To run this function it is required to upload the reshape package (http://had.co.nz/reshape/). Reshape is an R package for flexibly restructuring and aggregating data. It can be installed following this command:

```
> install.packages("reshape")
```

Then load the CalcLog2Ex function:

```
   > CalcLog2Ex=function(x,y,z,w)
{
require(reshape) #loads the library when the
function is executed
dt=x[,c("SampleName","GeneName","CtValue")]
mtable=cast(dt, SampleName ~ GeneName)
if (z=="ON")
{
write.table(mtable,file=paste(y,"_fail-
Removed_matrix.csv",sep=""), sep=",", dec=".",
quote=FALSE, row.names=F)
}
log2Ex=data.frame(mtable[,-1], row.
names=mtable[,1])
LOD=w
r = nrow(log2Ex)
c = ncol(log2Ex)
for(i in 1:r)
```

```
{
      for(j in 1:c)
      {
            if(LOD-log2Ex[i,j]>0)
            {
                  log2Ex[i,j]=LOD-log2Ex[i,j]
#log2Ex=LoD Ct-Ct
}
            else
            {
                  log2Ex[i,j]=0 #log2Ex=0 if Ct
> LoD
}
}
}
if (z=="ON")
{
write.table(log2Ex,file=paste(y,"_log2Ex_
matrix.csv",sep=""), sep=",", dec=".",
quote=FALSE, row.names=T)
}
return(log2Ex)
}
```

In "CalcLog2Ex" function "x" is the data frame object from RemoveFail, "y" is the name of the two files that will be saved with the suffixes: "_failRemoved_matrix.csv" and "_log2Ex_matrix.csv," "z" is ON or OFF, depending on if the user wants to save (ON) or not (OFF) the final files, and "w" is the number indicating the LOD (Limit of Detection). Running this function will return an object with data.frame structure:

```
> MyMatrix1=CalcLog2Ex(MyPlate1NoFail,"MyMatri
x1","ON",24)
```

The "CalcLog2Ex" function creates a data.frame shaped as a matrix with genes as columns, cells as rows, and Log2Ex values at the intersections. For each plate we need to be able to eliminate outlier cells that might show a drastically different gene expression average profile compared with the other cells. The function "makeBoxplot" uses the Log2Ex matrix to make boxplots.

```
> makeBoxplot=function(x,y)
{
require(reshape)
mlog2Ex=melt(as.matrix(x))
colnames(mlog2Ex)=c("SampleName","GeneName","L
og2Ex")
mlog2Exss=subset(mlog2Ex, mlog2Ex[,3] >0)
mlog2Exss$TYPE < with(mlog2Exss,
reorder(SampleName,-Log2Ex, median))
```

```
jpeg(file=paste(y,"_log2Ex_boxplots.
png",sep=""), quality=100, width=2000,
height=700,pointsize = 18)
mar.orig <- par()$mar # save the original val-
ues
par(mar = c(10,4,4,4)) # set new values
boxplot(Log2Ex ~ TYPE, data=mlog2Exss, las
= 2, col="blue",ylab="Expression(Log2Ex)",
varwidth=T, main="Outlier identification")
par(mar = mar.orig) # put the original values
back
dev.off()
}
```

In this case "x" is the matrix from CalcLog2Ex, "y" is the name of the file that will be saved with the suffix: "_log2Ex_boxplots.png." This function does not return an object but it saves the boxplot image in the working directory (Fig. W).

```
> makeBoxplot(MyMatrix1,"MyMatrix1")
```

This function will create the boxplot in Fig.6 using the data from Plate1. Representing the data as boxplots is useful since it helps to identify cells behaving as outliers. In the example in Fig. 6 we can immediately spot the cells that will have to be excluded by the analysis, being the ones expressing few or no genes (All "nocell" samples and 1-SampleA-89) and having the median lower than average (i.e., 1-positive-14).

Once the matrix is created, before proceeding to further analysis we need to eliminate the outlier cells. For the analysis of FAPs we eliminate cells that do not express Ly6a (Sca1, that is the FAPs marker) or that express Pecam1 (CD31, endothelial marker) or Itga7 (satellite cells marker) during this step. At this point having "no cell" controls, i.e., cDNA synthetized from a spot in the C1 IFC chip where there was no cell captured, is important to be able to detect "faulty" gene assays that may produce false-positive signals.

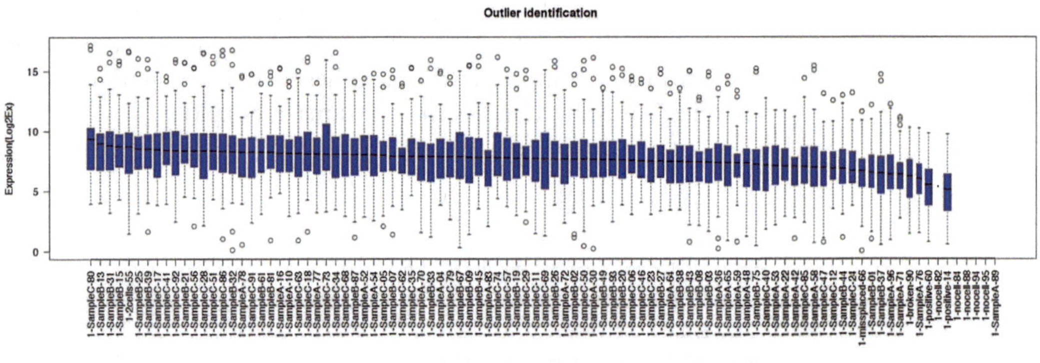

Fig. 6 Example of gene expression boxplots from the example Plate1, representing the gene expression distribution in each cell

We disregard "no cell" controls and other "faulty" cDNAs, i.e., the ones coming from C1 capture sites having two cells captured, identified during the visual inspection of C1 IFC after cell capture. To do this, we extract from the matrix only cells we previously labeled as "Sample" (that are the spots on the IFC C1 where only one cell was captured during the visual inspection, *see* **Note 13**):

```
> MyMatrixSamples1=MyMatrix1
[c(grep("SampleA", rownames(MyMatrix1)),grep("
SampleB", rownames(MyMatrix1)),grep("SampleC",
rownames(MyMatrix1))),]
```

We also eliminate outlier samples identified from the boxplot inspection:

```
> MyMatrixSamples1= MyMatrixSamples1[grep("1-
SampleA-89", rownames(MyMatrixSamples1)),]
```

Then filter matrix for control genes expression: Ly6a=Gene32 (FAPs marker); Pecam1=Gene11 (endothelial marker); Itga7=Gene19 (satellite cells marker):

```
> MyMatrixClean1=MyMatrixSamples1
[MyMatrixSamples1[,"Gene32"]!=0&
MyMatrixSamples1[,"Gene11"]<=0& MyMatrixSample
s1[,"Gene19"]<=0,-97]
```

At this point, we will have a matrix for each BioMark run plate looking as follows:

```
> head(MyMatrixClean1)
                Gene1Gene10Gene11Gene12Gene13Gene14
1-SampleA-04    7.551524  8.651576  08.990779
7.650140  0.000000
1-SampleA-22    4.490049  7.492508  08.320329
0.000000  0.000000
1-SampleA-36    8.811958  6.963579  09.085393
0.000000  0.000000
1-SampleA-42    0.000000  6.768611  07.267674
0.000000  0.000000
1-SampleA-48    0.000000  7.279961  07.651894
0.000000  7.625663
1-SampleA-52    8.871920  7.298877  08.186965
6.244493  0.000000
```

To proceed to secondary analysis, process each of three files (Supplementary Tables Plate1.csv, Plate2.csv, Plate3.csv) the same way (and eliminate cell "3-SampleA-86" from the Plate3 to have the same results shown in the next steps of the analysis). Join the matrices from each plate using rbind (making sure that the column names are the same and in the same order):

```
AllData=rbind(MyMatrixClean1, MyMatrixClean2,
MyMatrixClean3)
```

3.5.3 Introduction to Data Visualization with R

The secondary data analysis for single cell gene expression is very much data-driven since it is contingent to the particular experimental setting and tailored to answer to specific biological questions. Here, we will describe four basic visualization tools that can provide users with useful information about data quality, sample homogeneity, and gene expression profiles: PCA, heatmap, bar plots, and violin plots (*see* **Note 16**).

First, labeling of samples and plates has to be specified in R. Two vectors of the size of the data frame row count have to be created (basically one label each cell), one with the sample names (Sample A, Sample B, etc.) and the other with the plate identifiers (Plate 1, Plate 2, etc.). These vectors will be used as labels for the single cells in the data representation and the order has to respect the same order as in the matrix:

```
> Plate=c(rep("Plate1",65), rep("Plate2",67),
rep( "Plate3",71))
> Sample=vector(length=nrow(AllData))
```

With the "rep" function we repeat the label as many times as there are cells in that plate (Refer to "rep" function in R documentation). We can use this command to count how many cells are, for example, in plate 1 (Refer to "grep" function in R documentation):

```
> length(grep("1-",rownames(AllData)))
```

At this point, we can attach the label to the matrix (*see* **Note 17**).

```
> AllDataLabeled=cbind(AllData,Plate,Sample)
```

"grep" function returns a vector of the indices of the elements of the object AllData that yielded a match:

```
> SampleA=grep("SampleA",rownames(AllData))
> SampleB=grep("SampleB",rownames(AllData))
> SampleC=grep("SampleC",rownames(AllData))
```

In the Sample column we write the sample identity according to the row name (cell name):

```
> for(i in 1:nrow(AllDataLabeled)){
if(i %in% SampleA){
      AllDataLabeled[i,"Sample"]="SampleA"
}
if(i %in% SampleB){
      AllDataLabeled[i,"Sample"]="SampleB"
}
if(i %in% SampleC){
      AllDataLabeled[i,"Sample"]="SampleC"
}
}
```

This way the last two columns of the file will contain the specific label for each cell.

Now we can save the plate in our working folder:

```
write.table(AllDataLabeled,file="AllDataLabel
ed.csv", sep=",", dec=".", quote=FALSE, row.
names=T)
```

Principal Component Analysis (PCA)

The PCA is a statistical procedure that transforms the set of values from related variables (i.e., individual expression of 96 genes in single cells) into a set of unrelated variables. The highest variance in the data is represented in the first components of the data. Plotting the first two (or three) components will position the individual cells in the space in relation to each other on the basis of the similarity between their gene expression profiles. In order to plot a PCA we will need to download the "ggbiplot" package:

```
> install.packages("devtools")
> library(devtools)
> install_github("vqv/ggbiplot")
and then load it in the workspace:
> library(ggbiplot)
```

(*see* **Note 18**).

Now we can run the PCA analysis eliminating all columns that sum up to 0 (we use the file with no labels for that):

```
> pcaAll=prcomp(AllData[,colSums(AllData)!=0],
scale. = TRUE)
```

(*see* **Note 19**).

Using the analysis just performed we can now plot our bidimensional PCA plot with ggbiplot, where the cells are represented as dots. Each dot will be colored according to the Plate column (i.e., red for Plate 1, blue for Plate 2, green for Plate 3).

```
> pdf("Results_PCA_loadings_byPlate.pdf",
width = 8, height=6)
> g <- ggbiplot(pcaAll, obs.scale = 1, var.
scale = 1, groups=as.character(AllDataLabeled$
Plate), var.axes=F)
> g <- g + scale_colour_manual(values
= c("red","blue","green")) +scale_fill_
manual(values = c("red","blue","green"))
> g <- g + theme(legend.direction = 'horizon-
tal', legend.position = 'top')
> print(g)
> dev.off()
```

The same plot can be done coloring the cells (dots) according to the Sample column (i.e., white for Sample A, gray for Sample B, black for Sample C).

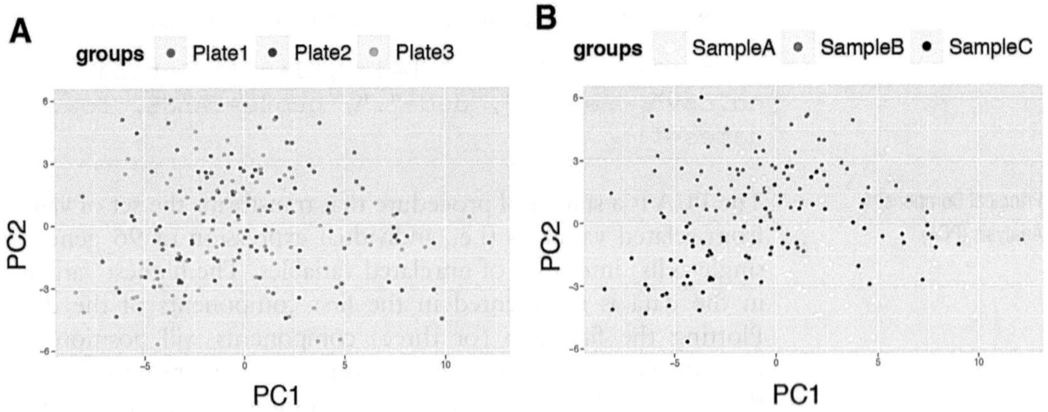

Fig. 7 Example of Principal Component Analysis (PCA) visualization of single cell gene expression. (**a**) Cells are colored according to the BioMark Array (plate) their gene expression was tested on. (**b**) Cells are colored according to the biological sample they belong to

```
> pdf("Results_PCA_loadings_bySample.pdf",
width = 8, height=6)
> g <- ggbiplot(pcaAll, obs.scale = 1, var.
scale = 1, groups=as.character(AllDataLabeled$
Sample), var.axes=F)
> g <- g + scale_colour_manual(values =
c("white","grey50","black")) +scale_fill_
manual(values = c("white","grey50","black"))
> g <- g + theme(legend.direction = 'horizon-
tal', legend.position = 'top')
> print(g)
> dev.off()
```

The resulting pdf files visualizing PCA on our data set will be saved in the working directory (Fig. 7a, b). The most interesting information we can extrapolate by looking at these graphs is that the cells will not cluster according to the BioMark run plates. Instead, the cells show a tendency to aggregate according to their sample nature. This is the expected result of such analysis, showing that there is no plate-related bias influencing the biological information and showing that there is no need for data normalization.

Heatmap with Hierarchical Clustering

Heatmaps are a popular way of large datasets visualization. The hierarchical clustering shows how cells cluster together according to their gene expression profile, and the heatmap visualization allows the association of gene expression profiling to cell clusters (Fig. 8). We start creating a vector with the colors we want to use for the plates and the samples:

```
> PlateCol=c(rep("Blue",65), rep("Red",67),
rep("Green",71))
```

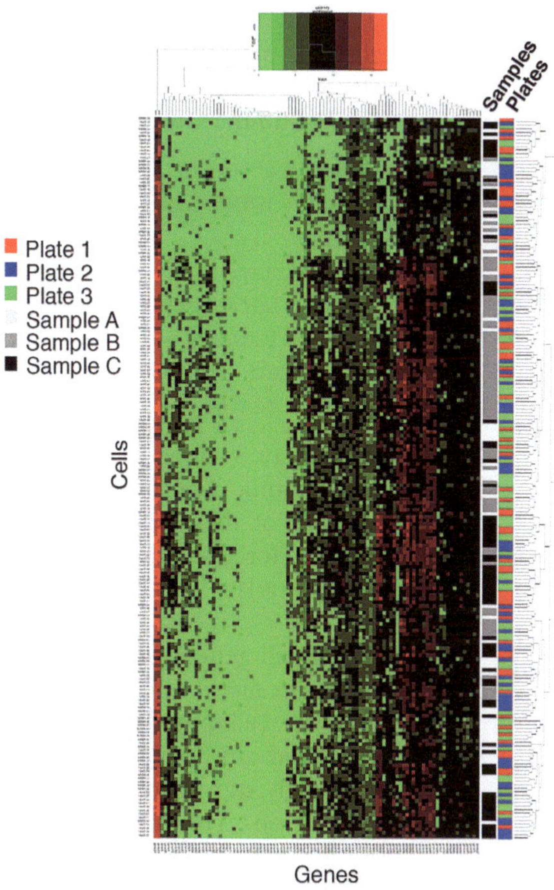

Fig. 8 Example of evaluation of single cell gene expression using heatmap with hierarchical clustering. *Colored bars* on the right side label the identity of each cell in terms of biological sample or BioMark Array (plate) run

```
> SampleCol=c(rep("Grey95",55),
rep("Grey50",75), rep("Black",73))
```

We install the gplots package and load it in the workspace:

```
> install.packages("gplots")
> library("gplots")
```

We create the heatmap with a colored column on top labeling with the proper color each cell and save it as a jpg file in the working directory:

```
> jpeg("Heatmap_ColorByPlate.jpg", width=3000,
height=1500, quality=100, pointsize = 18)
> heatmap.2(t(as.matrix(AllDataLabeled[order(A
llDataLabeled$Plate),-c(ncol(AllDataLabeled),n
col(AllDataLabeled)-1)])), trace="none", col=g
reenred(10),ColSideColors=PlateCol).
> dev.off()
```

Note that we sort the data according to the Plate column to allow a perfect match with the color vector uploaded before (PlateCol).

We do the same to add a colored column for the samples:

```
> jpeg("Heatmap_ColorBySample.jpg",
width=3000, height=1500, quality=100, point-
size = 18)
> heatmap.2(t(as.matrix(AllDataLabeled[order(A
llDataLabeled$Sample),-c(ncol(AllDataLabeled),
ncol(AllDataLabeled)-1)])), trace="none", col=
greenred(10),ColSideColors=SampleCol)
> dev.off()
```

Figure 8 shows the heatmap resulting from these commands, highlighting again that the cells cluster according to biological samples of their origin, rather than according to BioMark run plates. Moreover, this graph shows how the gene expression patterns correlate with clusters of cells.

Bar Plots

Single cell gene expression analysis is an extremely valuable technique because it can define subpopulations of cells within a heterogeneous cell population. Genes that we normally consider as expressed can be highly expressed in some cells in the population and lowly expressed or absent in other cells within the same cell population. This is why it can be interesting to visualize the percentage of cells that expresses each given gene and how it varies among samples. Here, we will show how to plot bar plots for each gene, with the three samples in different colors (Fig. 9).

First, we prepare a matrix 96 × 3 with genes as rows and samples as columns:

```
M=matrix(data=0, nrow=length(unique(AllDataLab
eled[,98])),ncol=ncol(AllData))
rownames(M)=unique(AllDataLabeled[,98])
colnames(M)=colnames(AllData)
```

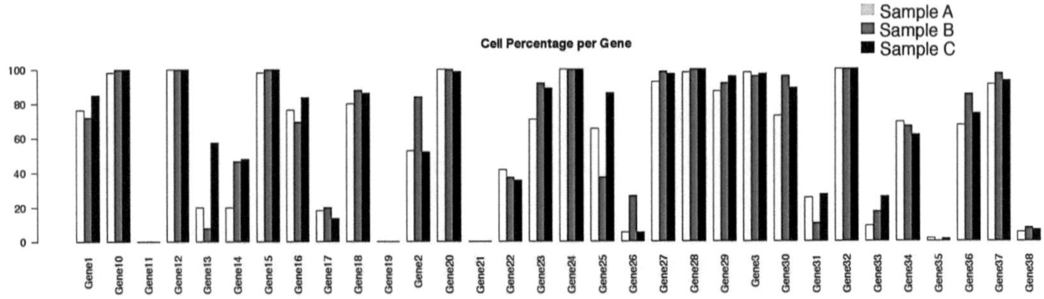

Fig. 9 Example of bar plot showing the percentage of cells expressing the genes in the different samples (expression higher than 0)

Then we count how many cells in each sample express each gene and write it in the matrix M:

```
for (j in 1:ncol(AllData)){
for(i in SampleA){
      if(AllData[i,j]!=0){
            M[1,j]=M[1,j]+1
}
}
}
for (j in 1:ncol(AllData)){
for(i in SampleB){
      if(AllData[i,j]>0){
            M[2,j]=M[2,j]+1
}
}
}
for (j in 1:ncol(AllData)){
for(i in SampleC){
      if(AllData[i,j]>0){
            M[3,j]=M[3,j]+1
}
}
}
```

Then we calculate the percentage of each sample expressing the gene:

```
Mnorm=M
Mnorm[1,]=M[1,]/length(SampleA)*100
Mnorm[2,]=M[2,]/length(SampleB)*100
Mnorm[3,]=M[3,]/length(SampleC)*100
```

Finally, we plot the percentage of cells expressing each gene for the first 32 genes:

```
jpeg("BarPlot_CellCountPerGene_all.jpg", width
= 2000, height=500, quality=100, pointsize =
18)
par(las=2)
barplot(Mnorm[,1:32], main="Cell Percentage
per Gene",col=c("white","grey50","black"),besi
de=TRUE)
dev.off()
```

Figure 9 results from this analysis and shows, for example, that gene 2 is expressed in a higher percentage of cells in sample B than in the others, while the other genes do not show differences in terms of percentage expression in the samples.

Violin Plots

While bar plots only discriminate between genes expressed and not expressed, violin plots finely visualize the expression levels within each sample. Violin plots are smoothed versions of histograms represented sideways and are used to show the distribution of single cell gene expression within a certain population (gene expression is on *y* axis and population density is on *x* axis). Figure 10 demonstrates the visual outcome of such representation of gene-by-gene expression of the first 43 genes in each sample population (*see* **Note 20**).

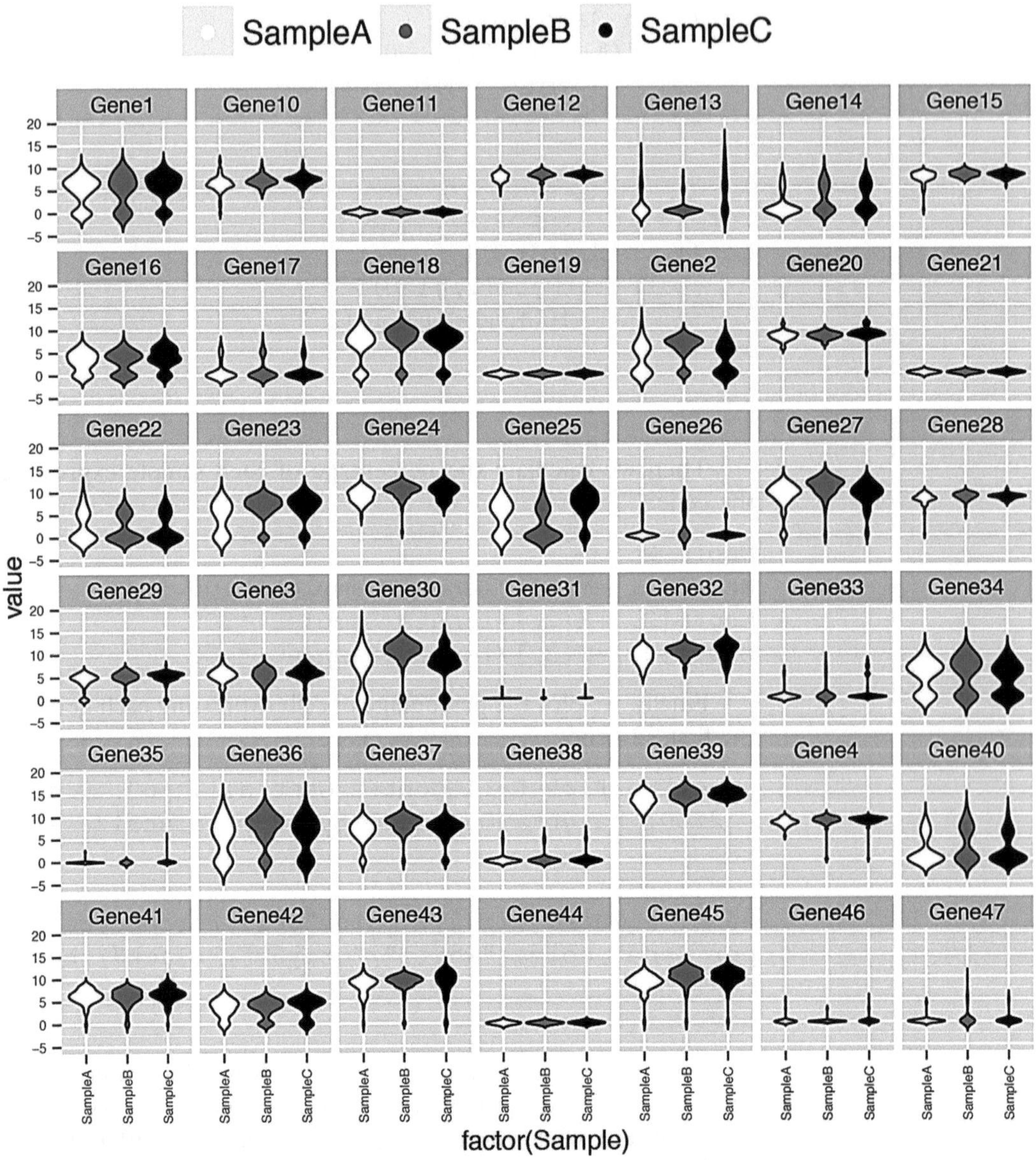

Fig. 10 Example of evaluation of single cell gene expression with violin plots. See text for detailed explanation

This plot shows how genes can have monophasic and biphasic distribution of expression and that the distribution can be compared across populations to identify signature genes.

First, install and load the necessary packages, as follows:

```
> install.packages("ggplot2")
> install.packages("scales")
> library(ggplot2)
> library(scales)
```

Then, format the matrix and plot the violin plots with ggplot:

```
> mdat=melt(AllDataLabeled
[1:nrow(AllDataLabeled),-c(43:97)],
id.vars="Sample")
> pdf("ViolinPlots_Samples.pdf")
> ggplot(mdat, aes(factor(Sample), value))
+ geom_violin(aes(fill = factor(Sample)),
trim = FALSE) + scale_fill_manual(values =
alpha(c("white","grey50", "black"), 1)) +
theme(legend.position="none", axis.text.x
= element_text(vjust=0.5, size=6, an-
gle = 90, hjust = 1), axis.text.y = el-
ement_text(vjust=0.5, size=6))+ facet_
wrap(~variable)
> dev.off()
```

From Fig. 10 we can clearly see how some genes show a sample-specific expression profile, which would probably be missed on a bulk cell population gene expression analysis. For example, bimodal expression of genes 23 and 30 in sample A becomes more uniform in samples B and C and genes 2 and 25 fluctuate a lot between high and low expression within the same sample.

4 Notes

1. Keep the buffer on ice all the time to keep it cold. FACS Buffer can be stored at 4 °C for couple of weeks. 200 ml FACS Buffer is sufficient for one cell sorting procedure (one mouse).
2. Keep the Digestion Solution on ice. Always prepare fresh before use.
3. Using DAPI instead of Propidium Iodide to monitor dead cells may improve the background in flow cytometry analysis.
4. Inverted microscope may be sufficient to observe cells captured on C1 chip without Live/Dead staining. FAPs are robust and viable cells, it is sufficient to monitor their capture on C1 array using an inverted microscope without using live/dead staining.

5. An alternative source of primers can be used. Forward and reverse primers need to be mixed together at the concentration 100 μM each. TaqMan® gene expression assays with TaqMan PCR system are an alternative to DELTAgene assays used with EvaGreen.
6. Different sizes of Dynamic Arrays are available according to the experimental need (https://www.fluidigm.com/faq/ge-20).
7. For detailed information see https://www.fluidigm.com/faq/ge-20.
8. Semisterile work may be sufficient. If isolated cells will not be cultured, single cell gene expression profiling does not require cells to be sterile.
9. Multiple 15 ml tubes can be taped together. Tubes in the water bath can be kept stable by sandwiching them between two heavy weights.
10. If smaller amount of muscle is processed resulting in smaller cell yield, scale down accordingly, avoid 100 μm cell trainer, and use directly 40 μm cell strainer to avoid cell loss.
11. If the cell suspension is too concentrated, dilute cells with cold FACS buffer.
12. If cells are too concentrated and need to be diluted to a higher volume, adjust the amount of antibodies accordingly to keep the antibodies concentration constant.
13. We generally name the cells like this: plateNumber_SampleName_CellNumber (1_SampleA_1). Spots where no cells or more than one cell was detected do not have the "Sample" nomenclature, instead they have either "NoCell" or "X cells": plateNumber_NoCell_CellNumber (1_NoCell_1).
14. During this step it is possible to pre-amplify the single cell cDNA with more than 96 DELTAgene Assays (in our experience we have had good results with up to 109 assays).
15. We find it useful to collect the 3 μl of C1 cDNA released by the C1 IFC chip by resuspending it with 10 μl of the C1 DNA dilution reagent from the plate using a multichannel pipette to assist with collection of such small volumes.
16. For data visualization in R a simple guide can be found here: http://www.cookbook-r.com/Graphs/.
17. Be careful! At this stage we are simply "attaching" the labels, so it is of extreme importance that the labels and the cells have the same order.
18. For more information about the ggbiplot, refer to its source https://github.com/vqv/ggbiplot and to http://www.r-bloggers.com/computing-and-visualizing-pca-in-r/.
19. For more information about the prcomp analysis refer to the R manual https://stat.ethz.ch/R-manual/R-devel/library/stats/html/prcomp.html.

20. Here you can find a reliable reference where to find more information about violin plots: http://docs.ggplot2.org/current/geom_violin.html.

Acknowledgment

This work was supported by NIH P30 pilot grant P30AR061303 and CIRM training grant TG2-01162 to B.M., and NIH grants R01AR056712, R01AR052779, and P30 AR061303 to P.L.P. We thank Thomas C. Roberts and Alessandra Dall'Agnese for critical reading of the manuscript.

Electronic Supplementary Materials

References

1. Joe AWB, Yi L, Natarajan A et al (2010) Muscle injury activates resident fibro/adipogenic progenitors that facilitate myogenesis. Nat Cell Biol 12:153–163
2. Heredia JE, Mukundan L, Chen FM et al (2013) Type 2 innate signals stimulate fibro/adipogenic progenitors to facilitate muscle regeneration. Cell 153:376–388
3. Mozzetta C, Consalvi S, Saccone V et al (2013) Fibroadipogenic progenitors mediate the ability of HDAC inhibitors to promote regeneration in dystrophic muscles of young, but not old Mdx mice. EMBO Mol Med 5:626–639
4. Saccone V, Consalvi S, Giordani L et al (2014) HDAC-regulated myomiRs control BAF60 variant exchange and direct the functional phenotype of fibro-adipogenic progenitors in dystrophic muscles. Genes Dev 28:841–857
5. Uezumi A, Fukada S-I, Yamamoto N et al (2010) Mesenchymal progenitors distinct from satellite cells contribute to ectopic fat cell formation in skeletal muscle. Nat Cell Biol 12:143–152
6. Uezumi A, Ito T, Morikawa D et al (2011) Fibrosis and adipogenesis originate from a common mesenchymal progenitor in skeletal muscle. J Cell Sci 124:3654–3664
7. Dulauroy S, Di Carlo SE, Langa F et al (2012) Lineage tracing and genetic ablation of ADAM12(+) perivascular cells identify a major source of profibrotic cells during acute tissue injury. Nat Med 18:1262–1270
8. Lemos DR, Babaeijandaghi F, Low M, et al (2015) Nilotinib reduces muscle fibrosis in chronic muscle injury by promoting TNF-mediated apoptosis of fibro/adipogenic progenitors, Nature medicine.
9. Pannerec A, Formicola L, Besson V et al (2013) Defining skeletal muscle resident progenitors and their cell fate potentials. Development 140:2879–2891
10. B.Malecova and P.L.Puri (2012) "Mix of Mics"- Phenotypic and Biological Heterogeneity of "Multipotent" Muscle Interstitial Cells (MICs), Journal of Stem Cell Research & Therapy.
11. Judson RN, Zhang R-H, Rossi FMA (2013) Tissue-resident mesenchymal stem/progenitor cells in skeletal muscle: collaborators or saboteurs? FEBS J 280:4100–4108
12. Natarajan A, Lemos DR, Rossi FMV (2010) Fibro/adipogenic progenitors: a double-edged sword in skeletal muscle regeneration. Cell Cycle 9:2045–2046
13. Giordani L, Puri PL (2013) Epigenetic control of skeletal muscle regeneration: Integrating genetic determinants and environmental changes. FEBS J 280:4014–4025
14. Yi L, Rossi F (2011) Purification of progenitors from skeletal muscle. Journal of visualized experiments, JoVE
15. Sacco A, Doyonnas R, Kraft P et al (2008) Self-renewal and expansion of single transplanted muscle stem cells. Nature 456:502–506
16. Livak KJ, Wills QF, Tipping AJ et al (2013) Methods for qPCR gene expression profiling applied to 1440 lymphoblastoid single cells. Methods 59:71–79
17. Core Team R (2013) R: a language and environment for statistical computing. R Foundation for Statistical Computing, Vienna, Austria http://www.R-project.org

Part IV

Assays for Stem Cell Functionality

Chapter 11

Engraftment of FACS Isolated Muscle Stem Cells into Injured Skeletal Muscle

Matthew Tierney and Alessandra Sacco

Abstract

Skeletal muscle stem cell (MuSC) isolation and transplantation are invaluable tools to assess their capacity for self-renewal and tissue repair. Significant technical advances in recent years have led to the optimization of these approaches, improving our ability to assess MuSC regenerative potential. Here, we describe the procedures for Fluorescent Activated Cell Sorting (FACS)-based isolation of MuSC, their intramuscular transplantation, and analysis of their engraftment into host tissues.

Key words FACS-based cell isolation, Skeletal muscle stem cells, Transplantation, Engraftment, Muscle regeneration

1 Introduction

The development of fluorescence-activated cell sorting (FACS)-based strategies to isolate muscle stem cells (MuSC) based on cell surface markers has made a significant impact in the skeletal muscle field. Previous approaches included the isolation of myogenic progenitors through tissue enzymatic digestion and cell culture based on adhesion properties, which presented inherent limitations such as the selection of cells in culture, extensive passaging in vitro, and subsequent changes in composition of the cell population [1]. In subsequent years, the isolation and transplantation of single myofibers from muscle tissue enabled the study of MuSC behavior in their satellite cell niche [2, 3]. This approach also allowed MuSC transplantation without removal from their native microenvironment [4]. However, this procedure is restricted to the limited and variable number of MuSC contained per myofiber, preventing the analysis of single cells in vivo. With the use of FACS-based isolation strategies, several combinations of cell surface molecules have been utilized to prospectively isolate MuSC from skeletal muscle tissue, including CD34, α7-integrin, β1-integrin, CXCR4, Vcam1, syndecan-4 [5–11]. Intramuscular transplantation of freshly

Eusebio Perdiguero and DDW Cornelison (eds.), *Muscle Stem Cells: Methods and Protocols*, Methods in Molecular Biology, vol. 1556, DOI 10.1007/978-1-4939-6771-1_11,

isolated MuSC revealed their robust capacity for muscle repair and ability to colonize the satellite cell niche. Finally, we have conclusively shown by single cell transplantation that freshly isolated MuSC are indeed capable of self-renewal in vivo [10], generating more copies of themselves while also giving rise to committed progenitors. These studies have also demonstrated the deleterious impact of in vitro propagation on MuSC function, as only a few days in culture significantly reduce their ability to engraft and contribute to host muscles [8, 10, 12]. Here, we provide details on the procedure to freshly isolate and transplant MuSC and assess their function in vivo.

2 Materials

Prepare and store all reagents and antibodies at 4 °C (unless indicated otherwise). Follow all waste disposal regulations when disposing waste materials.

2.1 Skeletal Muscle Digestion

1. Isoflurane vaporizer, supply gas (oxygen), flowmeter, and induction chamber.
2. Isoflurane (catalog number SC-363629RX) (Santa Cruz).
3. Razor blades, surgical scissors, and forceps for tissue dissection and mincing.
4. Shaking water bath, heated to 37 °C.
5. Media: Ham's F-10 media containing 10 % horse serum. Add 450 ml Ham's F-10 media and 50 ml horse serum to a 500 ml filtration unit, 0.2 μm pore size. Filter and store @ 4 °C.
6. Digestion solution I: Media (*see* **step 5**) containing 700 units/ml collagenase type II (catalog number 17101-015) (Gibco). Weigh after determining the correct amount to match activity units (varies with each batch) and suspend immediately prior to use (*see* **Note 1**).
7. Digestion solution II: Media (*see* **step 5**) containing 100 units/ml collagenase type II and 2 units/ml dispase II (catalog number 04942078001) (Roche). Weigh enzymes as in the previous step and suspend immediately prior to use (*see* **Note 2**).
8. Syringes, 10 ml with 20G × 1″ needles (catalog number 309644) (BD Biosciences).
9. Cell strainers, 70 μm (catalog number 22363548) (Fisher Scientific).

2.2 Cell Staining and FACS Sorting

1. FACS buffer: PBS, pH7.4 (1×) containing 2.5 % normal goat serum, 1 mM ethylenediaminetetraacetic acid (EDTA). Add 486.5 ml PBS, 12.5 ml goat serum, and 1 ml 0.5 M EDTA to

a 500 ml filtration unit, 0.2 μm pore size. Filter and store @ 4 °C.

2. FACS round-bottom tubes, 5 ml (catalog number 352063) (BD Falcon).
3. FACS antibodies.
   (a) Biotin CD45, clone 30-F11, 0.5 mg/ml (catalog number 553078) (BD Biosciences).
   (b) Biotin CD31, clone 390, 0.5 mg/ml (catalog number 13-0311-85) (eBiosciences).
   (c) Biotin CD11b, clone M1/70, 0.5 mg/ml (catalog number 553309) (BD Biosciences).
   (d) Biotin Ly-6A/E-Sca1, clone E13-161.7, 0.5 mg/ml (catalog number 553334) (BD Biosciences).
   (e) Alpha7-integrin/PE or FITC, clone R2F2, 1 mg/ml (catalog number 53-0010-01) (AbLabs).
   (f) CD34/Alexa Fluor 647, clone RAM34, 0.2 mg/ml (catalog number 560230) (BD Biosciences).
   (g) Streptavidin APC-Cy7, 0.2 mg/ml (catalog number 554063) (BD Biosciences).
4. Streptavidin microbeads (catalog number 130-048-101) (Miltenyi Biotec).
5. LS columns (catalog number 130042401) (Miltenyi Biotec).
6. MACS magnetic separator and multi-stand (catalog numbers 130-042-302, 130-042-303) (Miltenyi Biotec).
7. FxCycle Violet stain, 0.5 mg/ml (catalog number F10347) (Life Technologies).
8. Flow cytometer, FACSAria II (BD Biosciences) (*see* **Note 3**).

2.3 Intramuscular Transplantation and Tissue Collection

1. Isoflurane vaporizer, supply gas (oxygen), flowmeter, and induction chamber.
2. Small animal hair clippers (catalog number CLP-22965) (Braintree Scientific).
3. Insulin syringes, 0.3 ml with 29G × 0.5″ ultra-fine needles (catalog number 309301) (BD Biosciences).
4. Fixative solution: PBS, pH7.4 (1×) containing 0.5 % paraformaldehyde (*see* **Note 4**).
5. PBS, pH7.4 (1×) containing 20 % sucrose.
6. Biopsy cryomolds, 10 × 10 × 15 mm (catalog number 4565) (VWR).
7. O.C.T. compound (catalog number 4583) (VWR).

2.4 Tissue Cryosectioning and Immunostaining

1. Research cryostat, CM3050 S (Leica).
2. Microscope slides, Superfrost Plus, 25 × 75 × 1.0 mm (catalog number 12-550-15) (Fisher Scientific).
3. Hydrophobic PAP pen (catalog number 00-8899) (Life Technologies).
4. Slide staining humidity box, black cover (catalog number 71397-B) (Electron Microscopy Sciences).
5. Fixative solution: PBS, pH7.4 (1×) containing 1.5 % paraformaldehyde.
6. Blocking buffer: PBS, pH7.4 (1×) containing 20 % normal goat serum, 0.1 % Triton X-100 (catalog number H5142) (Promega).
7. Antigen retrieval solution: deionized water containing 100× citrate-based antigen unmasking solution, pH6.0 (catalog H-3300) (Vector). Add 0.5 ml antigen unmasking solution to 49.5 ml deionized water and store @ 4 °C (*see* **Note 5**).
8. Antibodies.
   (a) Rabbit anti-GFP, 2 mg/ml (catalog number A11122) (Life Technologies).
   (b) Rabbit anti-RFP, 2 mg/ml (catalog number 600-401-379) (Rockland).
   (c) Rat anti-laminin B2, clone A5, 0.5 mg/ml (catalog number 05-206) (Millipore).
   (d) Mouse anti-Pax7 concentrate (catalog number Pax7) (Developmental Studies Hybridoma Bank).
   (e) Alexa Fluor secondary antibodies (Life Technologies).
9. Hoechst 33258, 10 mg/ml (catalog number H3569) (Life Technologies).
10. Fluoromount-G mounting medium (catalog number 0100-01) (Southern Biotech).
11. Microscope cover glass, 24 × 50 mm (catalog number 12-545-F) (Fisher Scientific).

3 Methods

Carry out all procedures at room temperature unless otherwise specified. For a detailed description of FACS isolation of MuSC, see also [13].

3.1 Skeletal Muscle Digestion

1. Warm media to 37 °C and prepare digestive solution I, also stored at 37 °C until after muscle harvest.

2. Turn on the flowmeter to 200 ml/min oxygen and the isoflurane vaporizer to 2 %. Anesthetize the mouse by placing inside the induction chamber and wait 2–3 min until breathing has slowed and the mouse appears asleep. Remove the mouse from the induction chamber, confirm it is nonresponsive to a foot or tail pinch, and sacrifice by cervical dislocation.
3. Place the mouse on a surgical bench and soak the hind limbs in 70 % ethanol. Also soak the forceps, surgical scissors, and razor blade in 70 % ethanol to disinfect. Remove the skin covering the hind limb muscles by making an incision near the hip and slowly pull the skin down the limb and over the ankle.
4. To remove the gastrocnemius and soleus muscles, sever the Achilles tendon and grasp with forceps. Slowly pull up the limb until fully separated from the anterior muscle groups. Use a razor blade to remove from the surrounding hamstring muscles. Use a razor blade to sever the distal tendons of the anterior compartment, including the tibialis anterior and extensor digitorum longus. Group these tendons with forceps and slide a razor blade between these muscles and the tibialis bone until fully separated. Similarly, sever the quadriceps tendon and use a razor blade to separate the quadriceps from the femur. Place all muscles in a 10 cm tissue culture dish containing 5 ml media. Repeat this process with the contralateral limb.
5. Bring the two 10 cm tissue culture dishes to a sterile laminar flow hood. Using two razor blades, mince the muscles into small pieces (*see* **Note 6**). Once all muscles have been minced, use forceps to transfer the muscle slurry from each hind limb into a 15 ml centrifuge tube. Collect all media in the plate and wash plate with ready-made digestion media I to collect all remaining muscle pieces, reaching a final volume of 10 ml per hind limb (*see* **Note 7**). Place samples in a shaking water bath, warmed to 37 °C, at 200 rpm for 90 min.
6. Remove samples from the water bath and centrifuge at 300 × *g* for 5 min. Return samples to the laminar flow hood, aspirate the supernatant, and resuspend in ready-made 10 ml digestive solution II (*see* **Note 8**). Briefly vortex samples and place in a shaking water bath, warmed to 37 °C, at 200 rpm for 30 min.
7. Remove samples from the water bath, disinfect with 70 % ethanol, and return to the laminar flow hood. Pipet several times and pass each sample through a 10 ml syringe equipped with a 20G × 1" needle to mechanically break down remaining muscle pieces. Repeat this process 10 times to be sure all cells have been released into suspension (*see* **Note 9**). Transfer samples to a 50 ml centrifuge tube, washing the original tube twice to reach a final volume of 40 ml media. Centrifuge samples at 300 × *g* for 5 min.

8. Return samples to the laminar flow hood, aspirate the supernatant, and resuspend in 10 ml media. Pass each sample through a 70-μm cell strainer placed over a 50 ml centrifuge tube to filter any remaining debris, including muscle tendons. Wash the original tube twice to reach a final volume of 40 ml media and centrifuge the samples at 300 × *g* for 5 min.

3.2 Cell Staining

1. Return samples to the laminar flow hood, aspirate the supernatant, and transfer the cell suspension to a 15 ml centrifuge tube. Combine samples from both legs at this step, washing the original tubes once each to reach a final volume of 2 ml media.
2. To prepare single stain or single color controls for proper compensation during FACS sorting, add 10 μl cell suspension to 500 μl media contained in 5 ml FACS round-bottom tubes. One sample should be prepared for each antibody or stain used in the FACS sort and an additional sample as an unstained, negative control (*see* **Note 10**). Add the following volumes of antibody to each single color control: 1 μl α7-integrin/PE or FITC, 4 μl CD34/Alexa Fluor 647, 1 μl biotin CD45, and 1 μl FxCycle Violet stain. Add the following volumes of antibody to the main sample: 5 μl biotin CD45, 10 μl biotin CD31, 10 μl biotin CD11b, and 10 μl biotin Ly-6A/E-Sca1. Briefly vortex each sample and incubate for 20 min at 4 °C.
3. Following the primary incubation, add 10 ml chilled media to the main sample or 3 ml FACS buffer to each single color control sample and centrifuge at 300 × *g* for 5 min at 4 °C.
4. Return samples to the laminar flow hood, aspirate the supernatant, and resuspend in 1.5 ml media for the main sample or 500 μl FACS buffer for the single color control samples. Add 1 μl streptavidin APC-Cy7 to the biotin CD45 single color control. Add 150 μl streptavidin microbeads and the following volumes of antibody to the main sample: 10 μl α7-integrin/PE or FITC, 30 μl CD34/Alexa Fluor 647, and 10 μl streptavidin APC-Cy7 (*see* **Note 11**). Briefly vortex and incubate for 20 min at 4 °C.
5. Following the secondary incubation, add 10 ml chilled media to the main sample or 3 ml FACS buffer to the biotin CD45/APC-Cy7 single color control sample and centrifuge at 300 × *g* for 5 min at 4 °C. During centrifugation, set up the MACS magnetic separator with LS column in the laminar flow hood and prepare the LS column for depletion by preloading with 3 ml media.
6. Return samples to the laminar flow hood, aspirate the supernatant, and resuspend in 1 ml media for the main sample or 500 μl FACS buffer for the biotin CD45/APC-Cy7 single color control sample. Load the main sample onto the LS column placed

on the MACS magnetic separator and allow it to pass through the column. Wash the original centrifuge tube twice with 3 ml media and pass each wash through the column (*see* **Note 12**). Centrifuge the collected cell suspension after depletion at 300 × *g* for 5 min at 4 °C.

7. Return samples to the laminar flow hood, aspirate the supernatant, and transfer the depleted sample to a 5 ml FACS round-bottom tube. Resuspend the sample in 2 ml FACS buffer and add 2 μl FxCycle Violet stain to the sample. Keep the main sample and single color control samples at 4 °C until ready for sorting.

3.3 FACS Sorting

1. Bring the main sample and single color control samples to the flow cytometer and prepare the machine according to the facility instructions.
2. Run the unstained control sample, collecting >10,000 events. Adjust the forward-scattered light (FSC) and side-scattered light (SSC) detector gain and area scaling factor to ensure all signals are on scale (*see* **Note 13**). Wash with deionized water for 1 min to remove debris trapped within the fluidics system.
3. Run each single color control sample, collecting >10,000 events per sample (*see* **Note 14**), washing between each sample deionized water for 1 min. Determine the correct voltage for each fluorophore and rerun samples if necessary. Compensation should be performed using these control samples to avoid spectral overlap leading to "false positive" detection of one fluorophore into another. If prepared, FMO controls should also be run to aid proper gating in the next step.
4. Once proper voltage determination and compensation has been performed, wash with 10 % bleach for 1 min and deionized water for 2 min or until no events are observed. Load the main sample in the sample chamber and run to collect >20,000 events. The following steps should be taken to set up the proper gating scheme for MuSC isolation (Fig. 1):

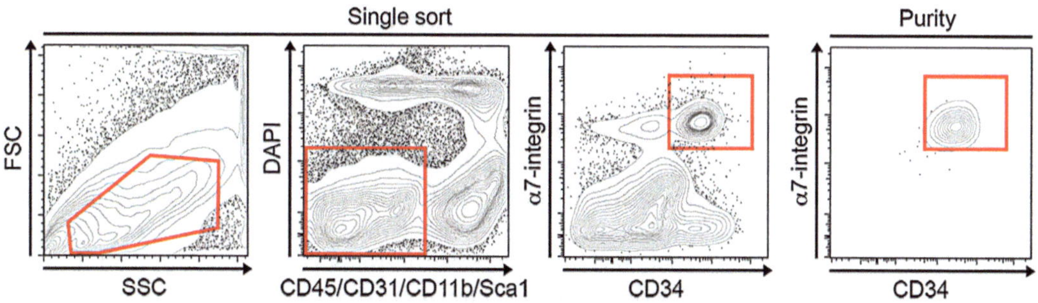

Fig. 1 Fluorescent-based cell sorting (FACS) of MuSC from adult hind limb skeletal muscle. Gating schemes to purify living (DAPI-negative) satellite cells negative for CD45, CD31, CD11b, and Sca1 and expressing a7-integrin and CD34 (*left*). This FACS strategy yields a population of MuSC with a purity of >95 % (*right*)

(a) To discriminate single cells and remove debris or cell doublets, gate conservatively for cells of low-to-medium cell size and low granularity. Additional pulse geometry gates can be established (ex. FSC-A vs. FSC-H) to further eliminate cell doublets.

(b) To remove dead cells (DAPI) and hematopoietic, endothelial or mesenchymal cell types (CD45, CD11b, CD31, and Ly-6A/E-Sca1 conjugated to APC-Cy7), gate for cells negative for all markers.

(c) To select for MuSC, gate for cells positive for both α7-integrin (conjugated to PE) and CD34 (conjugated to APC).

5. Begin sorting into a FACS round-bottom tube containing 0.5 ml PBS, pH7.4 and placed in the collection block. If possible, keep the sample chamber and collection block at 4 °C to preserve cell viability and prevent premature activation of MuSC, especially if presumed quiescent from undamaged muscle.
6. When FACS sorting has been completed, collect a small aliquot of sorted cells and wash with 10 % bleach for 1 min and deionized water for 2 min or until no events are observed. Transfer the aliquot of sorted cells to the sample chamber and run briefly until >5000 events have been collected to assess the purity of the sorted sample (*see* **Note 15**).
7. Centrifuge the sorted sample at 300 × *g* for 5 min at 4 °C and resuspend in a volume of PBS, pH7.4 conservatively less than the total volume of all transplantations (typically less than 150 μl). Count the cells using a hemocytometer and adjust the volume to the desired concentration (*see* **Note 16**). Keep the cell suspension at 4 °C until ready for transplantation.

3.4 Intramuscular Transplantation

1. Prepare the desired volume of cell suspension to be transplanted in 0.3 ml insulin syringes with a 29G × 0.5" ultra-fine needle (*see* **Note 17**). Set aside until recipient mice have been prepared for transplantation.
2. Turn on the flowmeter to 200 ml/min oxygen and the isoflurane vaporizer to 2 %. Anesthetize the recipient mouse by placing inside the induction chamber and wait 2–3 min until breathing has slowed and the mouse appears asleep. Remove the mouse from the induction chamber, divert flow from the flowmeter, and vaporizer to an anesthesia mask on a surgical bench and quickly place this mask on the mouse. Wait 2–3 min and confirm it is nonresponsive to a foot or tail pinch.
3. Remove hair from the transplantation site at the anterior hind limb with small animal hair clippers. Use an alcohol wipe to disinfect the transplantation site.

4. Insert the needle at the center of the tibialis anterior muscle, maintaining an angle of 45 degrees between the needle and the length of the tibialis (*see* **Note 18**). Slightly withdraw, maintaining the needle tip still in the center of the tibialis anterior muscle, to allow space for the cell suspension. Inject the cell suspension, holding the needle in place for 2–3 s to ensure the volume will not leak out from the muscle once the needle has been removed.
5. Remove the anesthesia mask and return the recipient mouse to its cage. Allow for up 5 min for the mouse to awaken. Reflect on the animal's health status to confirm normal activity and proper ambulation after an adjustment period of up to 15 min (*see* **Note 19**).

3.5 Tissue Harvest and Cryosectioning

1. At the desired time of harvest following cell transplantation, collect the recipient mouse for sacrifice, and prepare the fixative solution.
2. Turn on the flowmeter to 200 ml/min oxygen and the isoflurane vaporizer to 2 %. Anesthetize the mouse by placing inside the induction chamber and wait 2–3 min until breathing has slowed and the mouse appears asleep. Remove the mouse from the induction chamber, confirm it is nonresponsive to a foot or tail pinch, and sacrifice by cervical dislocation.
3. Place the mouse on a surgical bench and soak the hind limbs in 70 % ethanol. Remove the skin covering the hind limb muscles by making an incision near the hip and slowly pull the skin down the limb and over the ankle. Use a razor blade to sever the distal tendon of the tibialis anterior muscle. Hold this tendon with forceps, gently pulling toward the knee while sliding a razor blade between the muscle and the tibialis bone until fully separated. Place the muscle directly in fixative solution and incubate for 4 h at 4 °C (*see* **Note 20**).
4. Transfer the muscle from fixative solution to PBS, pH 7.4 containing 20 % sucrose (*see* **Note 21**). Incubate overnight at 4 °C.
5. Prepare for tissue freezing by gathering dry ice and liquid nitrogen. Chill a volume of 2-methylbutane sufficient to submerge the biopsy cryomold in liquid nitrogen until it begins to solidify. Transfer the recipient muscle to biopsy cryomolds containing O.C.T. compound, positioned longitudinally so cross-sections can be achieved by sectioning from top to bottom. Place the bottom of the cryomold on the top of the chilled 2-methylbutane until it has completely frozen over and then submerge for 1 min. Remove the cryomold and place on dry ice. Frozen samples can be stored at −80 °C until ready to be cryosectioned.

6. Set the cryostat chamber temperature to −20 °C, place samples into the cryostat, and allow them to come to temperature for 30 min. Cut and place two 10 μm sections serially on 4–8 slides, subject to the number of immunostainings that will be performed. Once two sections have been placed on each slide, eliminate 200 μm of tissue and resume normal cryosectioning. Repeat this process until the core 80 % of the muscle has been sectioned so as to represent the entire muscle and allow for analysis of the entire transplant area. Once completed, label and store slides at −20 °C.

3.6 Immunostaining

1. Remove slides from storage at −20 °C and let come to room temperature. Encircle sections with a hydrophobic PAP pen to retain solution on the mounted tissue. Place slides in a humid incubation box (dark, UV safe), where all subsequent steps should be performed.
2. Fix sections in PBS, pH 7.4 containing 1.5 % paraformaldehyde for 15 min.
3. Aspirate fixative and wash in PBS, pH 7.4 for 5 min. Block and permeabilize sections in blocking buffer for 1 h.
4. Incubate sections with the following primary antibodies, diluted in blocking buffer for 2 h (or overnight): rat anti-laminin B2 (1:200 dilution) and either rabbit anti-GFP (1:500 dilution) or rabbit anti-RFP (1:200 dilution), dependent upon the endogenous fluorophore present in the transplanted cells.
5. Aspirate primary antibodies and perform three washes in PBS, pH 7.4 for 5 min each. Incubate sections with the appropriate secondary antibodies (*see* **Note 22**), diluted in blocking buffer for 1 h.
6. Aspirate secondary antibodies and perform three washes in PBS, pH 7.4 for 5 min each. Postfix sections PBS, pH 7.4 containing 1.5 % paraformaldehyde for 15 min to preserve the signal specific to all antibodies used thus far. Antigen retrieval procedures are somewhat harsh and can greatly reduce signal without this step.
7. Aspirate fixative and wash in PBS, pH 7.4 for 5 min. Immerse slides in pre-warmed antigen retrieval solution contained in either a plastic or ceramic Coplin jar and incubate in a heated water bath maintained at >80 °C for 15 min.
8. Remove Coplin jar containing slides from the heated water bath and let cool for 5 min. Wash in PBS, pH 7.4 for 5 min. Reblock and permeabilize sections in blocking buffer for 30 min. Incubate sections with mouse anti-Pax7 (1:100 dilution) in blocking buffer overnight.

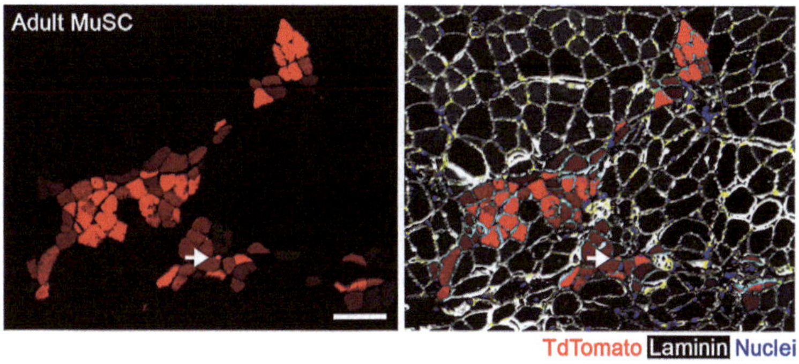

Fig. 2 Histological analysis and immunostaining of 1000 TdTomato$^+$ freshly isolated MuSC transplanted into the tibialis anterior muscle and harvested after 15 days. Donor cells have engrafted to both repair damaged myofibers and colonize the satellite cell niche (*arrow*). Scale bar 100 μm

9. Aspirate primary antibodies and perform three washes in PBS, pH 7.4 for 5 min each. Incubate sections with the appropriate secondary antibodies and Hoechst 33258 (1:1000 dilution) in blocking buffer for 1 h.
10. Aspirate secondary antibodies and wash in a Coplin jar with PBS, pH 7.4 for 30 min. Mount slides with Fluoromount-G water-based mounting medium and microscope cover glasses. Once the cover glass has been placed on the slide, utilize forceps to gently push visible air bubbles to the side of the slide as they can dry the sample and generate significant autofluorescence. Store slides at 4 °C until ready for imaging (*see* **Note 23**) (Fig. 2).

4 Notes

1. Stock concentrations of collagenase type II may be prepared prior to skeletal muscle digestion, aliquoted and stored @ −20 °C for up to 2 months.
2. Remaining stock solution of collagenase type II intended for the second digestion step may be used if stored @ 4 °C to minimize loss of enzymatic activity. We recommend that dispase type II is weighed and added to the digestion media immediately prior to use.
3. The FACS sorter must be equipped with 488, 405, and 633 nm lasers, given the antibodies listed. However, color combinations can be adjusted to match the laser combinations available.
4. This concentration of paraformaldehyde is appropriate for the tibialis anterior muscle, or other skeletal muscles of similar size. If fixing muscles considerably larger or smaller, both fixative concentration and time should be adjusted to optimize results.

5. Antigen retrieval solution at working concentration can be reused up to five times if stored at 4 °C in between uses.
6. The mincing of skeletal muscle into small pieces is a key step. Mincing into small pieces will artificially lower yield, ostensibly exposing the cells to prolonged enzymatic digestion. However, mincing into large pieces will result in an incomplete digestion. Mincing should be performed with only one to two plates kept at room temperature at once; if sorting hind limb muscles from more than one mouse, keep all other samples at 37 °C.
7. If 5 ml media is used during tissue mincing, prepare digestive solution I as a 2x concentrated stock so that it can be directly added to the muscle slurry once collected in a 15 ml centrifuge tube.
8. Prepare digestive solution II near the end of the first incubation to minimize the time the muscle slurry is kept at room temperature between digestive steps.
9. No visible muscle pieces should be present once completed; however, some remaining debris, including muscle tendons, is expected to remain until filtration through cell strainers in the next step.
10. Additional Fluorescence Minus One (FMO) controls can be prepared at this step to properly determine gating boundaries when sorting. Single color controls only allow for proper compensation of each fluorophore prior to sorting and should not be strictly used to determine gating boundaries.
11. Add the streptavidin microbeads before the streptavidin APC-Cy7 antibody to give it an advantage in binding to its target cell types. This improves the efficiency of depletion while still allowing the elimination of cells expressing CD45, CD31, CD11b, or Ly-6A/E-Sca1 during the sort.
12. If the positive cellular fraction expressing CD45, CD31, CD11b, or Ly-6A/E-Sca1 is desired following the depletion step, the plunger packaged with the LS column can be used to recover these cells.
13. Forward-scattered light is proportional to cell surface area or size and side-scattered light is proportional to cell granularity.
14. Cells from skeletal muscle preparations should be run through the flow cytometer at 20 psi through a 100 μm nozzle and at not more than 5000 events per second to avoid erratic streaming of the cells through the fluidics system.
15. The purity of living cells in the sorted sample should be >95 %; however, cell viability is negatively affected by the sorting process and only 50–70 % of the sorted cells are expected to remain negative for DAPI upon resorting. If the purity is less than 90 %, the sample can be resorted to eliminate cellular contaminants.

16. The number of cells transplanted largely depends on the assay being performed and the expected effect of any experimental manipulation on muscle stem cell behavior. We recommend transplanting 1000 cells, if freshly isolated, in any loss-of-function studies. If the cells are to be plated prior to transplantation as part of the experimental strategy, >5000 cells should be transplanted to achieve similar levels of engraftment to regenerating myofibers. Plating muscle stem cells in standard culture conditions, even overnight, dramatically reduces their engraftment potential to the satellite cell niche, while longer periods in culture will reduce both niche engraftment and myofiber repair.
17. The ideal transplantation volume is 10–15 μl, resulting in a localized donor cell engraftment site. Up to 40 μl may be injected into the tibialis anterior muscle, but the donor cells should be expected to be more sparely located and increasing the risk of leakage outside the muscle.
18. An angle of 45 degrees relative to the length of the tibialis increases the probability of successful and consistent transplantation into the tibialis anterior muscle, given its elongated shape.
19. In the event transplantation should be performed into injured muscle, the tibialis anterior muscle should be injected with 50 μl of notexin, cardiotoxin or 1.2 % barium chloride solution in PBS pH7.4 one day prior to cell transplantation.
20. Fixation concentration and time has been optimized for the tibialis anterior muscle. The fixation of smaller or larger muscle groups may require shorter or longer fixation times to successfully perfuse the tissue.
21. Incubation in sucrose solution following fixation protects the tissue from damage associated with freezing by increasing cellular solute concentration. Muscles incubated in fixative should float when initially placed in sucrose but eventually sink to the bottom of the tube. An additional change of sucrose solution is optional if significant freezing damage is observed.
22. Preference for any secondary antibody matched with either Pax7 or laminin is dependent upon the endogenous fluorophore identifying donor cells and available or preferred spectra given the researcher's imaging microscope.
23. To quantify donor cell contribution to regenerating myofibers following stable engraftment, the entire slide should be scanned by eye and a single section closest to the core of the transplantation area should be selected for quantification. If the researcher is unsure, several sections can be chosen for analysis and the section containing the largest number of donor myofibers should be chosen. Analysis of several sections should be avoided, as donor myofibers are likely to span a significant portion of the muscle but cannot be distinguished. To quantify

donor cell contribution to the satellite cell niche, each section on the slide should be imaged and quantified. As mononucleated cells are not expected to span the >40 μm length given four serial slides, there is little risk of scoring the same cell twice. Although tedious, this method enables a more accurate estimate of the total number of engrafted satellite cells in the recipient tissue and reduces the risk of nonrepresentative totals scored in any single section.

Acknowledgments

This work was supported by the US National Institutes of Health (NIH) grants R01AR064873 and P30 AR061303 to AS, and US National Institutes of Health (NIH) grant F31 AR065923-01 to MT. We thank Amy Cortez, Yoav Altman, Kenny Venegas and Buddy Charbono for technical support.

References

1. Rando TA, Blau HM (1994) Primary mouse myoblast purification, characterization, and transplantation for cell-mediated gene therapy. J Cell Biol 125(6):1275–1287
2. Bischoff R (1986) Proliferation of muscle satellite cells on intact myofibers in culture. Dev Biol 115(1):129–139
3. Rosenblatt JD, Lunt AI, Parry DJ, Partridge TA (1995) Culturing satellite cells from living single muscle fiber explants. In Vitro Cell Dev Biol Anim 31(10):773–779
4. Collins CA, Olsen I, Zammit PS, Heslop L, Petrie A, Partridge TA, Morgan JE (2005) Stem cell function, self-renewal, and behavioral heterogeneity of cells from the adult muscle satellite cell niche. Cell 122(2):289–301
5. Sherwood RI, Christensen JL, Conboy IM, Conboy MJ, Rando TA, Weissman IL, Wagers AJ (2004) Isolation of adult mouse myogenic progenitors: functional heterogeneity of cells within and engrafting skeletal muscle. Cell 119(4):543–554
6. Fukada S, Higuchi S, Segawa M, Koda K, Yamamoto Y, Tsujikawa K, Kohama Y, Uezumi A, Imamura M, Miyagoe-Suzuki Y et al (2004) Purification and cell-surface marker characterization of quiescent satellite cells from murine skeletal muscle by a novel monoclonal antibody. Exp Cell Res 296(2):245–255
7. Cornelison DD, Wilcox-Adelman SA, Goetinck PF, Rauvala H, Rapraeger AC, Olwin BB (2004) Essential and separable roles for Syndecan-3 and Syndecan-4 in skeletal muscle development and regeneration. Genes Dev 18(18):2231–2236
8. Montarras D, Morgan J, Collins C, Relaix F, Zaffran S, Cumano A, Partridge T, Buckingham M (2005) Direct isolation of satellite cells for skeletal muscle regeneration. Science 309(5743):2064–2067
9. Cerletti M, Jurga S, Witczak CA, Hirshman MF, Shadrach JL, Goodyear LJ, Wagers AJ (2008) Highly efficient, functional engraftment of skeletal muscle stem cells in dystrophic muscles. Cell 134(1):37–47
10. Sacco A, Doyonnas R, Kraft P, Vitorovic S, Blau HM (2008) Self-renewal and expansion of single transplanted muscle stem cells. Nature 456(7221):502–506
11. Tanaka KK, Hall JK, Troy AA, Cornelison DD, Majka SM, Olwin BB (2009) Syndecan-4-expressing muscle progenitor cells in the SP engraft as satellite cells during muscle regeneration. Cell Stem Cell 4(3):217–225
12. Gilbert PM, Havenstrite KL, Magnusson KE, Sacco A, Leonardi NA, Kraft P, Nguyen NK, Thrun S, Lutolf MP, Blau HM (2010) Substrate elasticity regulates skeletal muscle stem cell self-renewal in culture. Science 329(5995):1078–1081
13. Gromova A, Tierney MT, Sacco A (2015) FACS-based satellite cell isolation from mouse hind limb muscles. Bio Protoc 5(16):e1558

Chapter 12

Transplantation of Skeletal Muscle Stem Cells

Monica N. Hall, John K. Hall, Adam B. Cadwallader, Bradley T. Pawlikowski, Jason D. Doles, Tiffany L. Elston, and Bradley B. Olwin

Abstract

Transplanting adult stem cells provides a stringent test for self-renewal and the assessment of comparative engraftment in competitive transplant assays. Transplantation of satellite cells into mammalian skeletal muscle provided the first critical evidence that satellite cells function as adult muscle stem cells. Transplantation of a single satellite cell confirmed and extended this hypothesis, providing proof that the satellite cell is a bona fide adult skeletal muscle stem cell as reported by Sacco et al. (Nature 456(7221):502–506). Satellite cell transplantation has been further leveraged to identify culture conditions that maintain engraftment and to identify self-renewal deficits in satellite cells from aged mice. Conversion of iPSCs (induced pluripotent stem cells) to a satellite cell-like state, followed by transplantation, demonstrated that these cells possess adult muscle stem cell properties as reported by Darabi et al. (Stem Cell Rev Rep 7(4):948–957) and Mizuno et al. (FASEB J 24(7):2245–2253). Thus, transplantation strategies involving either satellite cells derived from adult muscles or derived from iPSCs may eventually be exploited as a therapy for treating patients with diseased or failing skeletal muscle. Here, we describe methods for isolating dispersed adult mouse satellite cells and satellite cells on intact myofibers for transplantation into recipient mice to study muscle stem cell function and behavior following engraftment.

Key words Skeletal muscle, Satellite cell, Muscle myofiber, Adult muscle stem cell, Stem cell transplantation

1 Introduction

Skeletal muscle is comprised of large multinucleated cells referred to as myofibers, which contain hundreds of nuclei from cell fusion during development. Skeletal muscle is dynamic, capable of hypertrophy and atrophy depending on the demands placed on the tissue. A rapid regenerative response is provided by a stem cell, ensuring lifelong function [1, 2]. Myofiber nuclei are terminally differentiated and thus, require muscle stem cells, termed satellite cells, to replenish myonuclei for tissue repair [1, 2]. During repair, satellite cells, which are typically mitotically quiescent, activate and then enter the cell cycle expanding as myoblasts where a majority

Eusebio Perdiguero and DDW Cornelison (eds.), *Muscle Stem Cells: Methods and Protocols*, Methods in Molecular Biology, vol. 1556, DOI 10.1007/978-1-4939-6771-1_12, © Springer Science+Business Media LLC 2017

fuse to repair myofibers and a minority undergo self-renewal to replenish the stem cell pool [3, 4]. When injected into injured or diseased host muscles, donor satellite cells expand, fuse into myofibers, and engraft into the host satellite cell niche [4]. Thus, satellite cell transplantation may prove useful for treating muscle diseases and age-associated atrophy in humans. Transplantation of skeletal muscle cells typically involves isolation of muscle cells from dissected muscle that is enzymatically digested to free cells of tissue debris, enriched by differential plating on tissue culture substrates, density gradient sedimentation, or by FACS (fluorescence activated cell sorting), followed by injection into a recipient muscle.

The first experiments demonstrating engraftment of donor myogenic cells and incorporation into mouse recipient muscle were performed in the 1980s [5–7]. An initial study on myoblast transplantation observed low survival rates postinjection with a small minority of donor cells exhibiting stem cell-like properties that survived long-term and participated in muscle homeostasis [8]. The observation that a subset of muscle cells possesses stem cell potential enabled the development of purification protocols to identify and study satellite cell subpopulations. FACS is typically used to separate satellite cell subpopulations with differing engraftment capacities [4, 9–11]. Isolation of individual multinucleated myofibers as a vehicle for transplanting associated satellite cells is a successful strategy, where transplantation of a few myofibers increased engraftment efficiencies with donor nuclei found in myofibers and occupation of donor cells in the satellite cell niche following transplantations [4, 12]. A similar approach was successful for xenografts, involving transplantation of human myofiber fragments into the irradiated muscle of immunodeficient mice, where donor cells engrafted into the host satellite cell niche and fused into myofibers as myonuclei [13]. Bioengineered substrates enhance satellite cell engraftment as they recapitulate key biophysical and biochemical features of the satellite cell niche [14]. Methods for transplanting isolated satellite cells and for transplanting myofiber-associated satellite cells described in this Chapter involve standard procedures, cell culture techniques, and reagents that permit isolation of a donor satellite cell population that engrafts into the host satellite cell niche and produces myoblasts capable of participating in skeletal muscle maintenance and regeneration.

2 Materials

Prepare and store all reagents at room temperature unless indicated otherwise. Follow all institutional waste disposal regulations when disposing of waste materials. Cell culturing requires standard tissue culturing techniques, reagents, and consumables with the addition of a low oxygen (4–6 % O_2) incubator. A fluorescent activated cell sorter, MoFlo XDP or equivalent, will be required for fluorescent activated cell sorting. We routinely screen multiple lots of horse serum and collagenase for cell survival, growth, and differentiation. Sufficient variability exists between lots of serum and collagenase such that lots that do not support cell survival and growth will result in failed transplants despite identical isolation methods. We strongly advise advance screening and selection of horse serum and collagenase lots prior to attempting cell transplantations.

2.1 Cell Isolation and Purification

1. Concentrated Saline G (20×): To 600 ml distilled water, add 160 g/l sodium chloride, 8.0 g/l potassium chloride, 3.0 g/l monopotassium phosphate, 3.0 g/l sodium phosphate dibasic, 22.0 g/l dextrose, 100 mg/l phenol red. Stir until dissolved and bring volume to 1 l. Filter sterilize solution with 0.2 μM filter in a tissue culture hood. Store at 4 °C.
2. Collagenase Type I (Worthington Biochemicals, LS004197) 4000 U/ml: Add collagenase to an appropriate volume of Saline G (1× or a 1:20 dilution of 20× stock) to generate 4000 U/ml, 10× stock solution. Stir 3–5 h at 4 °C until collagenase is dissolved. Serial filter the solution with 0.40 μM filter followed by a 0.22 μM filter. Aliquot 1 ml collagenase into 1.5 ml Eppendorf tubes or 15 ml conical tubes and store at −20 °C.
3. F-12C: To prepare 2 l of media, add 1 l tissue culture water to an Erlenmeyer flask containing a clean stir bar. Slowly add 2.352 g/2l powdered F-12 Media (Life Technologies, 21700-075) with stirring until dissolved. Add 2.352 g/2l sodium bicarbonate. Adjust pH to 7.15–7.20. Add penicillin/streptomycin to 100 U/ml penicillin, 100 μg/ml streptomycin, and 8 ml 0.2 M calcium chloride. Add tissue culture water to bring volume to 2 l as needed. Filter the final solution with a 0.22 μM filter and store at 4 °C.
4. Collagenase digestion media: 9 ml F-12C media, 1 ml Type I Collagenase 4000 U/ml.
5. FGF-2: Basic human fibroblast growth factor-2 (Promega) 5 ng/ml stock.
6. Growth Media: F-12C, 15 % Horse Serum: Life Technologies, Lot No. 1312382, 1 % penicillin/streptomycin (100 U/ml penicillin, 100 μg/ml streptomycin), 2 nM FGF-2.
7. Red Blood Cell Lysing Buffer: Sigma R7757.
8. Hanks' Balanced Salt Solution: Sigma H6648.
9. FACS Buffer: Hanks' Balanced Salt Solution, 5 % Horse Serum: Life Technologies.
10. Antibodies for FACS: FITC conjugated rat anti mouse CD31 (MEC 13.3) 0.5 mg/ml, FITC conjugated rat anti mouse

CD45 (30-F11) 0.5 mg/ml, FITC conjugated rat anti mouse Ly6A/E (E13-161.7) 0.5 mg/ml, AbLab, anti-Alpha7 Integrin (R2F2) -647 (1:100). An anti-Sydecan4 antibody can be substituted for anti-Alpha7 Integrin and is available per request from Dr. Olwin, University of Colorado, Boulder.

11. SiteClick Antibody Labeling kits: Life Technologies (Thermo Fisher Brand) for direct conjugation of antibodies to fluorophores.

2.2 Cell and Myofiber Transplantation

1.2 % Barium Chloride in 0.9 % Saline: Fill clean beaker with 50 ml of distilled water and add stir bar. While stirring add 0.9 g sodium chloride and 1.2 g barium chloride dehydrate. After salts have dissolved bring final volume to 100 ml in a clean graduated cylinder. In a tissue culture hood, filter 100 ml solution through a 0.22 μm Steritop filter (bottle top style filter) into a sterile tissue culture bottle. Aliquot for single use and store at 4 °C.

3 Methods

The following involves transplanting mononucleated satellite cell cultures or myofiber-associated satellite cells. Mononucleated cells must be transplanted immediately upon isolation or following purification via Fluorescence Activated Cell Sorting (FACS) without intervening cell culture. All mice were housed in a pathogen-free facility, and the Institutional Animal Care and Use Committee at the University of Colorado approved all protocols. Small rodent procedures require standard animal anesthesia setup, medical supplies and dissection tools. Follow all institutional regulations and safety protocols for animal care and use.

3.1 Cell Isolation and Purification

1. Harvest skeletal muscle tissue from the hindlimb of humanely euthanized mice via isoflurane inhalation and subsequent cervical dislocation.
2. Mince skeletal muscle in a glass petri dish using scissors or scalpels until a slurry is achieved (typically 5–10 min thorough mincing is required). Add the slurry to 10 ml of collagenase digestion media in a 15 ml conical tube and place at 37 °C. Gently vortex the slurry every 10 min for 1 h. For skeletal muscle myofibers, isolate the EDL and peroneus muscles and immediately place in collagenase digestion media for 1.5 h at 37 °C and precede to Subheading 3.4.
3. Filter the digest solution through a series of cell-strainer filters: 100, 70, and 40 μM. Sediment cells by centrifugation 1000 × *g* for 5 min. Suspend cells into 500 μl of phosphate buffered saline and determine cell concentration with a standard hemacytometer. If no further purification of the cells is desired,

proceed to Subheading 3.3 Cell Transplantation. Alternatively, cells can be purified via FACS in Subheading 3.2 or pre-plated to enrich for satellite cells and myoblasts (*see* **Note 1**).

4. Suspend cells in 10 ml of growth media containing FGF-2 and plate onto a 100 mm gelatin-coated tissue culture plate. Place in 6 % low oxygen tissue culture incubator (*see* **Note 1**).

3.2 FACS Enrichment of Satellite Cells

1. Directly conjugate the anti-Alpha7 Integrin antibody to an appropriate fluorophore using SiteClick Antibody Labeling kits according to manufacturer's instructions.
2. For FACS, all steps are performed at 4 °C. One mouse hindlimb yields 10,000–200,000 sorting events.
3. Sediment cells by centrifugation for 5 min at 1400 × *g* at 4 °C and resuspend the cells in 2 ml Red Blood Cell Lysing Buffer. Incubate the slurry for 1 min in lysing buffer and quench with 20 ml of phosphate buffered saline (*see* **Note 2**). Filter cells through a 40 μM cell strainer.
4. Sediment cells by centrifugation for 5 min at 1000 × *g*. Examine the cells for the presence of red blood cells, if red blood cells are present repeat **steps 2** and **3** until blood cells are not detected.
5. Bring cell solution to an appropriate final volume of Hanks' Balanced Salt Solution taking into account multiple immunostaining conditions and controls. Divide the cell solution among 2 ml Eppendorf tubes for immunostaining. Minimal volumes recommended: controls 200 μl and experimental conditions 500 μl.
6. Add the directly conjugated Alpha7 Integrin antibody at 1:100 dilution and FITC conjugated CD31, CD45, and Ly6A/E each at a 1:500 dilution.
7. Incubate tubes for 1 h at 4 °C on a rotary-style shaker.
8. Wash cells by adding 1 ml FACS Buffer and incubate for 5 min at 4 °C with gentle agitation.
9. Sediment cells by centrifugation for 5 min at 1400 × *g*, carefully remove the FACS buffer, and resuspend the cells in 500 μl fresh FACS buffer (*see* **Note 3**).
10. Label dying cells with either propidium iodide or DAPI (4′,6-diamidino-2-phenylindole).
11. Perform FACs, sorting for Alpha7 (positive) or Syndecan-4 (positive) and against CD31, CD45 and Ly6A/E (negative). The FITC conjugated antibodies can be run on a single channel.
12. Sort cells into F-12C growth media and keep on ice.

3.3 Cell Transplantation

1. Sediment cells by centrifugation in a 1.5 ml Eppendorf tube and suspend in 50 μl 1.2 % barium chloride for intramuscular injection. Cells may be suspended in F-12C or saline if no muscle injury is desired. Keep cells on ice until injection.
2. Anesthesize the host mouse with isofluorane and locate the tibialus anterior (TA) muscle of the hindlimb. With the syringe parallel to the length of the TA muscle, insert the syringe just below the kneecap and push the syringe in until the tip has reached the bottom of the TA near the ankle. Slowly pull syringe out while injecting so that the injection volume is released along the entire length of the TA muscle. This ensures delivery of donor cells to the entire TA muscle and the most consistent TA muscle injuries.

3.4 Myofiber Transplantation

1. Place muscle slurry into 100 mm tissue culture plates containing growth media. To clean myofibers of debris, use a dissection microscope to identify shiny (live) myofibers, transferring them onto a fresh media plate using a fire-polished glass Pasteur pipet with attached latex rubber bulb (*see* **Note 4**).
2. Place myofiber cultures in a 6 % low oxygen tissue culture incubator. Refeed cultures with 2 nM FGF-2 every 24 h.
3. For transplantation, transfer donor myofibers to a 35 mm dish containing 2 ml of 1.2 % Barium chloride, or 0.9 % saline. Using a dissection scope, identify myofibers and draw up the desired number of myofibers into a 27 G1/2 cc tuberculin syringe (Becton-Dickinson Company). To clear excess volume from the syringe, hold the syringe needle up and allow the myofibers to settle near the plunger. Once the correct volume is obtained, rotate the syringe so myofibers are centered within the syringe's solution. Inject myofibers along the length of the TA muscle (*see* **Note 5**).

3.5 Analysis

Regeneration of skeletal muscle appears complete at 30 days post $BaCl_2$ injection where donor satellite cells have homed to the host satellite cell niche and are quiescent, the minimum feret's diameter is stabilized and there is no evidence of infiltrating immune cells. Therefore, all analyses of cell engraftment should include a 30 days postinjection time point. Donor cells are most readily differentiated by host cells if they are lineage marked prior to transplantation. An alternative includes allosome (Y chromosome) labeling from male donor mice in female hosts [15]. FACS provides a quantitative measure of donor versus host populations present in engrafted skeletal muscle, while immunostaining of muscle cross-sections will provide visual identification and quantification of both donor and host satellite cells but is time consuming. Satellite cells are defined as mononuclear cells located between the myofiber plasma membrane and the basal lamina of myofibers [2]. Satellite cells can be readily identified by immunoreactivity to Pax7 [16], Syndecan-4 [10], or Calcitonin Receptor [17]. Quantification of donor cells that have fused into the host myofiber is not possible as all proteins produced by donor cell nuclei diffuse throughout the host myofiber, but the number of myofibers containing fused donor cells is readily quantified. Quantification of myofibers containing donor nuclei should be reported as percent of the total myofibers scored so that data are readily compared between research groups.

4 Notes

1. For mononucleated cell cultures, significant debris will be present in tissue extracts. Pre-plating for the first 12 h in culture permits separation of satellite cells from debris and contaminating cell populations that adhere to tissue culture plates as satellite cells require laminin, gelatin, or collagen to adhere.
2. Removal of red blood cells from satellite cell preparations is required for cell sorting and will prevent cell sorter nozzle clogging and improve profiling and sorting. Whenever possible primary directly conjugated antibodies should be used for FACS, the use of secondary labeled antibodies is not recommended.
3. To reduce nozzle clogging during FACS isolation of satellite cells, horse serum is added to the FACS buffer. Additional measures could involve increasing the percentage of serum or treatment of cell solutions with DNAse.
4. Select myofibers that are phase bright indicating live myofibers. Avoid opaque damaged myofibers. Cultured myofibers will remain suspended and not attach to tissue culture plates.
5. Following injection of myofiber-associated satellite cells into host mouse, pull up buffer into syringe and flush out several times while viewing under a dissecting microscope to quantify the number of myofibers that were injected as myofibers can be found following injection in the syringe.

Acknowledgment

This work was supported by NIH grants R01AG040074, R01AG040074 and an Ellison Medical Senior Scholar Award to BBO.

References

1. Shi X, Garry DJ (2006) Muscle stem cells in development, regeneration, and disease. Genes Dev 20(13):1692–1708. doi:10.1101/gad.1419406
2. Mauro A (1961) Satellite cell of skeletal muscle fibers. J Biophys Biochem Cytol 9:493–495
3. Hawke TJ, Garry DJ (2001) Myogenic satellite cells: physiology to molecular biology. J Appl Physiol 91(2):534–551
4. Collins CA, Olsen I, Zammit PS, Heslop L, Petrie A, Partridge TA (2005) Stem cell function, self-renewal, and behavioral heterogeneity of cells from the adult muscle satellite cell niche. Cell 122. doi:10.1016/j.cell.2005.05.010
5. Watt DJ, Lambert K, Morgan JE, Partridge TA, Sloper JC (1982) Incorporation of donor muscle precursor cells into an area of muscle regeneration in the host mouse. J Neurol Sci 57(2–3):319–331
6. Watt DJ, Morgan JE, Partridge TA (1984) Use of mononuclear precursor cells to insert allogeneic genes into growing mouse muscles. Muscle Nerve 7(9):741–750. doi:10.1002/mus.880070908
7. Morgan JE, Watt DJ, Sloper JC, Partridge TA (1988) Partial correction of an inherited biochemical defect of skeletal muscle by grafts of normal muscle precursor cells. J Neurol Sci 86(2–3):137–147
8. Beauchamp JR, Morgan JE, Pagel CN, Partridge TA (1999) Dynamics of myoblast transplantation reveal a discrete minority of precursors with stem cell-like properties as the myogenic source. J Cell Biol 144(6):1113–1122
9. Montarras D, Morgan J, Collins C, Relaix F, Zaffran S, Cumano A, Partridge T, Buckingham M (2005) Direct isolation of satellite cells for skeletal muscle regeneration. Science 309(5743):2064–2067. doi:10.1126/science.1114758
10. Cornelison DD, Filla MS, Stanley HM, Rapraeger AC, Olwin BB (2001) Syndecan-3 and syndecan-4 specifically mark skeletal muscle satellite cells and are implicated in satellite cell maintenance and muscle regeneration. Dev Biol 239(1):79–94. doi:10.1006/dbio.2001.0416
11. Tanaka KK, Hall JK, Troy AA, Cornelison DD, Majka SM, Olwin BB (2009) Syndecan-4-expressing muscle progenitor cells in the SP engraft as satellite cells during muscle regeneration. Cell Stem Cell 4(3):217–225. doi:10.1016/j.stem.2009.01.016
12. Hall JK, Banks GB, Chamberlain JS, Olwin BB (2010) Prevention of muscle aging by myofiber-associated satellite cell transplantation. Sci Transl Med 2(57):57ra83. doi:10.1126/scitranslmed.3001081
13. Marg A, Escobar H, Gloy S, Kufeld M, Zacher J, Spuler A, Birchmeier C, Izsvák Z, Spuler S (2014) Human satellite cells have regenerative capacity and are genetically manipulable. J Clin Invest 124(10):4257–4265. doi:10.1172/JCI63992
14. Gilbert PM, Havenstrite KL, Magnusson KE, Sacco A, Leonardi NA, Kraft P, Nguyen NK, Thrun S, Lutolf MP, Blau HM (2010) Substrate elasticity regulates skeletal muscle stem cell self-renewal in culture. Sci Signal 329(5995):1078
15. Gussoni E, Wang Y, Fraefel C, Miller RG, Blau HM, Geller AI, Kunkel LM (1996) A method to codetect introduced genes and their products in gene therapy protocols. Nat Biotechnol 14(8):1012–1016. doi:10.1038/nbt0896-1012
16. Seale P, Sabourin LA, Girgis-Gabardo A, Mansouri A, Gruss P, Rudnicki MA (2000) Pax7 is required for the specification of myogenic satellite cells. Cell 102(6):777–786
17. Fukada S, Uezumi A, Ikemoto M, Masuda S, Segawa M, Tanimura N, Yamamoto H, Miyagoe-Suzuki Y, Takeda S (2007) Molecular signature of quiescent satellite cells in adult skeletal muscle. Stem Cells 25(10):2448–2459

Chapter 13

Simultaneous Measurement of Mitochondrial and Glycolytic Activity in Quiescent Muscle Stem Cells

James G. Ryall

Abstract

Cellular metabolism has recently been identified as an important regulator of cell identity, with several adult stem cell populations having been observed to undergo a shift in metabolism during important changes in cell state, such as during the shift from quiescence to proliferation. In this chapter, a method to characterize the metabolism of quiescent skeletal muscle stem cells is presented. This technique will allow for the comparison of quiescent muscle stem cells isolated from two or more different mouse models.

Key words Metabolism, Satellite cells, Glycolysis, Oxidative phosphorylation, Mitochondria

1 Introduction

Skeletal muscle possesses an incredible capacity for repair and regeneration following injury, a result conferred on this tissue by a specialized population of skeletal muscle stem cells (MuSCs, also termed satellite cells) [1, 2]. In healthy adult individuals, MuSCs exist in a quiescent state outside the cell cycle, termed G_0. A similar state of quiescence has been observed in several tissue-specific adult stem cell populations, including hematopoietic stem cells and hair-follicle stem cells among others [3].

Over the last decade a significant amount of research has focused on the regulation of stem cell quiescence, as any disruption to this important cell state can lead to depletion and/or exhaustion of the stem cell pool [4–6]. Recently, cellular metabolism has been identified as an important regulator of changes in cell state (quiescence-to-activation) and lineage progression [7, 8]. The study of stem cell energetics has exploded over the last decade, with a change in stem cell metabolism leading to a change in cell identity now termed "metabolic reprogramming" [9, 10]. Coupling of advanced cell sorting techniques with downstream analyses has greatly enhanced our understanding of the quiescent state [7, 11–13].

Eusebio Perdiguero and DDW Cornelison (eds.), *Muscle Stem Cells: Methods and Protocols*, Methods in Molecular Biology, vol. 1556, DOI 10.1007/978-1-4939-6771-1_13, © Springer Science+Business Media LLC 2017

The introduction of the Seahorse Bioscience extracellular flux (XF) bioanalyzer in 2006 has allowed for the simultaneous measurement of oxygen consumption rates (OCR, a measure of mitochondrial oxidative phosphorylation activity) and extracellular acidification rates (ECAR, a measure of glycolysis) in live cells, tissues, and organisms [14]. The further development of assays designed to exploit several metabolic activators/inhibitors has led to the ability to investigate several aspects of mitochondrial and glycolytic kinetics [15]. Using freshly isolated quiescent MuSCs, the method presented in this chapter is designed to allow the user to measure basal metabolism (mitochondrial and glycolytic activity), mitochondrial bioenergetics (oxygen consumed for the production of ATP and the maintenance of the mitochondrial proton gradient; maximum mitochondrial activity; spare respiratory capacity; and, non-mitochondrial respiration), and maximal (and spare) glycolytic activity.

2 Materials

2.1 MuSC Seeding

1. Freshly isolated MuSCs in ice-cold media (*see* **Note 1**).
2. Centrifuge (with plate adapters).
3. Seahorse XF24 cell culture microplate.
4. OPTIONAL: Cell Tak or matrigel (*see* **Note 2**).

2.2 XF Assay Media

1. Seahorse XF assay media (*see* **Note 3**).
2. 45 % D-glucose solution.
3. 100 mM sodium pyruvate.
4. 2 M hydrochloric acid (HCl) and 2 M sodium hydroxide (NaOH) for adjusting pH.

2.3 Analysis of Mitochondrial/Glycolytic Kinetics

1. Seahorse XF24Bioanalyzer (*see* **Note 4**).
2. Seahorse XF24 sensor cartridge.
3. Seahorse XF calibrant solution.
4. Seahorse XF assay media (*see* **Note 3**).
5. 25 mM oligomycin (*see* **Note 5**).
6. 25 mM carbonyl cyanide 4-(trifluoromethoxy)phenylhydrazone (FCCP, *see* **Note 5**).
7. 50 mM rotenone (*see* **Note 5**).
8. 25 mM antimycin A (*see* **Note 5**).
9. 2-deoxy-D-glucose (2-DG, *see* **Note 6**).

2.4 Post-assay Cell Counts

1. Trypsin-EDTA (0.25 %).
2. Phosphate buffered saline containing 2 % horse serum (PBS+2 % HS).
3. A cell counter.

3 Methods

Analysis of cellular metabolism during quiescence can be conducted on any adult stem cell population where cell isolation and purification can be completed without inducing cell activation. For MuSCs the recommended protocol is to use fluorescence activated cell sorting, as previously described [7]. It should be possible to isolate greater than 1.5×10^5MuSCs from the hindlimb musculature of an adult C57BL/6 male mouse. The isolation of MuSCs must be completed on the same day as the assay (so that analysis is complete while MuSCs are still in a state of quiescence).

3.1 Hydrating the Sensor Cartridge

1. This step should be completed the day prior to the assay (*see* **Note 7**). Open one Seahorse XF24 sensor cartridge and add 750 μl of the supplied calibrant solution to each well of the calibrant plate.
2. Replace the (green) XF24 sensor cartridge and lid (making sure the notches in the bottom left of the calibrant plate, sensor cartridge and lid match up) and wrap in parafilm.
3. Place the XF24 sensor cartridge in a non-CO_2 incubator at 37 °C overnight.

3.2 Designing the Assay Protocol

1. Prior to running an assay on the Seahorse XF24 bioanalyzer, it is necessary to design the assay protocol and configure the plate layout. As the plate layout is assay/user dependent it will not be presented here, although users are recommended to keep at least three wells blank for background normalization.
2. Utilizing the XF24 assay wizard, enter the assay protocol as indicated in Fig. 1.

3.3 Preparation of XF Assay Media and Mitochondrial/Glycolytic Inhibitors

1. MuSCs should be isolated and sorted on the day of the assay. This can be completed on the morning of the assay so the cells are ready to be analyzed in the afternoon. The XF assay media and drug solutions described in this section can be made while the cells are sorting.
2. Aliquot 49 ml of XF assay media into a 50 ml falcon tube and add 500 μl D-glucose (final concentration 25 mM) and 500 μl sodium pyruvate (final concentration 1 mM). Recap the falcon tube and place in a water bath set to 37 °C for at least 30 min.
3. Once the assay media has reached 37 °C remove the falcon tube and measure the pH of the solution (*see* **Note 8**). Carefully add HCl or NaOH as needed to obtain a pH of 7.40 ± 0.02.
4. Decant 20 ml of the assay media into 4 × 5 ml aliquots in 15 ml falcon tubes and replace the remaining assay media in the 37 °C water bath. Label each 15 ml falcon tube as "A," "B," "C," or "D."

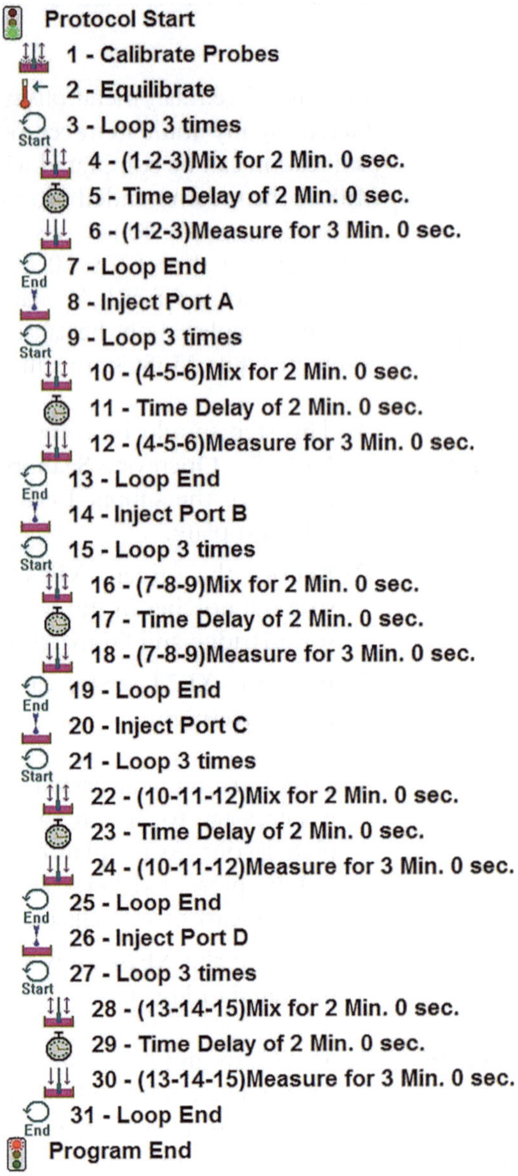

Fig. 1 Protocol design for the Seahorse assay, using XF24 software

5. Add 4 μl of oligomycin to tube A; 1 μl of FCCP to tube B; and 2 μl of rotenone and 4 μl of Antimycin A to tube C (*see* **Note 9**). Create a 1 M 2-DG solution by adding 2-DG directly to tube D (*see* **Note 6**). These tubes provide a 10× concentrate that will be added to each of the four Seahorse XF sensor plate ports. The final working concentration for each compound (following port injection) is as follows: 2 μM oligomycin; 500 nM FCCP; 1 μM rotenone; 1 μM antimycin A; and, 100 mM 2-DG.
6. Remove the hydrated sensor plate from the non-CO_2 incubator and remove the parafilm and plastic lid. Ensuring the green

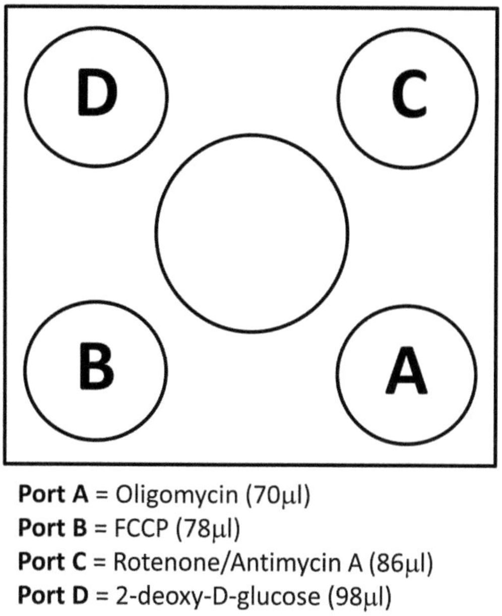

Fig. 2 Port layout for each well in an XF24 sensor cartridge. Before loading the ports, ensure the sensor cartridge is in the correct orientation. The optical probe is lowered into the center hole of the sensor cartridge, do not let any solution leak into this area

sensor plate is correctly aligned (notch in the bottom left) add the diluted compound from tubes A–D to every port as follows: port A, add 70 μl tube A; port B, add 78 μl tube B; port C, add 86 μl tube C; port D, add 96 μl tube D (*see* Fig. 2).

7. Start the equilibration and calibration phase of the assay by selecting "Start Assay" on the Seahorse computer and carefully placing the sensor cartridge (and calibrant plate) onto the loading tray, making sure the barcode on the sensor cartridge faces toward the back of the machine, and click "Start."

3.4 Seeding of Freshly Isolated MuSCs

1. Immediately following isolation of MuSCs, the total cell number should be accurately determined via cell counting (either via manual or automated techniques).
2. Once the total number of cells has been determined, centrifuge the MuSCs (5 min at 540 × *g*) and carefully remove the suspension buffer.
3. Slowly add ~1 ml of warmed XF assay media per 1×10^5 MuSCs and centrifuge for 5 min at 540 × *g*.
4. Carefully remove the XF assay media (without disturbing the cell pellet) and resuspend the cells (with slow trituration) in 630 μl warmed XF assay media per 7.5×10^4MuSCs.
5. Add 630 μl of the resuspended MuSCs to the desired number of wells in an XF24 cell plate by angling the tip of the pipette

toward the opposite wall of the well and ejecting the cells in a single constant motion. Ensure to leave at least three wells empty for background readings.

6. Immediately spin the XF24 cell plate on a centrifuge equipped with swing bucket plate holders (5 min at 200 × *g*). Following centrifugation, confirm even cell seeding on an inverted microscope (*see* **Note 10**).
7. Remove the calibration plate from the Seahorse Bioanalyzer and replace with the XF24 cell plate containing the MuSCs and start the measurement phase of the assay.

3.5 Post-assay Analysis

1. At the completion of the assay remove the XF cell plate and sensor cartridge from the Bioanalyzer and discard the used sensor cartridge.
2. Transfer the XF assay media from each well to a labeled 15 ml falcon tube, and then add 100 μl pre-warmed (37 °C) Trypsin-EDTA to each well of the XF cell plate. Following 5 min of incubation, remove the trypsin (and cells) from each well and transfer to the labeled tubes containing the XF assay media.
3. Centrifuge all 15 ml falcon tubes for 5 min at 540 × *g* and remove the supernatant. Resuspend the cell pellet in 100 ml of PBS+2 % HS and count the total number of cells in each well (either via manual or automated techniques). Oxygen consumption and extracellular acidification rates can be corrected to absolute cell numbers (pmol/min/10^5 cells and mpH/min/10^5 cells).
4. By calculating the area under the curve following the addition of each compound, it is possible to characterize several measure of mitochondrial and glycolytic function (Fig. 3a, b). In addition, by plotting the mean basal OCR (Y-axis) against ECAR (X-axis) it is possible to generate a metabolic phenogram (Fig. 3c). These analyses will allow a detailed investigation of cellular metabolism during quiescence and the comparison of metabolism following in vivo transgenic or pharmacologic manipulations.

4 Notes

1. MuSCs isolated via FACS are generally sorted in FACS buffer consisting of PBS + 2 % horse serum, and kept on ice. This buffer is appropriate for cell seeding if the experimenter wishes to let the cells settle and attach to the XF cell plate without centrifugation. In this case, cells should be left for several hours to attach.
2. While not essential for seeding of quiescent MuSCs, users may wish to pre-coat the XF cell plate with a thin layer of either Cell

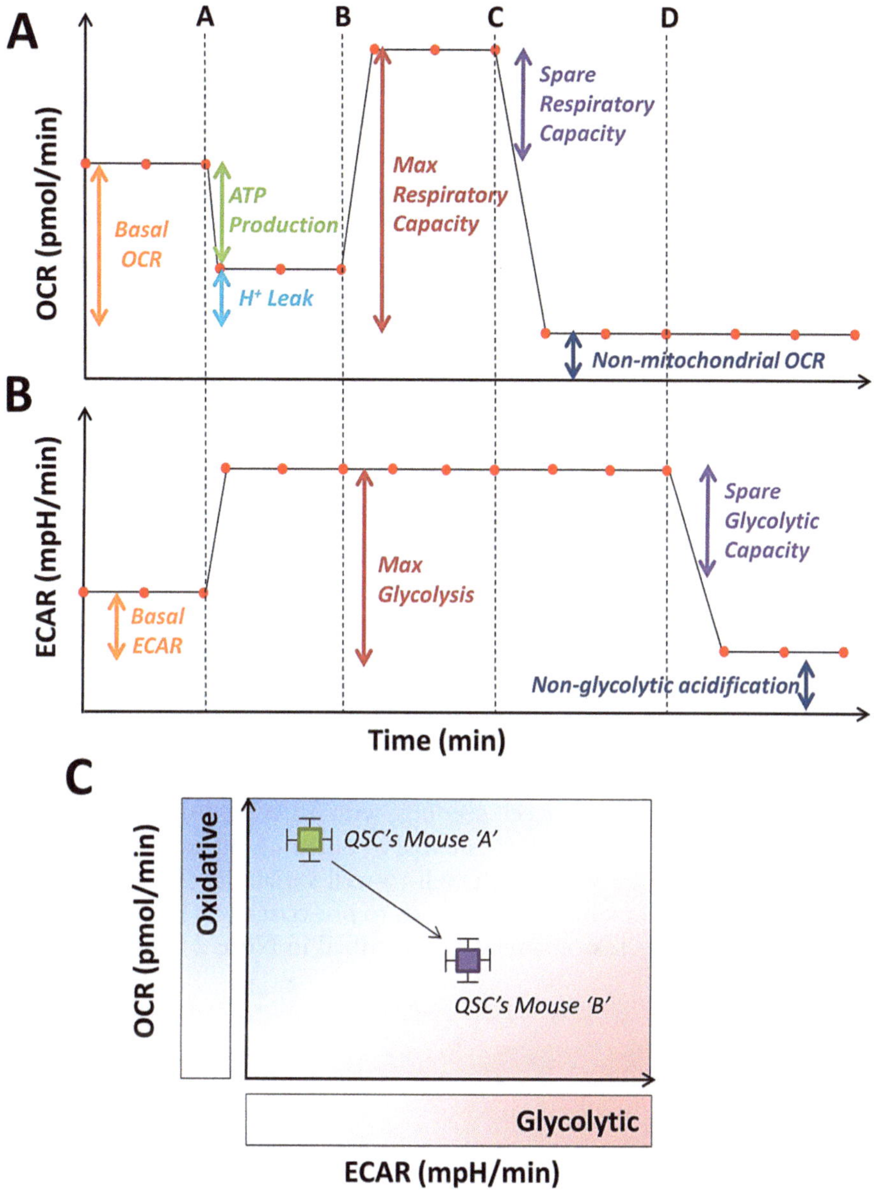

Fig. 3 (**a, b**) Representative Seahorse traces of the OCR and ECAR data generated from the described protocol. By determining the area under the curve of the indicated segments, it is possible to obtain values for several aspects of mitochondrial and glycolytic activity. (**c**) A metabolic phenogram can be constructed to directly compare the metabolic phenotype of two or more populations of quiescent MuSCs (QSCs) by plotting basal OCR against basal ECAR

Tak or matrigel. However, if a coating is to be applied, ensure that it does not contain serum or a buffering agent (as this will interfere with the measurement of OCR and ECAR).

3. While this protocol calls for the use of XF assay media, this can be substituted with XF base media which does not contain

GlutaMAX. This can be useful if you wish to replace GlutaMAX with L-Glutamine).

4. The protocol described has been optimized for the XF24; however, it should be possible to adapt cell numbers and volumes to the XF96 (this should be optimized by individual users). Additionally, users of XF*e* machines may find that cell numbers can be reduced to obtain similar values.
5. While Seahorse Bioscience sells these compounds as a kit, users may prefer to purchase each compound individual and make concentrated stock solutions by dilution in DMSO. Concentrated stocks should be stored at −20 °C in aliquots.
6. 2-DG should be diluted directly in Seahorse assay media, to a concentration of 1 M.
7. If necessary, calibration can be completed on the day of the assay. We have incubated flux plates for as little as 4 h with no observable decrease in recordings (although overnight calibration is preferable).
8. Typically at this point the XF assay media will be slightly acidic (pH: 7.00–7.25) and requires the addition of a small volume of NaOH.
9. Following the addition of the compounds to tubes A–D it is important to readjust the pH of each solution to 7.40 ± 0.02.
10. Correct cell seeding, with an even distribution of cells across each well, is essential to obtain good data. Uneven seeding will result in high well-to-well variability. To improve cell attachment users may wish to pre-coat the XF24 cell plate with Cell-Tak or similar (as described in **Note 2**).

References

1. Koopman R, Ly CH, Ryall JG (2014) A metabolic link to skeletal muscle wasting and regeneration. Front Physiol 5:32
2. Yin H, Price F, Rudnicki MA (2013) Satellite cells and the muscle stem cell niche. Physiol Rev 93(1):23–67
3. Cheung TH, Rando TA (2013) Molecular regulation of stem cell quiescence. Nat Rev Mol Cell Biol 14(6):329–340
4. Shea KL et al (2010) Sprouty1 regulates reversible quiescence of a self-renewing adult muscle stem cell pool during regeneration. Cell Stem Cell 6(2):117–129
5. Fukada S et al (2007) Molecular signature of quiescent satellite cells in adult skeletal muscle. Stem Cells 25(10):2448–2459
6. Fukada S et al (2011) Hesr1 and Hesr3 are essential to generate undifferentiated quiescent satellite cells and to maintain satellite cell numbers. Development 138(21): 4609–4619
7. Ryall JG et al (2015) The NAD(+)-dependent SIRT1 deacetylase translates a metabolic switch into regulatory epigenetics in skeletal muscle stem cells. Cell Stem Cell 16(2):171–183
8. Shyh-Chang N et al (2013) Lin28 enhances tissue repair by reprogramming cellular metabolism. Cell 155(4):778–792
9. Folmes CD, Terzic A (2014) Metabolic determinants of embryonic development and stem cell fate. Reprod Fertil Dev 27(1):82–88
10. Shyh-Chang N, Daley GQ (2015) Metabolic switches linked to pluripotency and embryonic stem cell differentiation. Cell Metab 21(3): 349–350
11. Cheung TH et al (2012) Maintenance of muscle stem-cell quiescence by microRNA-489. Nature 482(7386):524–528

12. Liu L et al (2013) Chromatin modifications as determinants of muscle stem cell quiescence and chronological aging. Cell Rep 4(1):189–204
13. Pallafacchina G et al (2010) An adult tissue-specific stem cell in its niche: a gene profiling analysis of in vivo quiescent and activated muscle satellite cells. Stem Cell Res 4(2):77–91
14. Ferrick DA, Neilson A, Beeson C (2008) Advances in measuring cellular bioenergetics using extracellular flux. Drug Discov Today 13(5–6):268–274
15. Nicholls DG et al (2010) Bioenergetic profile experiment using C2C12 myoblast cells. J Vis Exp 46

Chapter 14

Monitoring Autophagy in Muscle Stem Cells

Laura García-Prat, Pura Muñoz-Cánoves*, and Marta Martínez-Vicente*

Abstract

Autophagy is critical not only for the cell's adaptive response to starvation but also for cellular homeostasis, by acting as quality-control machinery for cytoplasmic components. This basal autophagic activity is particularly needed in postmitotic cells for survival maintenance. Recently, basal autophagic activity was reported in skeletal muscle stem cells (satellite cells) in their dormant quiescent state. Satellite cells are responsible for growth as well as for regeneration of muscle in response to stresses such as injury or disease. In the absence of stress, quiescence is the stem cell state of these cells throughout life, although which mechanisms maintain long-life quiescence remains largely unknown. Our recent findings showed that autophagy is necessary for quiescence maintenance in satellite cells and for retention of their regenerative functions. Importantly, damaged organelles and proteins accumulated in these cells with aging and this was connected to age-associated defective autophagy. Refueling of autophagy through genetic and pharmacological strategies restored aged satellite cell functions, and these finding have biomedical implications. In this chapter, we describe different experimental strategies to evaluate autophagic activity in satellite cells in resting muscle of mice. They should facilitate our competence to investigate stem cell functions both during tissue homeostasis as in pathological conditions.

Key words Stem cell, Satellite cell, Skeletal muscle, Autophagy, Quiescence, Aging

1 Introduction

All organisms undergo continuous tissue and cell renovation. This, in part, occurs to replace old components with fresh, better-quality ones. Such "cellular renovation" requires synthesis of new constituents but also degradation of pre-existing, often damaged materials, which can serve as building blocks for "the novo synthesis" [1]. Autophagy is a process whereby intracellular cytosolic components including both macromolecules, such as proteins, glycogens, lipids and nucleotides, and organelles, such as mitochondria, peroxisomes, ribosomes, and endoplasmic reticulum among others, are degraded within the lysosome [2–8]. Lysosomes are cellular organelles that contain acid hydrolases (proteases, lipases,

*These authors contributed equally to this work.

Eusebio Perdiguero and DDW Cornelison (eds.), *Muscle Stem Cells: Methods and Protocols*, Methods in Molecular Biology, vol. 1556, DOI 10.1007/978-1-4939-6771-1_14, © Springer Science+Business Media LLC 2017

nucleases, etc.) that break down macromolecules internalized inside this organelle. Once the macromolecules are degraded, their essential components can be transferred back to the cytosol for the building of new cellular material.

In mammals there are three types of autophagy: macroautophagy, microautophagy, and chaperone-mediated autophagy. Macroautophagy, which will be simply referred to here as autophagy, is mediated by the autophagosome. A small portion of the cytoplasm is engulfed by an isolated membrane or phagophore, leading to the formation of a double-membrane vesicle called autophagosome. Then, the autophagosome fuses with the lysosome, giving rise to an autolysosome, and materials inside the autophagosome are degraded by lysosomal hydrolases. Thus, autophagy is considered a dynamic process that comprises the following sequential steps: initiation, nucleation of the autophagosome membrane, recognition, and trapping of the substrate, transport of the loaded autophagosomes toward the lysosomes, fusion of both vesicles, and ultimately degradation of the cargo [9] (Fig. 1).

Although autophagy can be a nonselective degradation pathway where intracellular components are randomly engulfed, under some conditions and in response to certain signals, autophagy is a highly selective process by recognizing and eliminating components like mitochondria (mitophagy), peroxisomes (pexophagy) [10], ER portions (reticulophagy) [11], ribosomes(ribophagy) [12], lipid droplets (lipophagy) [13], pathogens (xenophagy) [14], or ubiquitinated-aggregates (aggrephagy) [15]. Most of these

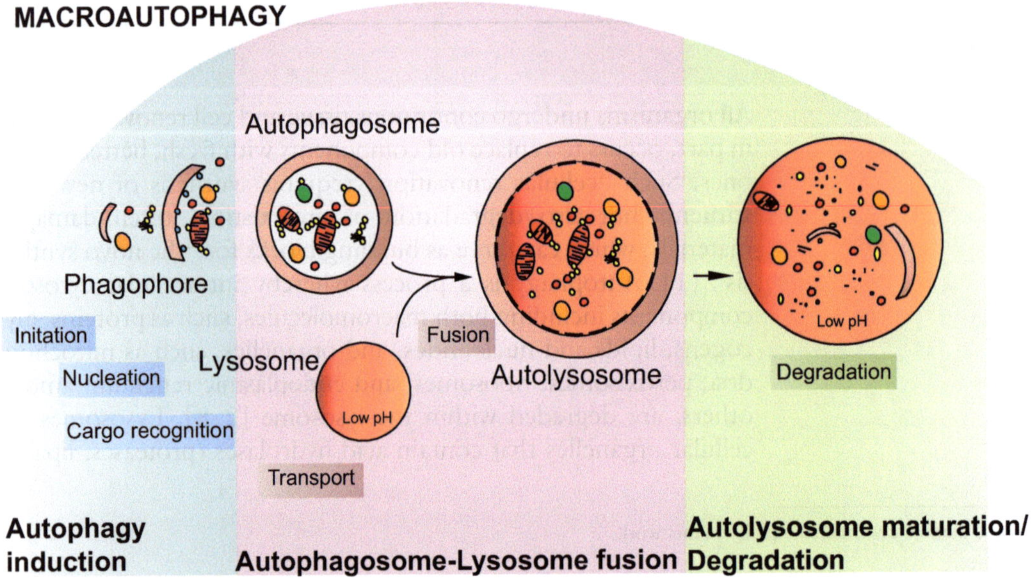

Fig. 1 The process of autophagic flux

pathways require the action of autophagy adaptors, like p62, that act as linkage between the autophagosome (interacting with LC3-II proteins) and the substrate (binding to its ubiquitin chain) [16–18].

The most fundamental and evolutionary conserved role of autophagy is the supply of amino acids in the "adaptation response" to starvation conditions for new protein synthesis and energy production [19, 20]. However, under conditions of stress, such as mitochondrial damage, ER stress, hypoxia, redox stress, accumulation of aggregated proteins, or pathogen-associated molecular patterns (PAMPs), autophagy might also be induced as part of the cellular stress-response mechanism [8].

Growing evidence points to the importance of basal autophagy, operating constitutively at low rate, even under a nutrient-rich environment, and to its key role in global turnover of cellular components including organelles. Consistent with this, constitutive autophagy acts as the quality-control machinery for cytoplasmic components, being crucial for cellular homeostasis. The importance of constitutive autophagy is specially critical in postmitotic cells-like neurons, which require an effective and basal autophagy activity to maintain cell homeostasis and survival [21, 22]. More recently, quiescent muscle stem cells (also called *satellite cells*) have also been shown to require constitutive autophagy to maintain their stemness status [23]. In this context, autophagy has been demonstrated to operate in two different scenarios. In the first one, a recent study reported that suppression of autophagy inhibits activation of young satellite cells from quiescence upon injury [24]. The authors proposed that autophagy is induced by Sirt1 during satellite cell activation to provide the nutrients necessary to meet bioenergetic demands during the transition from quiescence to activation [24]. Of note, this study proposed a relative lack of nutrient availability to induce autophagy during the satellite cell activation phase, mimicking starvation-induced autophagy, a process necessary for cellular adaptation to nutritional stress, similar to what was described in hematopoietic stem cells [25]. In a second scenario, we have demonstrated that young quiescent satellite cells (in resting muscle) display a basal autophagy flux, and that this basal activity preserves cell integrity and fitness over time. Through Atg7 (one of the essential genes for the formation of the autophagosomes) genetic deletion in young satellite cells, we have shown that constitutive autophagy, which declines during aging, functions as a cytoprotective and cellular quality control mechanism to balance protein and organelle homeostasis, which is essential to maintain both the population of satellite cells and their functional properties. These findings reflect the relevance of autophagy in stem cells homeostasis regulation.

Because of the low number of quiescent satellite cells, the peculiarities of their isolation in a resting state, and their particular morphology (very low proportion of cytoplasm), the study of the autophagic process in these stem cells and the consequences of its impairment, is therefore challenging. In this chapter, we review a variety of methods developed to study autophagy in quiescent muscle stem cells.

2 Materials

2.1 Isolation of Quiescent Satellite Cells by Fluorescence-Activated Cell Sorting (FACS)

DMEM (Dulbecco's Modified Eagle Medium) (Biowest, L0101-500)

Penicillin/Streptomycin (Biowest, L0022-100)

Fetal Bovine Serum (FBS) (Capricorn Scientific, FBS-12A)

Filters 100, 70, and 40 μM (SPL Lifescience, 93100/93070/93040)

Lysis Buffer (BD Pharm lyse, 555899)

FACS Buffer: PBS 1×, 5 %Goat Serum (Gibco, 16210-064)

Antibodies:

- PE/Cy7 anti-mouse/human Ly-6A/E (Sca-1) (Biolegend, 108114).
- PE/Cy7 anti-mouse/human CD45 (Biolegend, 101216).
- α-7 integrin R-Phycoerythrin (AbLab, 53-0010-05).
- Alexa Fluor 647 Rat anti-mouse CD34 (BD Pharmingen, 560230).

Digestion mix (four limb muscles of one mouse require 100 ml of digestion mix):

- Collagenase D, 0.8 % final concentration (Roche, 11088882001).
- Trypsin 2.5 %, 0.125 % final concentration (Gibco, 15090-046).
- DMEM, 1 %P/S.

DAPI, stock solution 1 mg/ml, final concentration 1 μg/ml (Invitrogen, D1306)

2.2 Autophagic Flux Determination

2.2.1 Turnover of LC3-II by Western Blotting of Isolated Quiescent Satellite Cells

FACS Buffer: PBS 1×, 5 % Goat Serum (Gibco, 16210-064)

Antibodies:

- PE/Cy7 anti-mouse/human Ly-6A/E (Sca-1) (Biolegend, 108114).
- PE/Cy7 anti-mouse/human CD45 (Biolegend, 101216).
- α-7 integrin R-Phycoerythrin (AbLab, 53-0010-05).
- Alexa Fluor 647 Rat anti-mouse CD34 (BD Pharmingen, 560230).

Bafilomycin A1, 10 nM final concentration (Sigma, B1793)

DMSO (Dimethyl sulfoxide) (Sigma, D2650)

Immunoprecipitation (IP) Buffer:

- 50 mM Tris–HCl.
- 150 mM NaCl.
- 1 % NP-40.
- 5 mM EGTA.
- 5 mM EDTA.
- 20 mM NaF.

IP Working Soultion (for 10 ml):

- One tablet of protease inhibidor (CompleteMini, Roche, 11836153001).
- 50 μl phosphatase inhibitor Cockatil 1 (Sigma, P2850).
- 50 μl phosphatase inhibitor Cocktail 2 (Sigma, P5726).
- NaVanadate, 0.1 mM final concentration.
- PMSF, 1 mM final concentration.
- Glicerophosphate, 10 mM final concentration.

Resolving gel 12 % (two gels):

- 8.6 ml ddH20.
- 6 ml 40 % acrylamide.
- 3 ml 1.5 M Tris–HCl, pH 8.8.
- 200 μl 10 % SDS.
- 200 μl 10 % APS.
- 8 μl TEMED.

Stacking gel 5 % (two gels):

- 5.9 ml ddH_20.
- 1 ml 40 % acrylamide.
- 1 ml 1.0 M Tris–HCl, pH 6.8.
- 80 μl 10 % SDS.
- 80 μl 10 % APS.
- 8 μl TEMED.

Running Buffer (pH 8.3):

- 0.025 M Tris–HCl.
- 0.192 M Glycine.
- 0.1 % SDS.

Transfer Buffer:

- 0.025 M Tris.
- 0.192 M Glycine.
- 20 % M ethanol.

Blocking Solution: 5 % milk, TBS-T

Antibody solution: 5 % BSA, TBS-T

PVDF membrane (Amersham Hybon PVDF, GE Healthcare LifeScience, 10600023)

Tris buffered sali.e (TBS, 10×): 1.5 M NaCl, 0.1 M Tris–HCl, pH 7.4.

TBS-T: TBS 1×, 0.05 % Tween

5× Laemmli Buffer:

- 100 mM Tris pH 6.8.
- 32 % Glycerol.
- 2 % SDS.
- 0.05 % β-mercaptoethanol.
- 0.1 % Bromopehnol Blue.
- ddH_20.

Antibodies:

- anti-LC3 (Novus Biologicals NB100-2331).
- Anti-Tubulin (Sigma, T-6199).
- Polyclonal Goat anti-rabbit Immunoglobulins/HRP (Dako, P0448).
- Alexa fluor 680 Goat anti-mouse IgG (Invitrogen, A21058).
- anti-p62/SQSTM1 antibody produced in rabbit (Sigma, P0067).

ECL reagent (Amersham ECL Western Blotting Detection Reagents, GE Healthcare LifeScience, RPN2209)

X ray-film (Amersham Hyperfilm ECL, GE Healthcare LifeScience, 28906837)

Odyssey Blocking Buffer (PBS) (LI-COR, 927-40000)

2.2.2 GFP-LC3 Fluorescence Microscopy of Isolated Quiescent Satellite Cells

FACS Buffer: PBS 1×, 5 % Goat Serum (Gibco, 16210-064)

Antibodies:

- PE/Cy7 anti-mouse/human Ly-6A/E (Sca-1) (Biolegend, 108114).
- PE/Cy7 anti-mouse/human CD45 (Biolegend, 101216).
- α-7 integrin R-Phycoerythrin (AbLab, 53-0010-05).
- Alexa Fluor 647 Rat anti-mouse CD34 (BD Pharmingen, 560230).

Bafilomycin A1, 10 nM final concentration (Sigma, B1793)

DMSO (Dimethyl sulfoxide) (Sigma, D2650)

Eight-well glass slides (Thermo Scientific, 177402)

Poly L-lysin (Sigma, P8920)

PBS 1×, 0.02 % azide

Fixation: 4 % PFA

Mowiol

2.2.3 Immunostaining Analysis of Freshly Isolated Quiescent Satellite Cells

Blocking Solution: 10 % Goat serum (Gibco, 16210-064), 10 % FBS (Capricorn Scientific, FBS-12A) in PBS 1×

Washes: PBS 1×, 0.01 % Tween (Sigma, P1379)

Permeabilization: PBS 1×, 0.5 % Triton (Sigma, T8787)

DAPI, stock solution 1 mg/ml, final concentration 1 μg/ml (Invitrogen, D1306)

Antibodies:

- anti-p62/SQSTM1 antibody produced in rabbit (Sigma, P0067).
- mouse monoclonal antibody to LC3 (NanoTools, 5F10).
- LAMP-1 (Santa Cruz Biotechnology, sc-19992).

2.2.4 Immunostaining Analysis of GFP-LC3 Quiescent Satellite Cells in Tissue Sections

Cassettes 15 × 15 × 5 mm (Tissue-Tek Cryomold, Sakura)

OCT (Tissue-Tek, Sakura)

Blocking Solution 1: 1/50 MOM mouse IgG blocking reagent (Vector, MKB-2213), 0.01 % Triton, PBS 1×

Blocking Solution 2: 10 % Goat Serum, 10 % FBS, PBS 1×

Washes: PBS 1×, 0.01 % Triton (Sigma, T8787)

Permeabilization: PBS 1×, 0.5 % Triton (Sigma, T8787)

Mowiol

Fixation: 4 % PFA

DAPI, stock solution 1 mg/ml, final concentration 1 μg/ml (Invitrogen, D1306)

2.2.5 Flow Cytometry Analysis of GFP-LC3 Quiescent Satellite Cells

FACS Buffer: PBS 1×, 5 % Goat Serum (Gibco, 16210-064)
Antibodies:

- PE/Cy7 anti-mouse/human Ly-6A/E (Sca-1) (Biolegend, 108114).
- PE/Cy7 anti-mouse/human CD45 (Biolegend, 101216).
- α-7 integrin R-Phycoerythrin (AbLab, 53-0010-05).
- Alexa Fluor 647 Rat anti-mouse CD34 (BD Pharmingen, 560230).

Bafilomycin A1, 10 nM final concentration (Sigma, B1793)

DMSO (Dimethyl sulfoxide) (Sigma, D2650)

DAPI, stock solution 1 mg/ml, final concentration 1 μg/ml (Invitrogen, D1306)

2.2.6 Fluorescence Microscopy Analysis of Cell Expressing the mRFP-GFP-LC3 Plasmid

Growth medium: HAM'S F-12 (Biowest, l0140-500), 20 % FBS (Capricorn Scientific, FBS-12A), 1 % L-Glutamine (Biowest, X0550-100), 1 % P/S (Biowest, L0022-100)

Bafilomycin A1, 10 nM final concentration (Sigma, B1793)

DMSO (Dimethyl sulfoxide) (Sigma, D2650)

DAPI, stock solution 1 mg/ml, final concentration 1 μg/ml (Invitrogen, D1306)

Coating Solution: 0.1 mg/ml of Collagen (Becton Dickinson, 354236), 0.02 N acetic acid (Sigma, 320099)

Fixation: 4 % PFA

Mowiol

Lipofectamine 3000 (Life Technologies, L3000001)

mRFP-GFP-LC3 plasmid[26]

Eight-well glass slides (Thermo Scientific, 177402)

2.3 Transmission Electron Microscopy (TEM)

Fixation: 2 % paraformaldehyde/2.5 % glutaraldehyde in phosphate buffer (0.1 M; pH 7.4)

2.4 Transcriptional Regulation

q-RT-PCR

RNeasy Micro kit (Qiagen, 74004)

SuperScript III Reverse Transcriptase (Invitrogen 1674043)

LightCycler 480 System using Light Cycler 480 SYBR Green I Master reaction mix (Roche Diagnostic Corporation)

Primers (Sigma) (*see* Table 1)

Global gene expression analysis of quiescent satellite cells

RNeasy Micro kit (Qiagen, 74004)

Agilent SurePrint G3 Mouse GE 8 × 60 K (Agilent Technologies, G4852A)

2.5 Autophagy Modulation: (In Vivo and In Vitro Treatments)

2.5.1 Activation

Rapamycin (LC Laboratories, R-5000)

Spermidine (Sigma, S2626)

LV-Atg7 (ref)

Lentivirus infection:

- Retronectin (Takara, T100A).
- Hank's Balanced Salt Solution (HBSS) (Gibco, 14025-050).
- Bovine Serum Albumins (BSA) (Sigma, A7906).
- Hepes (Sigma, H3375).

2.5.2 Inhibition

Bafilomycin A1 (Sigma, B1793)

Chloroquine (Sigma, C-6628)

E64d (Sigma, E3132)

Table 1
Primer sequences used for real-time PCR of autophagy-related mouse genes

Gene	Forward primer	Reverse primer
Atg7	TCTGGGAAGCCATAAAGTCAGG	GCGAAGGTCAGGAGCAGAA
Map1lc3	TTATAGAGCGATACAAGGGGGAG	CGCCGTCTGATTAT
Beclin1	ATGGAGGGGTCTAAGGCGTC	TCCTCTCCTGAGTTAGCCTC
Bnip3l	TTGGGGCATTTTACTAACCTTG	TGCAGGTGACTGGTGGTACTAA
Atg12	TTCGCTCCACAGCCCATTTC	TCCCCGGAACGAGGAACTC
Atg4b	ATTGCTGTGGGGTTTTTCTG	AACCCCAGGATTTTCAGAGG
Cathepsin	GTGGACTGTTCTCACGCTCAAG	TCCGTCCTTCGCTTCATAGG
p62	CCCAGTGTCTTGGCATTCTT	AGGGAAAGCAGAGGAAGCTC
Bnip3	TTCCACTAGCACCTTCTGATGA	GAACACCGCATTTACAGAACAA
Vps34	TGTCAGATGAGGAGGCTGTG	CCAGGCACGACGTAACTTCT
Gabarapl1	GGACCACCCCTTCGAGTATC	CCTCTTATCCAGATCAGGGACC

Pepstatin A (Sigma, P-5318)

Leupeptin (Sigma, L2884)

2.6 Mouse Models Useful for Monitoring Autophagy in Quiescent Satellite Cells

Tamoxifen (Sigma, T5648)

Corn Oil

3 Methods

3.1 Isolation of Quiescent Satellite Cells by FACS

Muscle stem cells (satellite cells) are located surrounding the basal lamina and outside the myofiber plasma membrane in a quiescent state. These small stem cells were first described in adult muscle in 1961 by Alexander Mauro [27], using electron microscopy. This technique also revealed other morphological attributes: large nuclear-to-cytoplasmic ratio, few organelles, and condensed interphase chromatin. This morphology is in harmony with the notion that most satellite cells, in healthy unstressed muscles, are mitotically quiescent (arrested in the G0 phase of the cell cycle) and transcriptionally inactive. The steps for isolation of these cells by FACS (*see* Fig. 2a) are described below:

1. Muscles from fore and hind limbs are collected in cold DMEM 1 %P/S into 50 ml falcon tubes.
2. Decant all the muscles collected in a petri dish and remove DMEM 1 %P/S completely.

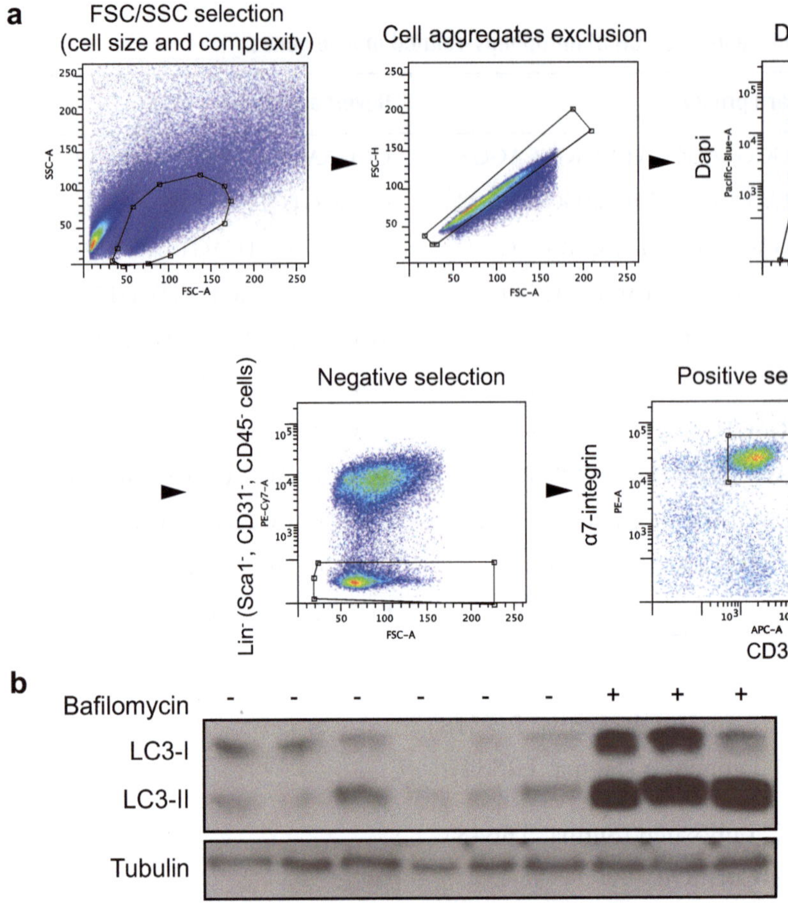

Fig. 2 (**a**) Representative example of the FACS strategy and gating scheme to isolate quiescent satellite cells from resting muscles of wild type (WT) mice. (**b**) LC3 and Tubulin protein quantity by western blot analysis of young WT satellite cells treated with Bafilomycin A1 or vehicle for 4 h prior to analysis

3. Mince muscles with scissors.
4. Mince muscles further with razor blades.
5. Collect minced muscles into a 50 ml falcon tube and add cold DMEM 1 %P/S. Leave muscle sediment and remove DMEM 1 %P/S, discarding floating fat pieces. Repeat this step to further clean the sample from non-muscle pieces.
6. In the last wash, remove DMEM 1 %P/S as much as possible and split the minced muscle into two 50 ml falcon tubes.
7. Add 10 ml of the prepared digestion mix (collagenase/trypsin) to each tube.
8. Incubate 25 min at 37 °C in a shaking water bath.
9. At the end of the digestion, leave the tube for 5 min on ice to let the sample sediment.
10. Transfer the digestion supernatant to a new 50 ml falcon tube on ice.

11. Add 10 ml of the digestion mix again and incubate 25 min at 37 °C.
12. Collect the new digestion supernatant and pool it with the supernatant from the first digestion round.
13. **Steps 6–12** are repeated two more times.
14. If some pieces of muscle still remain, decant the sample on a petri dish for a new mincing process until no more are seen, and repeat one round of digestion. At the end of the fourth round of digestion, all muscle tissue should be digested.
15. Add 5 ml of fetal bovine serum (FBS) to each 50 ml falcon tube to block the digestion.
16. Filter the supernatant with 100 and 70 μM filters.
17. Centrifuge the supernatant 10 min at 50*g* and at 4 °C.
18. Collect the supernatant and discard the pellet (optional: the pellet can be washed and the supernatant collected and pooled with the previous supernatants).
19. Centrifuge at 670×g for 15 min at 4 °C, repeat three times. The supernatant is discarded at each round and the pellet is resuspended gently in cold DMEM 1 %P/S.
20. After three centrifugations at 1700 rpm, pellets from both tubes are pooled together in cold DMEM 1 %P/S, and passed through a 40 μM filter.
21. Centrifuge at 1700 rpm for 15 min at 4 °C.
22. Resuspend the pellet in 3 ml of 1× Lysis Buffer and incubate at 4 °C for 10 min (protect from light). Before centrifugation add cold DMEM 1 %P/S up to 50 ml.
23. Centrifuge at 1700 rpm for 15 min at 4 °C.
24. Discard the supernatant and resuspend the pellet in 1 ml of cold DMEM 1 %P/S.
25. Count the number of cells for each sample.
26. Centrifuge at 1700 rpm for 15 min at 4 °C and resuspend the pellet at 1×10^4cells/μl (1×10^6 cells in 100 μl) in FACS Buffer.
27. Staining with antibodies:
    (a) Negative selection: Sca1-PECy7 and CD45-PECy7 (0.5/100 μl FACS Buffer)
    (b) Positive selection: α-7 integrin-PE (1/100 μl FACS Buffer) and CD34-APC (1.5/100 μl FACS Buffer)
    (c) Controls: Single staining and FMO controls are required to set up the gates
28. Incubate the cells with antibodies for 20 min at 4 °C, protected from light.

29. Centrifuge at 1700 rpm for 15 min at 4 °C.
30. Discard the supernatant and resuspend the cell bulk in 1 ml of FACS Buffer for sample sorting.
31. Add DAPI (final concentration 1 μg/ml) 5 min prior FACS to detect and exclude dead cells.
32. Collect Sca1$^-$/CD45$^-$/CD34$^+$/α7-integrin$^+$ satellite cells in Eppendorf tubes with 500 μl of FACS Buffer at 4 °C.

3.2 Autophagic Flux Determination

The term "autophagic flux" refers to the dynamic process of autophagy, including all the steps described before (*see* Fig. 1). The simple determination of numbers of autophagosomes is insufficient for an overall estimation of autophagic activity. Given that the autophagosome is an intermediate structure in a dynamic pathway, the number of autophagosomes observed at any specific time point depends on the balance between the rate of their generation and the rate of their conversion into autolysosomes [28]. Thus, accumulation of autophagosomes may represent either autophagy induction or, alternatively, suppression (dysfunction) of steps in the autophagy pathway downstream of autophagosome formation (as observed in satellite cells of advanced age) (Laura García-Prat, *in press*). The main biological autophagy marker used is the microtubule-associated protein LC3-II. LC3-I is normally located in the cytoplasm but is cleaved and lipidated by phosphatidylethanolamine when incorporated into the autophagosome inner leaflet of the membrane as LC3-II [29]. Thus, for autophagic flux determination, it is important to use inhibitors of autophagosome clearance to discern between increased autophagy or impaired autophagosome clearance. Currently, several compounds are used. In particular, we use Bafilomycin A1 (a vacuolar H^+-ATPase inhibitor) (*see* Fig. 3a)—but other lysosomal inhibitors can be used, like chloroquine and NH_4Cl (both impairing lyosomal acidification) or protease inhibitors (like pepstatin A, E64d, or leupeptin).

Another widely used autophagy marker is p62, also called sequestrome 1 (SQSTM1), which binds directly LC3 and GABARAP (Atg8 orthologues) proteins via a short LC3 interaction region (LIR). This may serve as a mechanism to deliver selective autophagic cargo for degradation in lysosomes. The p62 protein is itself degraded by autophagy, acting as a marker of cargo degradation and is, consequently, useful to study autophagic flux as a complementary technique to LC3-II turnover [17, 19, 30, 31]. Thus, when autophagy is activated, p62 levels decrease, while autophagy inhibition leads to p62 accumulation [29].

3.2.1 LC3-II Turnover Analysis by Western Blotting

1. After **step 29** of the protocol for isolation of quiescent satellite cells, transfer and divide the bulk of cells (already stained with the antibodies) into two Eppendorf tubes (500 μl/each).
2. Add FACS Buffer up to 1 ml.

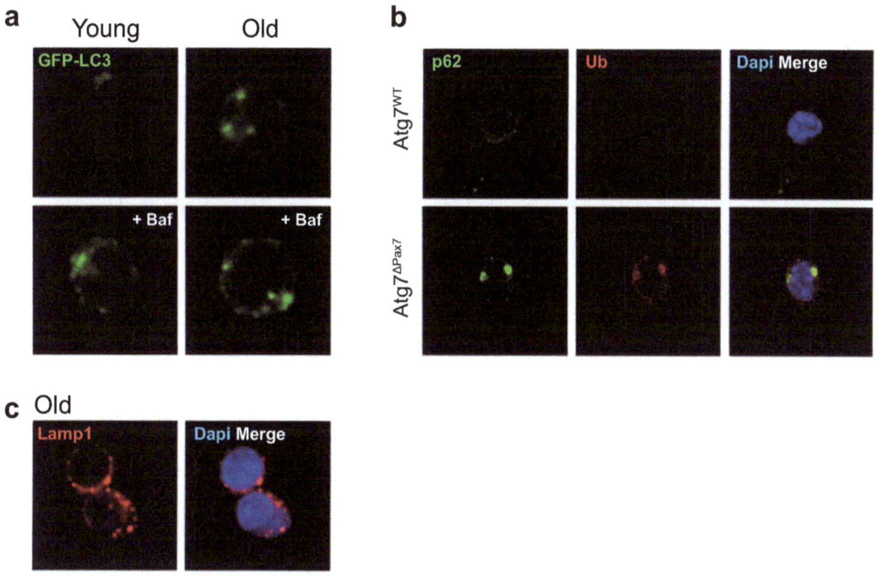

Fig. 3 (**a**) GFP-LC3 autophagosomes in quiescent satellite cells. Z projections. (**b**) p62 and Ub immunostaining in quiescent satellite cells freshly isolated from $Atg7^{WT}$ and $Atg7^{\Delta Pax7}$ mice. (**c**) LAMP-1 immunostaining in quiescent satellite cells freshly isolated from old mice

3. Treat the cells with BafilomycinA1 (10 nM) or DMSO in FACS Buffer prior to FACS analysis for 4 h at 37 °C in the incubator.
4. Add DAPI (final concentration 1 μg/ml) 5 min prior FACS to detect and exclude dead cells.
5. Collect 200,000 satellite cells ($Sca1^-/CD45^-/CD34^+/\alpha7$-integrin$^+$) from each sample in Eppendorf tubes with 500 μl of FACS Buffer at 4 °C.
6. Centrifuge at 14,000 × *g* for 5 min at 4 °C.
7. Remove the supernatant as much as possible. At this step, samples can be stored at −80 °C.
8. Add 100 μl of IP Working Solution in each Eppendorf tube for cell lysis.
9. Ensure complete cell lysis by incubating suspensions for 45 min at 4 °C with shaking.
10. Remove remaining cell debris by centrifuging at 14,000 × *g* for 15 min at 4 °C. The supernatant contains the proteins and can be quantified using the standard Bradford Assay. However, since quiescent satellite cells contain small amounts of proteins, and since the same number of satellite cells has been used for each sample, the protein quantification step can be omitted.
11. Prepare samples for gel loading: 40 μl of each sample with 5× Laemmli Buffer (10 μl).

12. Incubate samples for 5 min at 95 °C.
13. Prepare the stacking gel at 5 % and the resolving gel at 12 %, use 1.5 mm gel and 10-well comb to be able to load 50 μl of sample.
14. Load the maximum sample volume (around 50 μl).
15. Run the samples at 100 V with Running Buffer.
16. Transfer at 100 V for 1 h, using PVDF membrane and Transfer Buffer.
17. Block membrane with Blocking Solution for 1 h at RT. The membrane can be divided into different parts by cutting it horizontally at different levels, to allow Western blotting of proteins with different molecular weights, by using different antibodies.
18. Incubate O/N with the primary antibody (anti-LC3) in Antibody Solution. Dilution 1/200.
19. Wash 3× with TBS-T.
20. Incubate with the secondary HRP antibody for 2 h at RT, protected from light.
21. Wash 3× with TBS-T.
22. Detect HRP using ECL reagent and X-ray film. Ensure that the exposures are in the linear range.
23. Wash 3× with TBS-T.
24. Incubate O/N with the primary antibody used as loading control (Tubulin) in Antibody Solution.
25. Wash 3× with TBS-T.
26. Incubate with infrared secondary antibody for 2 h at RT, protected from light. Dilution 1/4000 in Odyssey Blocking Buffer (PBS).
27. Wash 3× with TBS.
28. Odyssey infrared imager (LI-COR) is used for antibody detection.
29. Quantify the bands using imageJ. LC3-I is usually detected on a gel at a molecular mass around 16kD, and LC3-II at approximately 14 kD (*see* Fig. 2b).

Note: Which is the best indicator of autophagy? The amount of LC3-II, the LC3-II/LC3-I ratio, or LC3-II/(LC3-I+LC3-II) ratio is now used. Levels of LC3-II should be compared preferably to tubulin or any other housekeeping protein instead of LC3-I.

3.2.2 GFP-LC3 Fluorescence Microscopy of Isolated Quiescent Satellite Cells

1. After **step 29** of the protocol for quiescent satellite cell isolation from muscles of GFP-LC3 mice, transfer and divide the bulk of cells (already stained with the antibodies) into two Eppendorf tubes (500 μl on each).
2. Add FACS Buffer up to 1 ml.

3. Treat the cells with BafilomycinA1 (10 nM) or DMSO in FACS Buffer prior to FACS analysis for 4 h at 37 °C in the incubator.
4. Add DAPI (final concentration 1 μg/ml) 5 min prior FACS to detect and exclude dead cells.
5. Collect $Sca1^{-}/CD45^{-}/CD34^{+}/\alpha 7$-integrin^{+} satellite cells in Eppendorf tubes with 500 μl of FACS Buffer at 4 °C.
6. Coat eight-wells glass slide with Poly L-lysin 30 min at room temperature.
7. Remove Poly L-lysin (it can be reused).
8. Add cell suspension (10,000–15,000 cells/well). Ensure that cell suspension is evenly distributed in the well; if necessary, add PBS1×.
9. Spin the slides 5 min at 50 × *g*.
10. Remove supernatant from each well.
11. Add 4 % PFA (200 μl) in each well. Incubate 10 min at RT.
12. Remove 4 % PFA and perform two washes with PBS1×. At this point, slides can be stored at 4 °C with PBS1× 0.02 % azide.
13. Mount slides with mowiol.
14. Analyze fluorescence in images obtained with Leica SPE confocal laser scanning microscope and Superresolution microscopy using stimulated emission depletion (STED), Leica TCS SP5 STED.
15. The number and percentage of cellular area occupied by GFP-LC3 puncta can be determined on digital images with Fiji and the cell image analysis software CellProfiler.

Note: Quiescent satellite cells have low cytoplasm content, and autophagosomes are hard to discern in such small cytoplasm. For this reason, high objectives with good resolution are necessary. In addition, doing several z-sections can improve their visualization. One needs to be able to see the increased number of autophagosomes in Bafilomycin-treated samples (*see* Fig. 3a). Video resconstructions of autophagosomes can be generated in Imaris software using full confocal z-stacks of each satellite cell. Z-stacks must be previously imported to Fiji software for background adjustments and then deconvolved using the blind-deconvolution wizard of Huygens software. Eventually, endogenous LC3 might also be detected by immunofluorescence with antibodies against LC3, but the sensitivity is much lower and detection in most of the cases is not feasible. Whenever possible it is recommended to use GFP-LC3 mice.

3.2.3 Immunostaining Analysis of Quiescent Satellite Cells

We can combine GFP-LC3 fluorescence analysis of quiescent satellite cells with other autophagy markers to assess for their colocalization (Fig. 3). For instance, impairment of autophagy in several cell types is accompanied by accumulation of p62 and Ubiquitin aggregates that stain positive for LC3 (Fig. 3b). In addition, LAMP-1 (lysosome marker) can be used to study the different phases of the autophagy process (Fig. 3c). Co-localization between p62 or ubiqitin and GFP-LC3 indicates a selective cargo of autophagy recognized by an autophagosome. Co-localization of GFP-LC3 with LAMP-1 (always in the presence of lysosomal inhibitors to avoid LC3-II degradation) is a marker of autolysosomes.

1. After **step 12** of the previous protocol, perform the following steps.
2. Permeabilize cells with PBS1× 0.5 % Triton for 15 min at RT.
3. Wash 3× with PBS1×.
4. Add Blocking Solution for 30 min at RT.
5. Incubate with primary antibody in Blocking Solution O/Nat 4 °C.
6. Wash 3× with PBS 1×, 0.01 %Tween
7. Incubate with secondary antibody in PBS 1×, 0.01 %Tween 1 h 30 min at RT (seat in the dark).
8. Wash 3× PBS 1×, 0.01 %Tween.
9. Add DAPI (final concentration 1 μg/ml) for 5 min at RT.
10. Wash 3× PBS 1×, 0.01 %Tween.
11. Mount with mowiol.
12. Analyze fluorescence, by doing confocal images taken using Leica SPE confocal laser scanning microscope system and Superresolution microscopy using stimulated emission depletion (STED), Leica TCS SP5 STED.

Note: Co-localizations can be determined on digital images Fiji, according to [32], with respect to the total cellular area. The Pearson's coefficient (r) is used to analyze the correlation of the intensity values of green and red pixels in dual-channels images. This coefficient measures the strength of the linear relationship between the intensities in two images calculated by linear regression and ranges from 1 to −1, with 1 standing for complete positive correlation.

3.2.4 Immunostaining Analysis of GFP-LC3 Quiescent Satellite Cells in Tissue Sections

1. Collect muscles from GFP-LC3 mice and put them in cassettes with OCT.
2. Freeze the muscles in isopentane cooled with liquid nitrogen and stored at −80 °C until analysis.
3. Cut the muscle samples in 10 μm sections with a cryostat.
4. Fix muscle sections with 4 % PFA 12 min on ice.
5. Wash 3× with PBS1× for 5 min.
6. Permeabilize with PBS 1×, 0.5 % Triton for 20 min at RT.

7. Wash 3× with PBS1× for 5 min.
8. Add Blocking Solution 1 for 1 h at RT.
9. Add Blocking Solution 2 for 20 min at RT.
10. Incubate primary antibody in Blocking Solution 2, O/N at 4 °C.
11. Wash 3× with PBS 1×, 0.01 % Triton for 5 min.
12. Incubate secondary antibody in Blocking Solution 2, for 2 h at RT (protected from light).
13. Wash 3× with PBS1×, 0.01 % Triton for 5 min.
14. Add DAPI (final concentration 1 μg/ml) for 5 min at RT.
15. Wash 3× with PBS 1×, 0.01 % Triton for 5 min.
16. Mount with mowiol.
17. Analyze fluorescence in images obtained with Leica SPE confocal laser scanning microscope system and Superresolution microscopy using stimulated emission depletion (STED) Leica TCS SP5 STED. Acquisition is performed using Leica Application or LAS AF software (Leica).

Note: For analysis of quiescent satellite cells in muscle section, we need to use a satellite cell marker, such as Pax7, for its identification (Fig. 4a). Analysis of autophagosomes in quiescent satellite cells on muscle sections is very difficult because of the small cytoplasm and the localization adjacent to the myofiber (Fig. 4a). This analysis would be an additional/complementary approach, but not a definitive one. Once more, to improve our analysis, we can use other antibodies as autophagy markers, such as LAMP-1, p62, and ubiquitin.

3.2.5 Flow Cytometry Analysis of GFP-LC3 Quiescent Satellite Cells

1. After **step 29** of the protocol for quiescent satellite cell isolation from muscles of GFP-LC3 mice, transfer and divide the bulk of cells (already stained with the antibodies) into two Eppendorf tubes (500 μl on each).
2. Add FACS Buffer up to 1 ml.
3. Treat the cells with BafilomycinA1 (10 nM) or DMSO in FACS Buffer prior to FACS analysis for 4 h at 37 °C in the incubator.
4. Add DAPI (final concentration 1 μg/ml) for 5 min prior FACS to detect and exclude dead cells.
5. Analyze GFP-LC3 fluorescence in FACS LSR Fortesa (Becton Dickinson) of 10,000 satellite cells ($Sca1^-/CD45^-/CD34^+/$ $\alpha7$-integrin$^+$) for each sample.
6. GFP-LC3 fluorescence signal is achieved by determining the mean fluorescence intensity (MFI) of the whole histogram signal for satellite cells and compared to the corresponding negative control sample (no GFP-LC3) in an overlaid histogram [33].

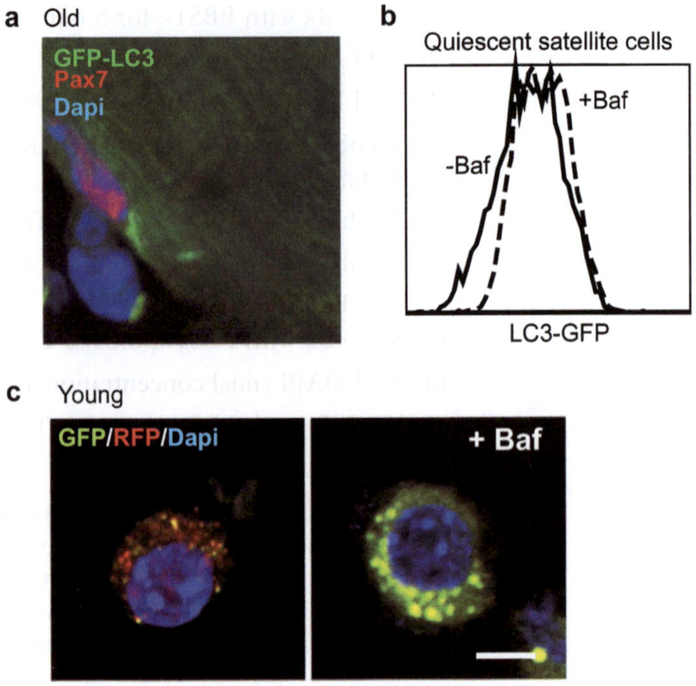

Fig. 4 (**a**) Pax7 and GFP immunostaining on tissue sections from resting TA muscles of old GFP-LC3 mice. (**b**) Autophagy flux analyzed by flow cytometry in quiescent satellite cells isolated by FACS from young mice. Satellite cells were treated with Bafilomycin A1 (+Baf) or vehicle for 4 h prior to analysis. (**c**) mRFP-GFP-LC3 plasmid was transfected into young satellite cells, for detecting autophagosomes (*yellow signal*) and their maturation into autolysosomes (*red signal*), and treated with Bafilomycin A1 or vehicle for 4 h prior to cell fixation. Three z projections are shown

7. For autophagy flux determination, relative changes in linear scaled MFI are compared between samples with or without Bafilomycin A1 (*see* Fig. 4b).

Note: MFI refers to the fluorescence intensity of each event (in average) of the selected cell population, in the chosen fluorescence channel.

3.2.6 Fluorescent Microscopy Analysis of mRFP-GFP-LC3 Plasmid

Measuring autophagy flux through this method is based on the concept of lysosomal quenching of GFP. GFP is a stably folded protein and relatively resistant to lysosomal proteases. However, the low pH inside the lysosome quenches the fluorescent signal of GFP, which makes it difficult to trace the delivery of GFP-LC3 to lysosomes. In contrast, RFP exhibits more stable fluorescence in acidic compartments, and mRFP-LC3 can readily be detected in autolysosomes. By exploiting the difference in the nature of these two fluorescent proteins (i.e., lysosomal quenching of GFP fluorescence versus lysosomal stability of RFP fluorescence), autophagic flux can be morphologically traced with an mRFP-GFP-LC3 tandem construct [26] (*see* Fig. 4c).

1. After **step 31** of the protocol for quiescent satellite cell isolation, satellite cells ($Sca1^-/CD45^-/CD34^+/\alpha7$-integrin$^+$) are collected in Eppendorf tubes with 500 μl of Growth Medium (GM).
2. Coat eight-well glass slides with Coating Solution (0.1 mg/ml collagen in sterile water and 0.02 N acetic acid).
3. Plate sorted satellite cells in coated glass slides and culture them in GM.
4. After 4 days in culture, transfect satellite cells with mRFP-GFP-LC3 plasmid [34] using Lipofectamine 3000.
5. After 48 h of plasmid transfection, add Bafilomycin A1 (10 nM) or DMSO in the culture medium and incubate cells for 4 h at 37 °C.
6. Fix cells with 4 %PFA for 10 min at RT.
7. Wash 3× with PBS1×.
8. Stain cell nuclei with DAPI (final concentration 1 μg/ml) in PBS1× for 5 min at RT.
9. Wash 3× with PBS1×.
10. Mount with mowiol.
11. Analyze fluorescence in images obtained with Leica SPE confocal laser scanning microscope system and Superresolution microscopy using stimulated emission depletion (STED) Leica TCS SP5 STED.
12. Co-localization of mRFP-LC3 and GFP-LC3 puncta is determined on the maximum projection of three Z-sections using Fiji automated macro pipeline calculating single and double-positive autophagosomes (Fig. 4c).

Note: With this tandem construct, autophagosomes and autolysosomes are labeled with yellow (i.e., mRFP and GFP) and red (i.e., mRFP only) signals, respectively (Fig. 4c). Results can be expressed as the percentage of yellow or red puncta from total (yellow + red) puncta. In Bafilomycin-treated samples, yellow puncta should be increased.

3.3 Transmission Electron Microscopy (TEM)

TEM is a valid and important method for both qualitative and quantitative analyses of changes in various autophagic structures. Autophagosomes have a double membrane and contain cytosol and/or organelles that look morphologically intact, while autolysosomes usually have only one limiting membrane, and contain cytoplasmic material and/or organelles at various stages of degradation. In quiescent satellite cells, as explained above, images with very high resolution will be required for proper analysis of the presence of autophagosomes/autolysosomes due to the reduced size and cytoplasmic cell content.

1. Intact muscles are collected and fixed with 2 % paraformaldehyde/2.5 % glutaraldehyde in phosphate buffer (0.1 M, pH 7.4), as fast as possible.
2. Process sample for TEM analysis following standard procedures.
3. Images were acquired using a Jeol 1010 microscope, working at 80 kv and equipped with a CCD Megaview III camera.
4. Identification of satellite cells in skeletal muscle by electron miscoscopy was based on cell size, content of heterochromatin, and position with respect to basal lamina.

3.4 Transcriptional Regulation

Analysis of the transcriptional regulation of the autophagy gene network can also provide valuable information about possible alterations in the autophagic system. We chose a list of autophagic-related genes involved in different steps of the autophagic machinery, from autophagic initiation to lysosomal clearance.

3.4.1 RT–qPCR

1. After **step 31** of the protocol for quiescent satellite cell isolation, satellite cells (Sca1$^-$/CD45$^-$/CD34$^+$/α7-integrin$^+$) are collected in Eppendorf tubes with 500 μl of FACS Buffer.
2. Centrifuge Eppendorf tubes at 14,000 × *g* for 5 min.
3. Remove supernatant.
4. Perform total RNA extraction using RNeasy Micro kit following manufacturer's protocol.
5. Complementary DNA (cDNA) is synthesized from total RNA using SuperScript III Reverse Transcriptase according to manufacturer's protocol.
6. Real-time PCR reactions are performed on a LightCycler 480 System using Light Cycler 480 SYBR Green I Master reaction mix and specific primers.
7. Thermocycling conditions: initial step of 10 min at 95 °C, then 50 cycles of 15 s denaturation at 94 °C, 10 s annealing at 60 °C, and 15 s extension at 72 °C.
8. Reactions must be run in triplicate, and automatically detected threshold cycle (Ct) values are compared between samples.
9. Transcripts of the ribosomal protein L7 or GAPDH housekeeping genes can be used as endogenous control, with each unknown sample normalized to L7 or GAPDH content.
10. Primers of autophagy-related genes (mouse) (Table 1).

3.4.2 Global Gene Expression Analysis of Quiescent Satellite Cells

1. After **step 31** of the protocol for quiescent satellite cell isolation, satellite cells (Sca1$^-$/CD45$^-$/CD34$^+$/α7-integrin$^+$) are collected Eppendorf tubes with Lysis Buffer.
2. Extract total RNA using RNeasy Micro kit according to manufacturer's protocol.
3. Analyze transcriptome with Agilent SurePrint G3 Mouse GE 8 × 60 K high-density microarray slides.
4. Raw data must be taken from the Feature Extraction output files and corrected for background noise using the normexp method.
5. Normalize data using cyclic loess, and identify differentially expressed genes using AFM 4.0 [35] for all pairwise comparisons.
6. Use quantile normalization to assure comparability across samples.
7. Analyze differential expression analysis on noncontrol probes with an empirical Bayes approach on linear models (limma).
8. Results must be corrected for multiple testing according to the False Discovery Rate (FDR) method.
9. Statistical analysis can be done with the Bioconductor project (http://www.bioconductor.org/) in the R statistical environment.
10. Gene ontology analysis functional annotation using DAVID [36].
11. Venn diagrams can be generated using BioVenn [37].

3.5 Autophagy Modulation: In Vivo and In Vitro Treatments

Autophagy is induced as a response to both extracellular stress conditions (nutrient deprivation, hypoxia, and oxidative stress) and intracellular stress conditions (endoplasmic reticulum stress, accumulation of damaged organelles, and aggregation of proteins). The large number of stimuli able to trigger autophagy implies the involvement of multiple signaling pathways in autophagosome formation.

3.5.1 Activation

Autophagy upregulation may have therapeutic benefits in a range of diseases. New research related to autophagy activators has become a hot topic owing to their potential clinical value [38].

Pharmacological

1. Rapamycin is a natural product with potent antifungal and immunosuppressive activities. It forms a complex with the immunophilin FK506-binding protein 12 (FKBP12), which then stabilizes the raptor-mTOR association and inhibits the kinase activity of mTOR. As an inhibitor of mTOR, rapamycin has been widely reported in the literature to induce autophagy both in vivo and in vitro [39].

For in vivo induction of autophagy in quiescent satellite cells:

Inject mice intraperitonally with Rapamycin (4 mg/kg) or vehicle every day for 2 weeks.

For in vitro induction of autophagy in satellite cells:

Add Rapamycin (100 ng/ml final concentration) or vehicle (DMSO) in cell media for 48 h.

2. Spermidine is a polyamine compound found in citrus fruit and soybean, which has recently been shown to increase lifespan of yeast, nematodes, and flies in an autophagy-dependent manner (Einsberg et al. 2009). Spermidine acts as an acetylase inhibitor that stimulates autophagy independently of SIRT1 and mTOR in human and yeast as well as in nematodes [39, 40].

   For in vivo induction of autophagy in quiescent satellite cells:

   Treat mice with Spermidine (3 mM final concentration) in drinking water for 2 weeks. Water should be changed every 2 days.

   For in vitro induction of autophagy in satellite cells:

   Add Spermidine (5 μM) or vehicle (DMSO) in cell media for 48 h.

Note: To assess autophagy induction, Bafilomycin A1 treatment (4 h at 37 °C) is necessary to observe the increase in autophagosome accumulation.

Genetic: LV-Atg7

The autophagy-related genes and their products are named ATG and Atg, respectively. Once the pagophore has been formed, the membrane structure expands to sequester materials to form autophagosome; this process is mediated by two ubiquitin-like conjugation systems, the Atg12-Atg5 and Atg8 conjugation systems. Of these Atgs, Atg7 is an E1-like activating enzyme that activates Atg12 for its conjugation with Atg5 as well as the Atg8 family proteins for their conjugation with phosphatidylethanolamine. Thus, Atg7 is essential for autophagosome formation [41].

Ex vivo infection of freshly isolated quiescent satellite cells with LV-ATG7 [42]

1. After **step 31** of the protocol for quiescent satellite cell isolation, satellite cells ($Sca1^-/CD45^-/CD34^+/\alpha7$-integrin$^+$) are collected in Eppendorf tubes with 500 μl of Growth Medium (GM).
2. Coat petri dishes for cell culture with Coating Solution (0.1 mg/ml of Collagen in sterile water and 0.02 N acetic acid).
3. Plate sorted satellite cells in coated petri dishes and culture them with GM.

4. Centrifuge petri dishes at 50 × *g* for 5 min.
5. After 22 h, coat new petri dishes with Retronectin (10 μg/ml) for 2 h at 37 °C.
6. Recover Retronectin and wash with 2 % BSA for 30 min at RT.
7. Wash with HBSS, 2.5 % Hepes 1 M.
8. Transfer satellite cells from Collagen to Retronectin-coated dishes.
9. Add LV-Atg7[42] or LV-control (10^5 pfu, final concentration) in cell media.
10. Centrifuge petri dishes at 50 × *g* for 5 min.
11. Culture cells with Lentivirus O/N.
12. Change cell media.
13. At that point, infected satellite cells can be transplanted in vivo or cultured in vitro for new experiments.

Note: To assess rapidly autophagy induction by LV-ATG7 infection, GFP-LC3 satellite cells can be infected, and the increase in GFP-LC3 fluorescence can be monitored by flow cytometry between samples with or without Bafilomycin A1 treatment. Alternatively, in other cell types, LV-Beclin 1 has also been used as a genetic activator of autophagy [43, 44].

3.5.2 Inhibition

Autohagy could potentially be suppressed at any stage of autophagic flux. To do so, many chemical inhibitors have been identified and used in various cells and animal models. However, most chemical inhibitors of autophagy are not entirely specific, and caution should be taken when interpreting the findings obtained with the use of these compounds, especially regarding their dose and incubation time [38].

Pharmacological

1. Bafilomycin A1. Vacuolar-type H (+)-ATPases (V-ATPases) are found within the membranes of many organelles including lysosomes, endosomes, and secretory vesicles, where they play a variety of roles crucial for organelle function. Bafilomycin A1 is a specific inhibitor of V-ATPases in cells, which inhibits the acidification of lysosomes and endosomes and consequently the activity of their enzymes that are active at acidic pH. Bafilomycin A1 was reported to prevent maturation of autophagic vacuoles by inhibiting the fusion between autophagosomes and lysosomes; however, contradictory results have also appeared; nevertheless, this alternative effect does not affect the interpretation of the autophagic flux [38].

   Bafilomycin A1 is used at 10 nM for 4 h at 37 °C.

2. Chloroquine. Chloroquine is a lysosomal lumen alkalizer and is used to block the autophagic progress by impairing lysosomes [38].
   Cholorquine is used at 50 mg/ml for 4 h at 37 °C.
3. Protease inhibitors. The lysosome is the ultimate degradative autophagic compartment in the cell. Protease inhibitors block autophagy at the step of degradation of cytoplasm enclosed in lysosomes, and cause accumulation of autolysosomes and/or many cytoplasmic inclusions in the central vacuoles. Leupeptin is a naturally occurring protease inhibitor that inhibits cysteine, serine, and threonine peptidase.

   Lysosomal cathepsins, which are enclosed in lysosomes, help maintain the homeostasis of the cell's metabolism by participating in the degradation of autophagic bodies. E64d and pepstatin A are two autophagy inhibitors that function by suppressing lysosomal proteases. E64d is a membrane-permeable inhibitor of cathepsin B, H, and L, whereas pepstatin A is an inhibitor of cathepsin D and E [38].

   Cocktail of protease inhibitors used: Pepstatin A 10 μM, Leupeptin 100 μM, E64d 10 μM, for 4 h at 37 °C.

3.6 Mouse Models Useful for Monitoring Autophagy in Quiescent Satellite Cells

3.6.1 GFP-LC3 Transgenic Mice

The GFP-LC3 transgenic mouse was first described by Mizushima et al. (2004) [45]. Detection of GFP-LC3 is not only simple but also generally highly specific. Specificity for GFP-LC3 puncta was previously determined by several studies. This transgenic mouse contains an enhanced GFP (EGFP)-LC3 cassette inserted between the CAG promoter (cytomegalovirus immediate-early (CMV) enhancer and chicken β-actin promoter) and the SV40 late polyadenylation signal [46]. GFP-LC3 fragment is randomly integrated into the mouse genome, but does not affect the functions of other genes [47].

For initial experiments, it is recommended to maintain GFP-LC3 mouse colony in heterozygosis, because to distinguish true GFP-LC3 signals from background autofluorescent signals, it is important to compare transgenic mice with wild-type (nonfluorescent) siblings. After finishing the initial experiments (or if littermate control is not necessary), GFP-LC3 mice can be maintained in homozygosis [46].

Note: Some cautions should be taken regarding the use GFP-LC3 transgenic mice. GFP-LC3 can be incorporated into protein aggregates independently of autophagy [48]. LC3 localization should be carefully interpreted in cells having protein aggregates or inclusion bodies, such as cells defective in autophagy machinery [21, 41].

3.6.2 Specific Deletion of Atg7 Gene in Muscle Stem Cells

1. Constitutive deletion. Atg7-floxed mouse ($Atg7^{f/f}$) was previously described in Komatsu et al., 2005 [41]. To examine the consequences of deleting Atg7 in satellite cells from the embry-

onic stage, we bred Atg7$^{f/f}$ mice with transgenic mice expressing the Cre recombinase under the control of the Pax7 promoter [49].

2. Inducible deletion. For inducible deletion of the Atg7 gene in Pax7-expressing cells (satellite cells), Atg7$^{f/f}$ mice were bred with transgenic mice expressing the Cre recombinase under the control of the Pax7 promoter, and in the presence of tamoxifen [50].

   Tamoxifen administration: one injection per day for 4 days of 5 mg/25 g body weight. (10 mg/ml in corn oil).

Acknowledgments

Work in the authors' laboratories has been supported by: MINECO, Spain SAF2012-38547, AFM, E-Rare/Eranet, Fundació Marató-TV3, MDA, EU-FP7 (Myoage, Optistem, and Endostem), and DuchennePP-NL, to PM-C; and ISCIII-FEDER, Spain (FIS-PS09/01267, FIS-PI13/02512, CP09/00184, PI14/01529) and CIBERNED to MM-V. L.G.-P. was supported by a Predoctoral Fellowship from Programa de Formación de Personal Investigador (Spain).

References

1. Mizushima N, Komatsu M (2011) Autophagy: renovation of cells and tissues. Cell 147(4):728–741
2. Cuervo AM (2004) Autophagy: many paths to the same end. Mol Cell Biochem 263(1 2):55–72
3. Levine B, Klionsky DJ (2004) Development by self-digestion: molecular mechanisms and biological functions of autophagy. Dev Cell 6(4):463–477
4. Shintani T, Klionsky DJ (2004) Autophagy in health and disease: a double-edged sword. Science 306(5698):990–995
5. Klionsky DJ (2005) Autophagy. Curr Biol 15(8):R282–R283
6. Mizushima N, Klionsky DJ (2007) Protein turnover via autophagy: implications for metabolism. Annu Rev Nutr 27:19–40
7. Klionsky DJ (2005) The molecular machinery of autophagy: unanswered questions. J Cell Sci 118(Pt 1):7–18
8. Kroemer G, Marino G, Levine B (2010) Autophagy and the integrated stress response. Mol Cell 40(2):280–293
9. Barth S, Glick D, Macleod KF (2010) Autophagy: assays and artifacts. J Pathol 221(2):117–124
10. Dunn WA Jr et al (2005) Pexophagy: the selective autophagy of peroxisomes. Autophagy 1(2):75–83
11. Bernales S, Schuck S, Walter P (2007) ER-phagy: selective autophagy of the endoplasmic reticulum. Autophagy 3(3):285–287
12. Kraft C et al (2008) Mature ribosomes are selectively degraded upon starvation by an autophagy pathway requiring the Ubp3p/Bre5p ubiquitin protease. Nat Cell Biol 10(5):602–610
13. Singh R et al (2009) Autophagy regulates lipid metabolism. Nature 458(7242):1131–1135
14. Bauckman KA, Owusu-Boaitey N, Mysorekar IU (2015) Selective autophagy: xenophagy. Methods 75:120–127
15. Lamark T, Johansen T (2012) Aggrephagy: selective disposal of protein aggregates by macroautophagy. Int J Cell Biol 2012:736905
16. Komatsu M et al (2007) Homeostatic levels of p62 control cytoplasmic inclusion body formation in autophagy-deficient mice. Cell 131(6):1149–1163
17. Pankiv S et al (2007) p62/SQSTM1 binds directly to Atg8/LC3 to facilitate degradation of ubiquitinated protein aggregates by autophagy. J Biol Chem 282(33):24131–24145

18. Lamark T et al (2009) NBR1 and p62 as cargo receptors for selective autophagy of ubiquitinated targets. Cell Cycle 8(13):1986–1990
19. Komatsu M, Ichimura Y (2010) Selective autophagy regulates various cellular functions. Genes Cells 15(9):923–933
20. Ichimura Y, Komatsu M (2010) Selective degradation of p62 by autophagy. Semin Immunopathol 32(4):431–436
21. Hara T et al (2006) Suppression of basal autophagy in neural cells causes neurodegenerative disease in mice. Nature 441(7095):885–889
22. Komatsu M et al (2006) Loss of autophagy in the central nervous system causes neurodegeneration in mice. Nature 441(7095):880–884
23. García-Prat L et al (2016) Autophagy maintains stemness by preventing senescence. Nature 529(7584):37–42
24. Tang AH, Rando TA. (2014) Induction of autophagy supports the bioenergetic demands of quiescent muscle stem cell activation. EMBO J 33(23):2782–2797.
25. Warr MR et al (2013) FOXO3A directs a protective autophagy program in haematopoietic stem cells. Nature 494(7437):323–327
26. Itoh N, Kimura T (2007) Cytokine-induced metallothionein expression and modulation of cytokine expression by metallothionein. Yakugaku Zasshi 127(4):685–694
27. Mauro A (1961) Satellite cells of skeletal fibers. JBiophysBiochemCytol 9:493–495
28. Mizushima N, Yoshimori T, Levine B (2010) Methods in mammalian autophagy research. Cell 140(3):313–326
29. Jiang P, Mizushima N (2015) LC3- and p62-based biochemical methods for the analysis of autophagy progression in mammalian cells. Methods 75:13–18
30. Komatsu M et al (2007) Constitutive autophagy: vital role in clearance of unfavorable proteins in neurons. Cell Death Differ 14(5):887–894
31. Komatsu M, Kageyama S, Ichimura Y (2012) p62/SQSTM1/A170: physiology and pathology. Pharmacol Res 66(6):457–462
32. Bolte S, Cordelieres FP (2006) A guided tour into subcellular colocalization analysis in light microscopy. J Microsc 224(Pt 3):213–232
33. Warnes G (2015) Flow cytometric assays for the study of autophagy. Methods 82:21–28
34. Kimura S, Noda T, Yoshimori T (2007) Dissection of the autophagosome maturation process by a novel reporter protein, tandem fluorescent-tagged LC3. Autophagy 3(5):452–460
35. Breitkreutz BJ et al (2001) AFM 4.0: a toolbox for DNA microarray analysis. Genome Biol 2(8):SOFTWARE0001
36. Huang da W, Sherman BT, Lempicki RA (2009) Systematic and integrative analysis of large gene lists using DAVID bioinformatics resources. Nat Protoc 4(1):44–57
37. Hulsen T, de Vlieg J, Alkema W (2008) BioVenn - a web application for the comparison and visualization of biological lists using area-proportional Venn diagrams. BMC Genomics 9:488
38. Yang YP et al (2013) Application and interpretation of current autophagy inhibitors and activators. Acta Pharmacol Sin 34(5):625–635
39. Bollon AP et al (1988) Human cytokines, tumor necrosis factor, and interferons: gene cloning, animal studies, and clinical trials. J Cell Biochem 36:353–367
40. Buttner S et al (2014) Spermidine protects against alpha-synuclein neurotoxicity. Cell Cycle 13(24):3903–3908
41. Komatsu M et al (2005) Impairment of starvation-induced and constitutive autophagy in Atg7-deficient mice. J Cell Biol 169(3):425–434
42. Crews L et al (2010) Selective molecular alterations in the autophagy pathway in patients with Lewy body disease and in models of alpha-synucleinopathy. PLoS One 5(2):e9313
43. Spencer B et al (2009) Beclin 1 gene transfer activates autophagy and ameliorates the neurodegenerative pathology in alpha-synuclein models of Parkinson's and Lewy body diseases. J Neurosci 29(43):13578–13588
44. Nascimento-Ferreira I et al (2011) Overexpression of the autophagic beclin-1 protein clears mutant ataxin-3 and alleviates Machado-Joseph disease. Brain 134(Pt 5):1400–1415
45. Mizushima N et al (2004) In vivo analysis of autophagy in response to nutrient starvation using transgenic mice expressing a fluorescent autophagosome marker. Mol Biol Cell 15(3):1101–1111
46. Mizushima N (2009) Methods for monitoring autophagy using GFP-LC3 transgenic mice. Methods Enzymol 452:13–23
47. Kuma A, Mizushima N (2008) Chromosomal mapping of the GFP-LC3 transgene in GFP-LC3 mice. Autophagy 4(1):61–62
48. Kuma A, Matsui M, Mizushima N (2007) LC3, an autophagosome marker, can be incorporated into protein aggregates independent of autophagy: caution in the interpretation of LC3 localization. Autophagy 3(4):323–328
49. Keller C et al (2004) Pax3:Fkhr interferes with embryonic Pax3 and Pax7 function: implications for alveolar rhabdomyosarcoma cell of origin. Genes Dev 18(21):2608–2613
50. Nishijo K et al (2009) Biomarker system for studying muscle, stem cells, and cancer in vivo. FASEB J 23(8):2681–2690

Part V

Cell culture, In Vitro, and In Silico Studies

Chapter 15

Mimicking Muscle Stem Cell Quiescence in Culture: Methods for Synchronization in Reversible Arrest

Reety Arora, Mohammed Rumman, Nisha Venugopal, Hardik Gala, and Jyotsna Dhawan

Abstract

Growing evidence supports the view that in adult stem cells, the defining stem cell features of potency and self-renewal are associated with the quiescent state. Thus, uncovering the molecular logic of this reversibly arrested state underlies not only a fundamental understanding of adult tissue dynamics but also hopes for therapeutic regeneration and rejuvenation of damaged or aging tissue. A key question concerns how adult stem cells use quiescence to establish or reinforce the property of self-renewal. Since self-renewal is largely studied by assays that measure proliferation, the concept of self-renewal programs imposed during non-proliferating conditions is counterintuitive. However, there is increasing evidence generated by deconstructing the quiescent state that highlights how programs characteristic of this particular cell cycle exit may enhance stem cell capabilities, through both cell-intrinsic and extrinsic programs.

Toward this end, culture models that recapitulate key aspects of stem cell quiescence are useful for molecular analysis to explore attributes and regulation of the quiescent state. In this chapter, we review the different methods used to generate homogeneous populations of quiescent muscle cells, largely by manipulating culture conditions that feed into core signaling programs that regulate the cell cycle. We also provide detailed protocols developed or refined in our lab over the past two decades.

Key words Quiescence, In vitro, G_0 phase, Stem cell, Satellite cell

1 Introduction

The defining characteristics of stem cells are captured by two primary features: self-renewal, the process by which stem cells generate more stem cells, and potency, the program whose expression allows stem cell progeny to have distinct fates. Self-renewal and potency are associated with different programs in embryonic and adult stem cells. In embryonic stem cells (ESC) both self-renewal and potency are linked to the highly proliferative state whose molecular regulation is different from the somatic cell cycle (Fig. 1) [1–3]. Consequently, any perturbation that slows the rate of cell division activates differentiation programs, and negatively impacts

Eusebio Perdiguero and DDW Cornelison (eds.), *Muscle Stem Cells: Methods and Protocols*, Methods in Molecular Biology, vol. 1556, DOI 10.1007/978-1-4939-6771-1_15, © Springer Science+Business Media LLC 2017

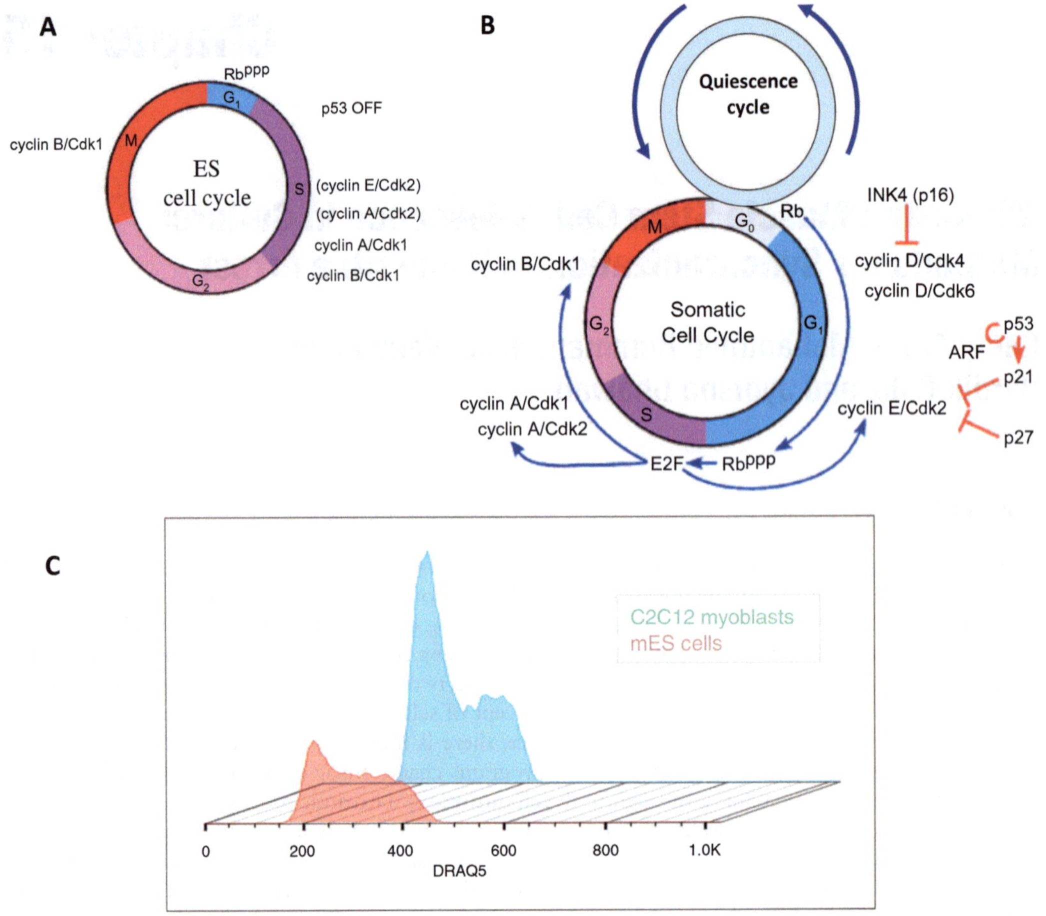

Fig. 1 Regulation of the cell cycle differs in embryonic stem cells and adult stem cells. The ESC cell cycle is characterized by a truncated G_1 phase, low Cyclin D1 expression, and persistent phosphorylation of Rb, such that entry into S phase is not a highly regulated step as it is in somatic cells. Consequently, ESC populations display a far greater proportion of cells in S and G_2/M as compared to somatic cells. The ASC cell cycle essentially follows the regulation of a typical somatic cell cycle showing oscillating phase-specific expression of the cyclins. The duration of G1 phase in somatic cells is extended, phosphorylation of Rb is tightly controlled at the restriction point, and CDKI inhibitory activity is integrated as a brake on the cell cycle. Most importantly, ASC can exit the cell cycle and enter into quiescence (G_0) in response to growth inhibiting signals or lack of growth promoting signals. The G_0 phase is visualized here as a quiescence cycle to underscore its reversible nature. Typical cell cycle profile comparing ES cells vs. C2C12 myoblasts, visualized using flow cytometric analysis of DNA content revealed by DRAQ5 staining. Note the greatly increased proportion of the ESC population in S phase of cell cycle compared to somatic cells, along with a decrease in the G_1 population.[Note: Panels **(a)** and **(b)** are modified from Berthet & Kaldis 2007; panel **(c)** depicts data from our lab]

both self-renewal and potency of the slower dividing progeny. Adult mammalian tissues also retain a small pool of resident stem cells, which contribute to homeostatic maintenance as well as repair and regeneration. Unlike ESCs, adult stem cells (ASC) are quiescent or nondividing lineage-committed precursors, embedded in the surrounding milieu of the fully differentiated functional tissue [4, 5].

Proliferative activity of these quiescent adult progenitors defines two types of tissue homeostatic mechanisms: ASC activation may occur in an episodic, programmed fashion as in the continually renewing bone marrow compartment to ensure a constant supply of the diverse circulating cells of the blood, immune and inflammatory system, or only in response to damage as in solid tissues such as epithelia or skeletal muscle [5]. In skeletal muscle, dormant stem cells break quiescence when activated by damage-induced signals, and reentry of a minor population of the activated stem cells into a quiescent state follows a period of expansion, differentiation, and restoration of stable functional tissue [5]. During differentiation, cell cycle withdrawal is coupled to the irreversible induction of the myogenic program. Thus, multinucleated myofibers are incapable of reentering the cell cycle when damage occurs, and they undergo necrosis or apoptosis, leaving the regeneration of the tissue to the activity of the quiescent stem cell.

Although its role in adult stem cell biology has long been recognized, the quiescent state in mammalian cells is still poorly understood. Earlier studies considered quiescence as a default state resulting simply from the depression of biosynthetic, signaling, and metabolic pathways as an adaptation to starvation for nutrients or mitogens [6]. However, there is mounting evidence for active control of entry into this reversibly arrested state, suggesting a specific program analogous to but distinct from other states characterized by mitotic inactivity, such as terminal differentiation, senescence, and apoptosis. It is now apparent that quiescent adult stem cells exercise active control to preserve reversibility of arrest and fulfill their regenerative role when activated. For example, dormant cells must simultaneously maintain their genomes in a mutation-free state, remember their identity as tissue-specific stem cells, conserve energy, as well as retain the ability to respond specifically to activation signals. A critical component of the quiescence program is the suppression of alternative nondividing programs, such as senescence, death, and differentiation. We have recently reviewed the evidence that the quiescence program is evolutionarily ancient, by comparing conserved pathways from yeast to mammals [7]. Despite the differences in complexity, broadly conserved conceptual frameworks appear to describe entry and exit into quiescence in these disparate systems, and studies from yeast have been instrumental for identifying key metabolic, signaling, and regulatory events that control entry into and exit from quiescence.

In this chapter, we briefly review the evidence linking the pathways controlling quiescence in adult muscle stem cells in the context of methods for generating homogeneous populations of reversibly arrested myoblasts.

Quiescence is a reversibly arrested cell cycle state characterized primarily by the absence of cell growth and proliferation. Proliferating cells in vitro will exit the cell cycle and enter G_0 or quiescence in response to growth-inhibiting signals or due to

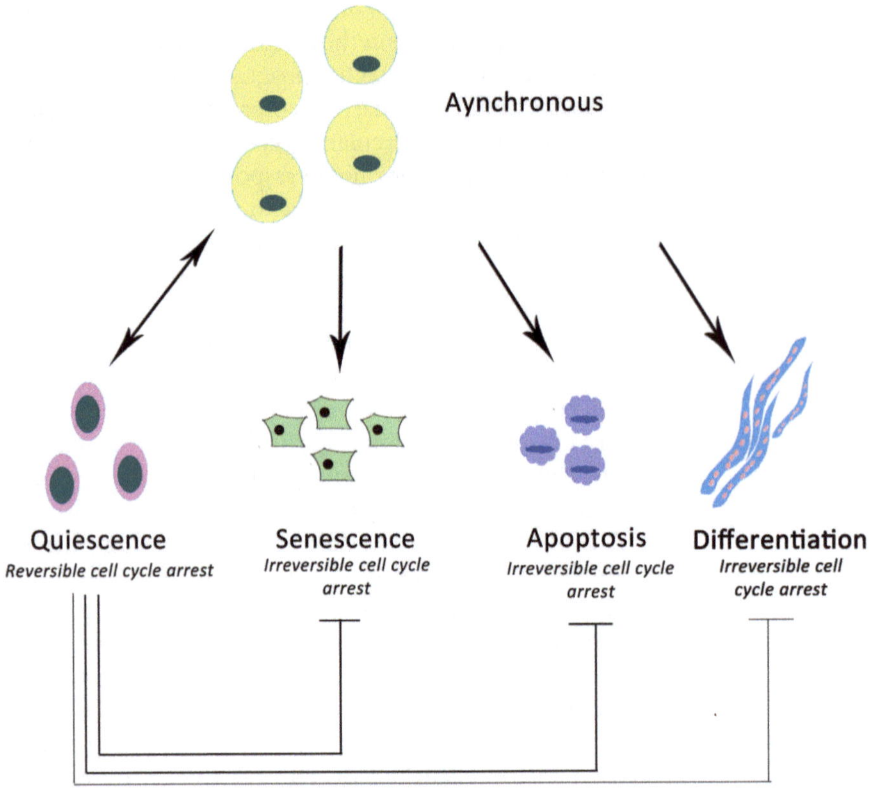

Fig. 2 The quiescence program is not only reversible but also inhibits other out-of-cycle states. Quiescence is reversible and resists differentiation, senescence, and apoptosis

absence of growth-promoting signals. Quiescent cells possess an un-replicated genome or G_1 DNA content, altered metabolism (such as increased autophagy), and increased nucleus to cytoplasm ratio [8–11]. Growth arrest associated with G_0 is fundamentally different from that observed in terminally differentiated cells or senescent cells as the former retains the ability to reenter cell cycle [4, 12, 13] (Fig. 2). This state of reversible arrest is achieved by active suppression of alternate pathways leading to permanent arrest [14, 15].

Much of our knowledge about quiescence has been derived from studies in cultured cells- from yeast to mammalian cells. In baker's yeast, *Saccharomyces cerevisiae,* quiescence is often induced by nutrient limitation [6], and induces a survival mechanism that maintains viability in adverse environmental conditions [16, 17].

In multicellular organisms, postnatal tissue homeostasis and regeneration after injury is dependent on the presence of a small population of cells known as Adult Stem Cells (ASCs) that have the capacity to self-renew and differentiate into lineage-specific cell types that comprise the tissue in which they are resident. ASCs reside in normal uninjured tissues, remain in a quiescent undifferentiated state, and are activated upon tissue damage by signals emanating from the damaged tissues [18–20]. Upon activation, ASCs proliferate, producing committed progenitors, which in turn differentiate to produce functionally mature cells that mediate tissue regeneration. A small minority of these activated ASCs exit the cell cycle and self-renew by reentering quiescence to maintain the stem cell reserve that can respond to future demands [21, 22]. Thus, quiescence ensures a steady number of ASCs available for tissue regeneration. Quiescence also plays a protective role in cells against proliferation-associated stresses [23–25].

In mammalian skeletal muscle, differentiated myofibers are permanently arrested (post-mitotic), but a rare population of stem cells called satellite cells enters an alternate cell cycle exit (quiescence or G_0), retaining the option to reactivate and repair damage (reviewed in [26]). With recent evidence it is becoming clear that rather than a state of passive hibernation, quiescence is a transcriptionally [8, 27] and epigenetically [28, 29] actively regulated program. From the perspective of the quiescent stem cell, failure to enter G_0 can lead to uncontrolled proliferation (underlying tumorigenesis), whereas failure to exit G_0 leads to a loss of progenitor function (leading to degenerative disease) [5]. Thus, understanding the mechanisms that control quiescence has implications beyond developmental tissue dynamics to disease.

In vitro, early indications of cells entering quiescence include a gradual drop in proliferative rate, and upregulation of genes responsible for cell cycle inhibition [8]. At a molecular level, the quiescent state is characterized by downregulation of positive regulators of cell proliferation such as cyclins and cyclin-dependent kinases (CDKs) and upregulation of negative regulators of cell cycle such as CDKIs [8, 30]. This feature is similar to cells that are entering differentiation, but a distinct family of negative growth regulators appears to accompany stable differentiation associated arrest [31, 32]. Quiescence is also associated with strong reversible suppression of global RNA and protein synthesis [9, 33]. In adult stem cells particularly, besides suppression of pro-cell cycle and induction of anti-proliferative genes, the G_0 state is also associated with active suppression of pro-differentiation genes and upregulation of anti-differentiation factors, preventing the cells from permanent cell cycle exit [32, 34].

Direct molecular analysis of quiescent stem cells in vivo poses a significant challenge, as they constitute a minor proportion of the tissue (satellite cells constitute ~2 % of adult muscle) and that the removal of these cells from their niche invariably leads to their activation. Freshly isolated primary stem cells are a possible source of studying these cells; however, the isolation procedures are complicated (invariably triggering activation) and the number of cells obtained is low [35], requiring enrichment by flow cytometry. Isolated single muscle fibers have served as an excellent model to

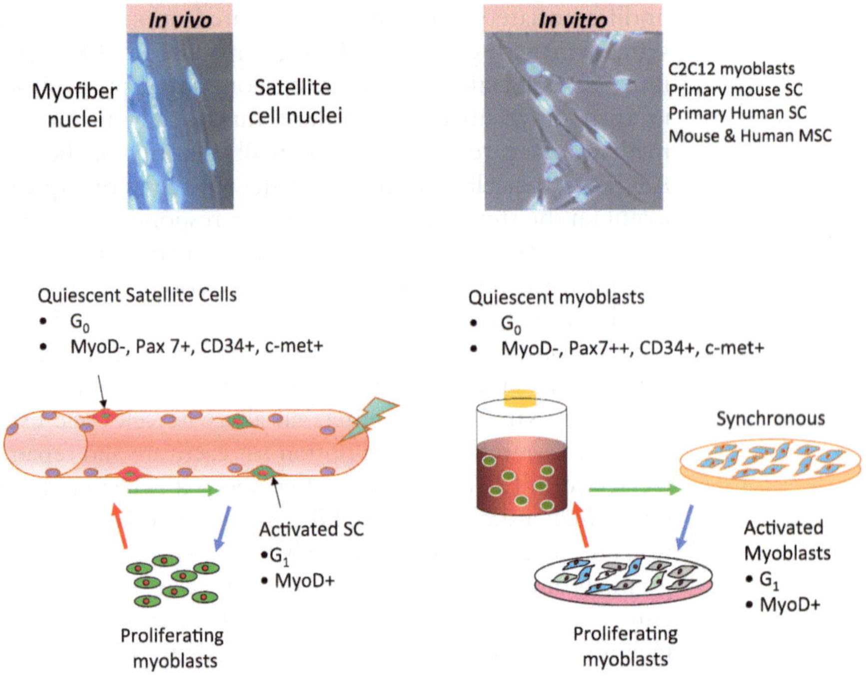

Fig. 3 Quiescence in culture mimics elements of quiescent satellite cells in vivo. The micrographs depict a section of an isolated myofiber showing the peripheral location of two satellite cell nuclei on a multinucleated muscle fiber (*left panel*) and adherent cultured myoblasts (*right panel*). Nuclei are visualized by DAPI staining. The schematic depicts satellite cell reactivation during damage-induced regeneration in vivo (*left panel*); events during suspension culture and reactivation in vitro (*right panel*)

study satellite cells in their niche particularly using immunohistochemistry [36–38]. Though robust and useful, these isolation techniques also activate the fiber-associated satellite cells rendering studies of quiescence difficult.

To study how cells enter, maintain, and exit the quiescent state we describe here an established and well-characterized culture method. By manipulating simple culture conditions (anchorage restriction in mitogen-rich semisolid medium), relatively homogeneous populations of viable quiescent myoblasts can be generated from muscle cell lines and primary muscle cells of both mouse and human origin (Figs. 3 and 4). Despite the caveats associated with the loss of the complex signaling environment of the tissue, the resultant G_0 cells in culture mimic key cellular attributes and behaviors of quiescent stem cells in vivo (Fig. 3). Importantly, this cell cycle arrest is not accompanied by activation of tissue-specific genes or the senescence/death pathways and the undifferentiated cells can then be synchronously activated back into the cell cycle upon restoration of surface contacts, by simply re-plating cells harvested from suspension culture. The advantage of this method is that a large number of cells can be arrested in G_0 and subsequently reactivated as synchronized homogenous populations, enabling robust

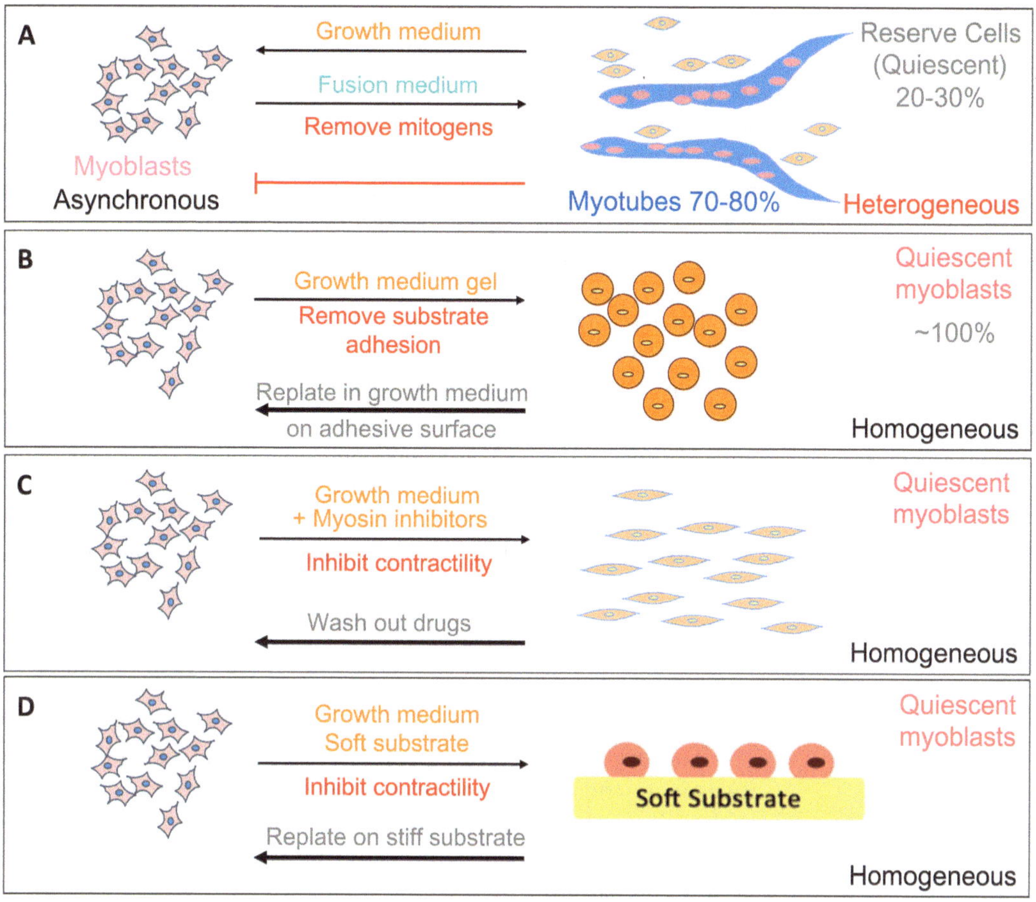

Fig. 4 Different protocols for achieving reversible arrest: abrogation of adhesion-dependent signals leads to homogeneous populations of quiescent cells while mitogen withdrawal leads to heterogeneous mixtures of differentiated (multinucleated) and quiescent (mononucleated) cells. Exposure of an asynchronously proliferating population of undifferentiated myoblasts to mitogen deprivation triggers fusion and terminal differentiation (irreversible) in the majority of cells but a minority enters the "reserve cell" pool and remains mononucleated and enters reversible arrest (**a**). By contrast, inhibition of adhesion (suspension culture, **b**), or contractility using myosin inhibitors (BDM, ML7; **c**), or plating on a soft substrate (**d**) leads to homogeneous populations of reversibly arrested myoblasts

genome-wide profiling of epigenetic states, transcriptomes, and proteomes, single cell analysis using flow cytometric or immunofluorescence studies, as well as targeted studies using RNAi. When used as an adjunct to in vivo studies, these culture methods allow considerable insights into aspects of quiescence biology that cannot be readily accessed in intact animals.

In Fig. 4 we compare synchronization by suspension culture [4, 39] with other methods used to generate quiescent myoblasts. These include the purification of a minor population of undifferentiated mononucleated cells or "reserve cells" [40] from heterogeneous cultures employed to generate differentiated multinucleated myotubes and inhibition of acto-myosin contractility by the use of pharmacological agents that target the non-muscle myosins [41] and restriction of amino acids in low mitogen media [42]. We have used the suspension culture protocol not only to model the quiescent state but also to uncover new programs associated with entry into and maintenance of reversible arrest [29, 32, 43–46]. The degree to which synchronization is achieved and maintained in populations released from G_0 can be seen by stage-specific gene expression and flow cytometric analysis (Fig. 5).

Importantly, synchronized C2C12 mouse myoblasts are able to effectively model some key molecular characteristics of muscle satellite cells (MuSCs) in vivo [4, 29, 43] (Fig. 3). These include arrest in G_0 accompanied by suppression of differentiation and appropriate regulation of genes such as *CD34*, *Pax7*, *MyoD*, *c-Met*, and *M-cadherin*, which are implicated in satellite cell arrest, commitment, and activation respectively [37, 47, 48]. In accordance with the induction of *MyoD* mRNA in activated satellite cells within 6 h of muscle injury [37, 48], suspension-arrested myoblasts also express *MyoD* RNA in G_1 by 6 h after replating [4, 44, 49] (Fig. 5a). Thus, the G_0 myoblasts in the suspension culture model mimic MuSCs in vivo. The morphology and gross cytoskeletal configuration of cells derived from serum deprivation (myotubes) and adhesion deprivation (quiescent myoblasts) are shown in Fig. 6.

The suspension culture protocol has been successfully used to investigate and understand important pathways that are associated with quiescence in muscle satellite cells (mouse and human) [4, 5, 39], and primary mesenchymal cells (mouse and human) (Rumman et al, in preparation).

2 Materials

Prepare and store all reagents at 4 °C (unless indicated otherwise). Be careful while weighing and dissolving solutions and diligently follow all waste disposal regulations.

2.1 Cell Lines and Cell Culture Reagents

1. **Cell lines:** This method was originally developed for C2C12 mouse myoblasts [45], and subsequently extended to mouse primary myoblasts [Subramaniam et al. unpublished], human myoblasts derived from patients (cc-2580) [46], and primary human mesenchymal stem cells [Rumman et al. in preparation]. The medium requirements are adjusted to provide conditions optimized for survival of each cell line.
2. **Proliferation medium:**

   C2C12 growth medium

   Dulbecco's Modified Eagle Medium (DMEM, high glucose) (Gibco # 11965-092), Fetal Bovine Serum—20 % (Gibco # 16000-044), Glutamax—1 % (Gibco # 35050-061), Penicillin-Streptomycin—1 % (Gibco #15140122).

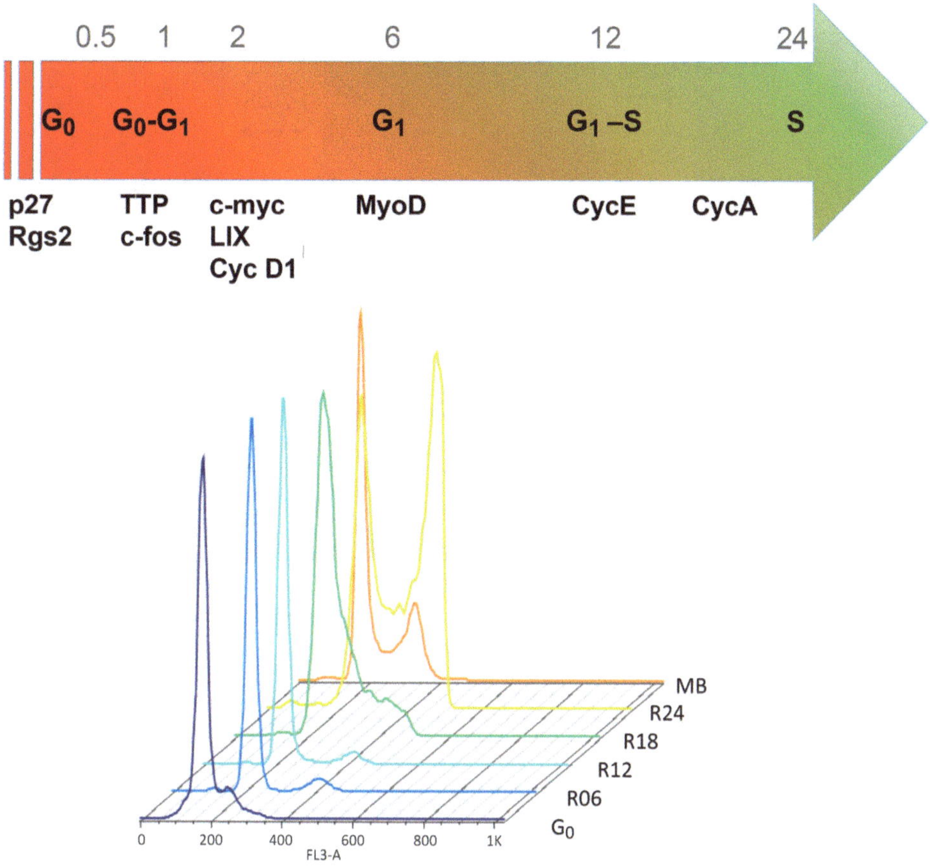

Fig. 5 G_0 myoblasts synchronously reenter the cell cycle upon replating and recapitulate satellite cell activation in vivo: A time-course of reactivation of G_0-synchronized myoblasts showing landmark events and associated genes. The CDKI p27 (cell cycle inhibitor) and regulator of G protein signaling RGS2 (associated with self-renewal) are highly induced during entry into quiescence and sustained in G_0, whereas MyoD expression is suppressed. Reactivation leads to a G_0-G_1 transition during which the expression of TTP, c-fos, c-myc, LIX, and Cyclin D1 is induced with 30 min-2 h of replating. An important landmark is the induction of MyoD expression at 4–6 h in the G_1 phase, marking a recoupling of lineage determination with cell cycle progression and the return of competence for differentiation. The G_1-S transition and entry into S phase are marked by expression of Cyclins E and A that drive DNA replication.

Reentry time course monitored by flow cytometry. Staining with DRAQ5, followed by quantitative flow cytometric estimation of DNA content in the population, clearly depicts the distribution of cells between G_1 (2N, unreplicated genome), S (2N-4N, replication to different extents in different cells in the population), and G_2/M (4N, completely duplicated genome) DNA content in asynchronous myoblasts (MB). After 48 h in suspension culture, the entire population is shifted to 2N DNA content (G_0). After 6 and 12 h of reactivation by restoration of surface contacts (R06, R12), the population has still largely remained in G_1, while at R18, a substantial "shoulder" of S phase cells appears, followed by the emergence of the G_2/M peak at R24 when cell cycle progression has moved beyond S phase. Note the increased enrichment of G_2/M cells in R24 compared to Mb, indicating that the population as a whole has a distinct profile from the asynchronous state

Primary human MSC growth medium

MEM Alpha (Gibco # A10490-01), Fetal Bovine Serum—20 % (Gibco # 16000-044), Glutamax—1 % (Gibco # 35050-061), Penicillin-Streptomycin—1 % (Gibco #15140122).

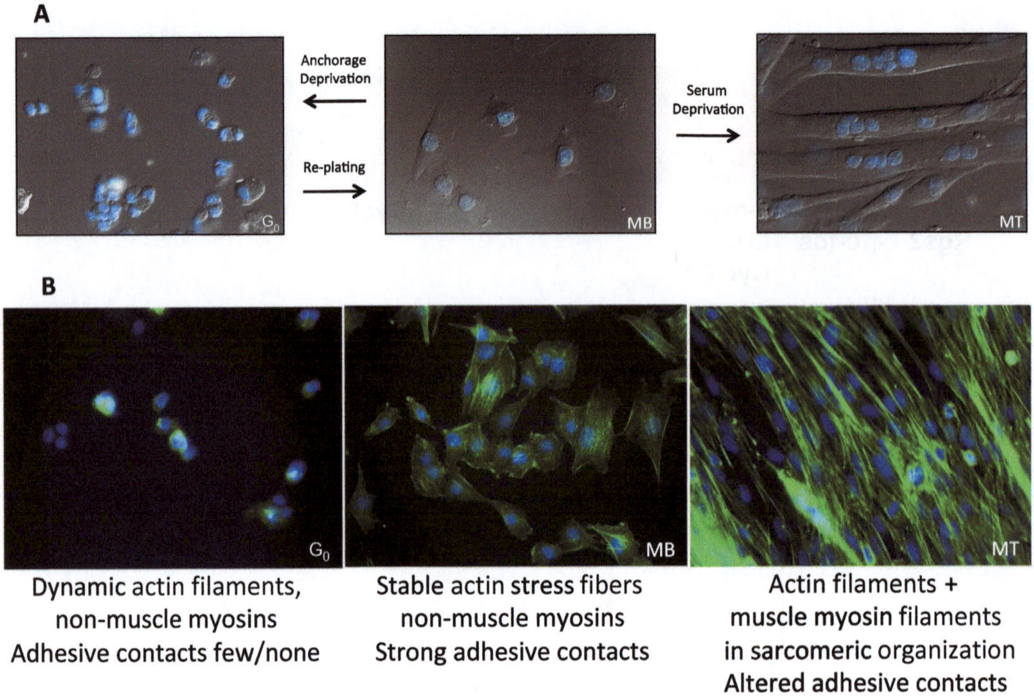

Fig. 6 Morphology of proliferating, quiescent, and differentiated myoblasts. (**a**) Phase contrast images, nuclei are stained with DAPI. (**b**) Immuno-staining to show altered actin cytoskeleton in the three states—actin is stained with phalloidin (*Green*) and the nucleus with DAPI (*blue*)

Primary human myoblast growth medium

Dulbecco's Modified Eagle Medium (DMEM), Fetal Bovine Serum—10 %, Penicillin-Streptomycin—1 %.

3. **Phosphate Buffer Saline (PBS)**—For 1 liter—Dissolve 200 mg of Potassium Chloride, 200 mg of Potassium biphosphate, 8 g of Sodium Chloride, 1.15 g of disodium hydrogen phosphate in deionized water. Make up the volume to 1 l with deionized water.
4. **1 M HEPES pH 7.4**—For 1 l dissolve 238.3 g HEPES (Sigma #H3375) in 800 ml of deionized water, Adjust the pH to the 7.4 with 10 N NaOH. Make up the volume to 1 l with deionized water.

2.2 Chemicals, Plastic-Ware, and Equipment

1. Methylcellulose—(Sigma # M0512).
2. Lymphoprep ™ (Stem Cell Technologies #07861).
3. Tissue Culture grade dishes.
4. Polypropylene conical bottom centrifuge tubes- 50 ml falcon tubes (Corning # 430829).
5. Polypropylene conical tubes with cap, 175 ml (Nalgene, 352076) (*see* **Note 1**).
6. Schott DURAN glass bottles (1 l with wide mouth).

7. Centrifuge and appropriate adaptors (Eppendorf #5810 R or Sorvall TC -7000, swinging bucket rotor).
8. 10 ml Syringes and 0.22 μm Syringe Filters (Millipore # SLGP033RS).
9. Autoclave.
10. Water bath—temperature adjustable (37–55°).
11. CO_2 Incubator for Tissue Culture.
12. Laminar Flow Hood (Sterile conditions for Tissue Culture).
13. Magnetic stir-bar and magnetic stirrer (with heating).
14. 4 °C Refrigerator.

3 Methods

3.1 Suspension Media (Methocel) Preparation Protocol (For 500 ml):

1. Add 10 g methylcellulose powder to an empty wide-mouth 1-liter media glass bottle. Add a clean large magnetic stir-bar to the bottle and autoclave (120 °C, 20 min). Cool the bottle to room temperature (*see* **Note 2**).
2. Heat DMEM to 55 °C for 10 min in a water bath (Do not warm media for longer as heat-sensitive components will denature).
3. In a laminar flow hood (sterile conditions) add 250 ml of pre-warmed (at 55 °C) DMEM into the 1-liter bottle containing autoclaved methylcellulose powder.
4. Place the bottle on a magnetic stirrer on a medium speed for 1 h at room temperature to allow the methylcellulose to dissolve.
5. In the laminar flow hood add the remaining 250 ml media, pre-warmed to 37 °C. Stir at room temperature on magnetic stirrer for 1 h. At this point the methylcellulose is well dispersed in the medium, and gelation has been initiated but not yet completed. The media looks frothy and fine white particles of methylcellulose can still be discerned if the stirring is stopped.
6. Stir on a magnetic stirrer for 12–16 h (overnight) at 4 °C, to dissolve completely. At this time, the methylcellulose solution should clarify and become a semisolid, viscous transparent gel.
7. Transfer the viscous medium to 50 cc centrifuge tubes by pouring in a continuous manner without tipping back media from the lip, in a laminar flow hood. You will find that the rate of pouring can be easily controlled due to the viscous nature of the gel. Cap the tubes tightly and centrifuge at 3200 × *g*, at 4 °C for 30 min in, to sediment all the un-dissolved fines of methylcellulose. Use only the top 80 % of the gel medium for further steps (*see* **Notes 3** and **4**)

8. Before use, loosen the cap of the tube and warm to 37 °C, in the CO_2 incubator.

3.2 Suspension Culture Protocol

1. Passage the cultured cells that you wish to suspend 1 day before starting the suspension protocol. Cells should be in the exponential growth phase when collected for suspension culture, i.e., should be grown till no greater than 70 % confluent (*see* **Note 5**).
2. On the day of the experiment, prepare the **suspension mix** (FBS, Glutamine, Penicillin-Streptomycin and HEPES) as mentioned in Table 1. Sterilize the suspension mix using a syringe and 0.22 micron filter (*see* **Note 6**).
3. Pour required volume of methocel into a 50 ml falcon tube (Refer to Table 2 for volumes for a 10 ml suspension culture) (*see* **Note 7**).
4. Add the filter-sterilized suspension mix to the pre-aliquoted methocel and mix thoroughly by capping the tube and rotating. Be careful not to allow the supplemented gel mix to touch the top of the tube. After adding the mix, the suspension culture media will change color (turns orange, due to HEPES which is added to ensure stable pH in the first hours of culture during equilibration in the CO_2 atmosphere of the incubator).

Table 1
Suspension mix

Cell culture reagents	Volume	Final concentration
FBS	2 ml	20 %
200 mM Glutamine	200 μl	4 mM
Penicillin-Streptomycin	100 μl	1×
1 M HEPES pH 7.4	100 μl	10 mM

(For addition to methocel generate 10 ml suspension culture)

Table 2
Volumes of components for standard suspension culture (10 ml)

	Component	Volume
1	Methocel (from Subheading 3.1, **step 8**)	6.6 ml
2	Suspension Mix (from Subheading 3.2, **step 2**)	2.4 ml
3	Single cell suspension (from Subheading 3.2, **step 6**)	1 ml (1×10^6 cells/ml)

5. Trypsinize and resuspend cells in pre-warmed (37 °C) proliferation medium, making sure to disperse trypsinized cells well by pipetting, to get a single cell suspension (check under microscope to ensure). If cells are clumped, they will not enter quiescence but proliferate as spheres and will not represent G0.
6. Count cells and adjust volume to ensure a cell density of 1×10^6 cells per ml. The final density of cells in suspension culture should be no greater than 1×10^5 cells per ml (*see* **Note 8**). For 10 ml methylcellulose suspension culture 1×10^6 cells are suspended. (For scaling up *see* **Note 9**)
7. Add trypsinized single cell suspension to the supplemented gel mixture, pre-aliquoted in the falcon tube. Rotate the falcon gently by hand for 3 min to mix media well. A homogenous color should be achieved. Remove a small aliquot by dipping in a Pasteur pipette and transferring to a microscope slide to examine under inverted microscope and ensure a single cell suspension has been achieved.
8. Loosen the cap of the falcon tube by one turn and place in a wire rack within the incubator at 37 °C, 5 % CO_2 for 48 h (*see* **Note 10**).
9. After 24 h close cap of tube, remove from incubator to hood, and rotate the tube gently for 5 min to mix media again and ensure that cells are still dispersed in suspension.

3.3 Cell Harvest

Suspended cell populations being to enter quiescence within 12–24 h of abrogation of the adhesion, and by 36–48 h greater than 99 % of cells do not incorporate BrdU and are still viable. Cultures at 48 h post suspension therefore show near complete synchronization in G0. Cells continue to show undiminished viability for another 24 h after which some cell death is initiated. Therefore, we normally harvest cells at either 48, or 72 h to represent quiescent populations.

3.3.1 By Density Gradient Medium Method Using Lymphoprep™ (see Note 11)

(For C2C12 myoblasts and primary human myoblasts, this method is critical particularly for FACS analysis where any clumps cause problems in the flow cell)

1. Pre-warm sterile PBS to 37 °C in the water bath.
2. In the hood, add pre-warmed PBS to the tube containing the suspension culture. Fill the tube (to the 50 cc mark) using PBS. Mix well by rotating the tube. (PBS should be about four times the volume of the culture).
3. Spin down the mix using a centrifuge at $1250 \times g$ for 25 min without brakes. (The brakes on the centrifuge should be set to 0 so that the centrifugation process is smooth and the loose cell pellet is not disturbed by juddering during braking).

4. Carefully decant the supernatant without dislodging the pellet, which is still very loose at this first collection step, leaving 2–3 ml of media/PBS mix in the tube.
5. Resuspend the pellet in the residual minimum volume (i.e., 2–3 ml) using a 1 ml pipetteman so as to generate a complete dispersal of the cells. At this point you can pool multiple samples into one tube, if needed, and add PBS to make a total volume of 20 ml.
6. Prepare 15 ml each of 70 and 40 % Lymphoprep™ in PBS. (70 % = 10.5 ml Lymphoprep™ and 4.5 ml PBS, 40 % = 6 ml Lymphoprep™ and 9 ml PBS).
7. Add 15 ml of the 70 % Lymphoprep™ solution to an empty 50 ml falcon tube.
8. Slowly and carefully overlay 15 ml of the 40 % solution on the top of this 70 % layer to create a density gradient (*see* **Note 12**).
9. Carefully and slowly add the 20 ml of cell-PBS mix to the 40 % layer.
10. Centrifuge at 290 × *g* for 15 min with brakes on. (The brakes of the centrifuge can be reactivated at this point).
11. After the spin, decant most of the PBS as above, without disturbing the pellet.
12. Resuspend with PBS and make up the volume to 50 ml.
13. Centrifuge again at 290 × *g* for 10 min with brakes on.
14. Aspirate all PBS and resuspend the cells in appropriate proliferation medium/buffer for replating/analysis of quiescent cells (Fig. 6).
15. A viability test for live cells by propidium iodide exclusion should reveal >95 % cell viability at this stage—trypan blue staining is misleading as it is taken up by residual methocel particles.
16. At this point the quiescent cells that have been harvested can be used for DNA, RNA, protein, immunofluorescence—please see Immunostaining section below), or replated for further cell cycle stage analysis.

3.3.2 PBS Washing Method

(For C2C12 myoblast, primary human myoblasts and primary human MSCs)

1. Pre-warm sterile PBS to 37 °C in a water bath.
2. Fill the cell suspension tube with pre-warmed PBS (up to the 50 ml mark of the Falcon tube). Mix well by rotating the tube and centrifuge at 1250 × *g* for 25 min (1800 × *g*, 30 min for MSCs) with brakes off. (The brakes on the centrifuge should be set to 0, so that the centrifugation process is smooth).

3. Carefully decant the supernatant (leaving about 15 ml in the tube) without dislodging the pellet.
4. Repeat **step 2**.
5. Decant the supernatant again (leaving about 10 ml in the tube) without dislodging the pellet.
6. Resuspend the cell pellet and fill the tube again with pre-warmed PBS.
7. Centrifuge at 290 × *g* for 10 min with breaks on. (The brakes of the centrifuge can be resumed to the highest setting (usually 9) at this point).
8. If necessary, give another PBS wash (**steps 6** and 7) to remove all methylcellulose.
9. Aspirate all PBS and resuspend the cells in appropriate proliferation medium/buffer for replating/analysis (*see* **Notes 13** and **14**).

3.4 Replating

1. Harvested cells should be resuspended in proliferation medium and replated at subconfluent densities (for C2C12 myoblasts, not more than 4×10^5 cells per 60 mm dish, 1×10^6 cells per 100 mm dish, 10,000 cells per 18 mm^2 coverslip or 3000 cells per 13 mm^2 coverslips in proliferation medium) (*see* **Note 15**).
2. After replating, cells rapidly and synchronously undergo G_0-G_1 transition and can be harvested at different time points (30 min, 1, 2, 6, 8, 12, and 24 h). After attachment, 6 h is required to enter G_1, and S phase peaks at 20–24 h (Fig. 5b).

3.5 Immunostaining for Suspended (Quiescent) Cells

For immunofluorescence assays, quiescent cells that have been harvested following protocols mentioned above should be fixed and stained while in suspension to avoid attachment-induced reactivation into cell cycle.

Fixing:

1. Resuspend cells into a minimal volume of PBS and add drop-wise into 5 ml of appropriate fix solution in a 50 cc falcon tube while continually mixing.
2. Wash the cells by filling up the falcon tube to the 50 ml mark with PBS to dilute out the fix solution, followed by centrifugation at 290 × *g* for 10 min.
3. Aspirate PBS carefully without disturbing the cell pellet and resuspend the fixed cells in 1 ml of PBS. These cells can be stored in 1.5 ml centrifuge tubes at 4 °C or immediately used for immunostaining following standardized protocols for immunostaining (permeabilization, blocking, antibody concentrations).

Immunostaining

1. Immunostaining for up to 1×10^5 cells can be performed efficiently in 200 μl volumes in 1.5 ml centrifuge tubes. All washes

should be performed by diluting the sample to 1.5 ml with blocking buffer followed by centrifugation to ensure proper recovery of cells (*see* **Notes 16** and **17**).

2. After staining is complete, resuspend cells in 200 μl of PBS and centrifuge onto a labeled slide using a cyto-centrifugation instrument such as Cytospin (Thermo Scientific), add mountant, coverslip, remove excess mountant by wicking off with Whatman #1 filter paper and proceed for imaging.

3.6 Biochemical Analysis

1. Isolation of RNA:

   Resuspend the final cell pellet obtained after harvesting in TRIzol® reagent and proceed as recommended by manufacturer.

2. Protein lysates for Western Blotting:
   (a) Wash the final cell pellet in ice cold PBS, making sure to aspirate out as much of PBS as possible.
   (b) Resuspend the cell pellet in Laemmli buffer, incubate on ice for 30 min.
   (c) Sonicate briefly and centrifuge at 20,000 × g for 10 min at 4 °C to clear insoluble debris.
   (d) Collect supernatant and heat at 95 °C for 5 min before proceeding for quantification using standard Amido Black protocols and use for western blotting.

4 Notes

1. For small-scale experiments involving analysis by qRT-PCR, western blots, or IFA, 50 ml Falcon tubes containing a maximum of 12.5 ml suspension culture are sufficient for each time point. However for larger-scale experiments involving replating time courses or genome-wide studies either 250 ml polypropylene flat bottom centrifuge bottles or 175 ml polypropylene tubes with conical bottom, from Nalgene, can be used. The maximum suspension culture mix to be incubated in these bottles is 60 ml (for 250 ml bottle) or 45 ml (for 175 ml bottle)-this is vital to maintain optimal gas exchange during culture. If using these bottles/tubes ensure you have the necessary centrifuge adaptors for them. A dedicated set of these tubes and bottles (if reusable) should be kept aside for suspension culture. Wash well to get rid of residual methocel gel *immediately after use*, prior to autoclaving for reuse.
2. After autoclaving the bottle with methylcellulose powder, it may take several hours for the bottle to cool down. You can choose to autoclave one day earlier and leave the bottle overnight at room temperature to cool down.

3. This centrifugation of the methylcellulose medium is very important as it sediments the undissolved fines. Make sure to keep the tubes upright after the centrifugation and discard the bottom 20 % of centrifuged medium, using only the top 80 %.
4. These aliquots can be kept for up to 1 month after preparation at 4 °C. Make sure that the tubes are tightly capped and parafilmed.
5. It is important that the cells (to be suspended) are growing at subconfluent conditions and in the exponential phase of growth. This ensures that the cells exit the cell cycle and enter into the G0 phase upon suspension.
6. A suspension mix (FBS, Glutamine, Penn-Strep and HEPES) can be made beforehand and stored at 4 °C for up to 1 month.
7. Pipetting methylcelullose medium is not recommended due to its highly viscous nature because of which you cannot dispense the correct volume from the pipette. It is useful to pipette the exact amount of PBS into the tube and mark the level with a marker before disposing the PBS; then pour in the methocel.
8. Cell density is a critical parameter in suspension culture experiment for optimal viability and entry into quiescence, as high cell density suspensions may not arrest efficiently. Greater density can compromise quiescence and cells will proliferate and will not represent G_0.
9. The suspension culture can be scaled up according to the number of cells that have to be suspended. The scaling up should be done in the same ratio as mentioned: 1×10^6 per 10 ml of suspension medium. Also the 10 ml is the maximum solution that should be kept in a 50 ml Falcon tube for a 48-h suspension experiment (i.e., one fifth the total volume). Hence if a 175 ml conical flask is being used, a maximum of 3.5×10^6 cells in 35 ml suspension mix should be used.
10. The incubation conditions used for suspension cultures are the same as that for adherent cells.
11. We have used Lymphoprep™ as a density gradient medium for isolation of quiescent myoblasts. However, either Percoll or Ficoll-Paque can be used for the cell harvest protocol. Both give similar efficiencies and cellular yields.
12. Set the pipetteman on slow-dispense setting and hold the tip of the serological pipette along the wall of the tube to dispense the liquid slowly.
13. The density gradient method is faster and gives a higher cellular yield than the PBS wash method. The density gradient method is preferred for replating experiments and harvest for protein. RNA can be extracted using Trizol after either method of cell harvest.

14. Tips on volumes of methylcellulose culture (number of cells) to use for various experiments:
    - Immunostaining (BrdU incorporation or endogenous antigen): We usually suspend 5×10^5 cells (5 ml culture) and replate cells (after 48 h in suspension culture) at 3000 cells per 13 mm cover slip placed in a 24-well dish. This cell number is therefore sufficient for ~30 time points/conditions).
    - RNA for Northerns, DNA for genomic southerns: you will need at least 50 ml methylcellulose culture (5×10^6 cells total) per condition-yield (Trizol) ~100–120 μg).
    - For FACS analysis: you may have to filter the fixed cells using a tube-top cell strainer (100 μ) to get rid of clumps.
15. For early time points in the cell cycle (2–12 h, i.e., prior to the onset of DNA synthesis) we often plate cells at 2×10^6 cells per 100 mm dish. For later time points (18–36 h) 1×10^6 cells per 100 mm dish is sufficient as cells spread and go through S phase.
16. Fixed cells in suspension do not sediment efficiently upon centrifugation. Hence, it is essential to add detergent and BSA/serum to ensure efficient recovery. For this purpose we have performed all washes in blocking buffer instead of wash buffer.
17. We have used Eppendorf tabletop centrifuges (MiniSpin or Eppendorf 5417R) for immunostaining. For this purpose, use low speed settings (not more than $600 \times g$) so that the pellet is collected but loose enough to permit ready dispersal, else cell damage will occur from pipetting).

Acknowledgments

We thank past and present members of the Dhawan lab, especially Chetana Sachidanandan, Ramkumar Sambasivan, and Sindhu Subramaniam for developing and refining the protocols presented here, and Lamuk Zaveri and Hardik Gala for the comparative analysis of the cell cycle in ESC and ASC. RA was supported by a DST Start up grant for young scientists and an NCBS-inStem Career Development Fellowship; MR and HG were supported by doctoral fellowships from CSIR and NV by a doctoral fellowship from DBT. The Dhawan lab is supported by core funds from the Dept. of Biotechnology to InStem, core funds from the Council of Scientific and Industrial Research to CCMB, and grants from the Dept. of Biotechnology Indo-Danish Strategic Fund, Indo-Australia Biotechnology Fund, and the Indo-French Center for the Promotion of Advanced Research.

References

1. Abdelalim EM (2013) Molecular mechanisms controlling the cell cycle in embryonic stem cells. Stem Cell Rev 9(6):764–773
2. Kapinas K et al (2013) The abbreviated pluripotent cell cycle. J Cell Physiol 228(1):9–20
3. Berthet C, Kaldis P (2007) Cell-specific responses to loss of cyclin-dependent kinases. Oncogene 26(31):4469–4477
4. Sachidanandan C et al (2002) Tristetraprolin and LPS-inducible CXC chemokine are rapidly induced in presumptive satellite cells in response to skeletal muscle injury. J Cell Sci 115(Pt 13):2701–2712
5. Dhawan J, Rando TA (2005) Stem cells in postnatal myogenesis: molecular mechanisms of satellite cell quiescence, activation and replenishment. Trends Cell Biol 15(12): 666–673
6. Gray JV et al (2004) "Sleeping beauty": quiescence in Saccharomyces cerevisiae. Microbiol Mol Biol Rev 68(2):187–206
7. Dhawan J, Laxman S (2015) Decodoing the stem cell quiescence cycle: lessons from yeast for regenerative biology. J Cell Sci 128(24): 4467–4474
8. Coller HA et al (2006) A new description of cellular quiescence. PLoS Biol 4(3):e83
9. Lemons JM et al (2010) Quiescent fibroblasts exhibit high metabolic activity. PLoS Biol 8(10):e1000514
10. Allen C et al (2006) Isolation of quiescent and nonquiescent cells from yeast stationary-phase cultures. J Cell Biol 174(1):89–100
11. Rubin H, Koide T (1973) Inhibition of DNA synthesis in chick embryo cultures by deprivation of either serum or zinc. J Cell Biol 56(3):777–786
12. Srivastava S et al (2010) Regulation of cellular chromatin state: insights from quiescence and differentiation. Organogenesis 6(1):37–47
13. Ellisen LW et al (2001) The Wilms tumor suppressor WT1 directs stage-specific quiescence and differentiation of human hematopoietic progenitor cells. EMBO J 20(8):1897–1909
14. Sang L et al (2008) Control of the reversibility of cellular quiescence by the transcriptional repressor HES1. Science 321(5892): 1095–1100
15. Sousa-Victor P et al (2014) Geriatric muscle stem cells switch reversible quiescence into senescence. Nature 506(7488):316–321
16. Herman PK (2002) Stationary phase in yeast. Curr Opin Microbiol 5(6):602–607
17. Brejning J et al (2003) Genome-wide transcriptional changes during the lag phase of Saccharomyces cerevisiae. Arch Microbiol 179(4):278–294
18. Cheshier SH et al (1999) In vivo proliferation and cell cycle kinetics of long-term self-renewing hematopoietic stem cells. Proc Natl Acad Sci U S A 96(6):3120–3125
19. Cotsarelis G et al (1990) Label-retaining cells reside in the bulge area of pilosebaceous unit: implications for follicular stem cells, hair cycle, and skin carcinogenesis. Cell 61(7): 1329–1337
20. Schultz E et al (1978) Satellite cells are mitotically quiescent in mature mouse muscle: an EM and radioautographic study. J Exp Zool 206(3):451–456
21. Schultz E (1996) Satellite cell proliferative compartments in growing skeletal muscles. Dev Biol 175(1):84–94
22. Fukada S et al (2011) Hesrl and Hesr3 are essential to generate undifferentiated quiescent satellite cells and to maintain satellite cell numbers. Development 138(21):4609–4619
23. Arai F et al (2004) Tie2/angiopoietin-1 signaling regulates hematopoietic stem cell quiescence in the bone marrow niche. Cell 118(2):149–161
24. Mourikis P et al (2012) A critical requirement for notch signaling in maintenance of the quiescent skeletal muscle stem cell state. Stem Cells 30(2):243–252
25. Rossi DJ et al (2007) Deficiencies in DNA damage repair limit the function of haematopoietic stem cells with age. Nature 447(7145): 725–729
26. Relaix F, Zammit PS (2012) Satellite cells are essential for skeletal muscle regeneration: the cell on the edge returns centre stage. Development 139(16):2845–2856
27. Subramaniam S et al (2014) Distinct transcriptional networks in quiescent myoblasts: a role for Wnt signaling in reversible vs. irreversible arrest. PLoS One 8(6):e65097
28. Evertts AG et al (2013) H4K20 methylation regulates quiescence and chromatin compaction. Mol Biol Cell 24(19):3025–3037
29. Sebastian S et al (2009) MLL5, a trithorax homolog, indirectly regulates H3K4 methylation, represses cyclin A2 expression, and promotes myogenic differentiation. Proc Natl Acad Sci U S A 106(12):4719–4724
30. Jin K et al (2009) Mirk regulates the exit of colon cancer cells from quiescence. J Biol Chem 284(34):22916–22925

31. Buttitta LA, Edgar BA (2007) Mechanisms controlling cell cycle exit upon terminal differentiation. Curr Opin Cell Biol 19(6):697–704
32. Subramaniam S et al (2013) Distinct transcriptional networks in quiescent myoblasts: a role for Wnt signaling in reversible vs. irreversible arrest. PLoS One 8(6):e65097
33. Ben-Ze'ev A et al (1980) Protein synthesis requires cell-surface contact while nuclear events respond to cell shape in anchorage-dependent fibroblasts. Cell 21(2):365–372
34. Fukada S et al (2007) Molecular signature of quiescent satellite cells in adult skeletal muscle. Stem Cells 25(10):2448–2459
35. Pallafacchina G et al (2010) An adult tissue-specific stem cell in its niche: a gene profiling analysis of in vivo quiescent and activated muscle satellite cells. Stem Cell Res 4(2):77–91
36. Bonavaud S et al (2002) Preparation of isolated human muscle fibers: a technical report. In Vitro Cell Dev Biol Anim 38(2):66–72
37. Cornelison DD, Wold BJ (1997) Single-cell analysis of regulatory gene expression in quiescent and activated mouse skeletal muscle satellite cells. Dev Biol 191(2):270–283
38. Rosenblatt JD et al (1995) Culturing satellite cells from living single muscle fiber explants. In Vitro Cell Dev Biol Anim 31(10):773–779
39. Milasincic DJ et al (1996) Anchorage-dependent control of muscle-specific gene expression in C2C12 mouse myoblasts. In Vitro Cell Dev Biol Anim 32(2):90–99
40. Yoshida N et al (1998) Cell heterogeneity upon myogenic differentiation: down-regulation of MyoD and Myf-5 generates 'reserve cells'. J Cell Sci 111(Pt 6):769–779
41. Dhawan J, Helfman DM (2004) Modulation of acto-myosin contractility in skeletal muscle myoblasts uncouples growth arrest from differentiation. J Cell Sci 117(Pt 17):3735–3748
42. Kitzmann M et al (1998) The muscle regulatory factors MyoD and myf-5 undergo distinct cell cycle-specific expression in muscle cells. J Cell Biol 142(6):1447–1459
43. Sambasivan R et al (2009) The small chromatin-binding protein p8 coordinates the association of anti-proliferative and pro-myogenic proteins at the myogenin promoter. J Cell Sci 122(Pt 19):3481–3491
44. Sambasivan R et al (2008) A gene-trap strategy identifies quiescence-induced genes in synchronized myoblasts. J Biosci 33(1):27–44
45. Sellathurai J et al (2013) A novel in vitro model for studying quiescence and activation of primary isolated human myoblasts. PLoS One 8(5):e64067
46. Cheedipudi S et al (2015) A fine balance: epigenetic control of cellular quiescence by the tumor suppressor PRDM2/RIZ at a bivalent domain in the cyclin a gene. Nucleic Acids Res 43(13):6236–6256
47. Seale P et al (2000) Pax7 is required for the specification of myogenic satellite cells. Cell 102(6):777–786
48. Beauchamp JR et al (2000) Expression of CD34 and Myf5 defines the majority of quiescent adult skeletal muscle satellite cells. J Cell Biol 151(6):1221–1234
49. Gopinath SD et al (2007) The RhoA effector mDiaphanous regulates MyoD expression and cell cycle progression via SRF-dependent and SRF-independent pathways. J Cell Sci 120(Pt 17):3086–3098
50. Yaffe D, Saxel O (1977) A myogenic cell line with altered serum requirements for differentiation. Differentiation 7(3):159–166

Chapter 16

Methods for Observing and Quantifying Muscle Satellite Cell Motility and Invasion In Vitro

Dane K. Lund*, Patrick McAnulty*, Ashley L. Siegel*, and DDW Cornelison

Abstract

Motility and/or chemotaxis of satellite cells has been suggested or observed in multiple in vitro and in vivo contexts. Satellite cell motility also affects the efficiency of muscle regeneration, particularly in the context of engrafted exogenous cells. Consequently, there is keen interest in determining what cell-autonomous and environmental factors influence satellite cell motility and chemotaxis in vitro and in vivo. In addition, the ability of activated satellite cells to relocate in vivo would suggest that they must be able to invade and transit through the extracellular matrix (ECM), which is supported by studies in which alteration or addition of matrix metalloprotease (MMP) activity enhanced the spread of engrafted satellite cells. However, despite its potential importance, analysis of satellite cell motility or invasion quantitatively even in an in vitro setting can be difficult; one of the most powerful techniques for overcoming these difficulties is timelapse microscopy. Identification and longitudinal evaluation of individual cells over time permits not only quantification of variations in motility due to intrinsic or extrinsic factors, it permits observation and analysis of other (frequently unsuspected) cellular activities as well. We describe here three protocols developed in our group for quantitatively analyzing satellite cell motility over time in two dimensions on purified ECM substrates, in three dimensions on a living myofiber, and in three dimensions through an artificial matrix.

Key words Muscle satellite cells, Cell motility, Myofiber culture, Cell invasion

1 Introduction

The sequence of events involved in satellite cell activity in vivo is generally listed as activation, proliferation, migration, differentiation, and self-renewal (reviewed in [1–4]). Some of these, such as proliferation,

The original version of this chapter was revised. The correction to this chapter is available at https://doi.org/10.1007/978-1-4939-6771-1_21

$These authors contributed equally to this work.

Electronic supplementary material: The online version of this chapter (doi:10.1007/978-1-4939-6771-1_16) contains supplementary material, which is available to authorized users.

Eusebio Perdiguero and DDW Cornelison (eds.), *Muscle Stem Cells: Methods and Protocols*, Methods in Molecular Biology, vol. 1556, DOI 10.1007/978-1-4939-6771-1_16, © Springer Science+Business Media LLC 2017

are relatively easy to quantify in vitro and in vivo using commonly accepted methods. Others are more difficult to observe and quantify, and few standardized techniques exist, and thus the degree, function, and even occurrence of these are somewhat more controversial. In particular, the questions of whether, how much, how, and why satellite cells translocate through the tissue as a component of damage-induced muscle regeneration are not yet fully resolved.

The strongest argument for the ability of satellite cells to leave their original location in vivo in response to more distal trauma has been the successful replacement of muscle tissue after injuries (such as experimental freeze-crush) that would be expected to kill effectively all local cells including stem cells [5–7], although some cases were described in which repeated freezing successfully prevented muscle regeneration [8]. Similarly, several groups described results suggesting that satellite cells could cross the basal laminae of their host myofibers in vivo (as they are routinely observed to do in single myofiber culture) and/or move between myofibers in the interstitial space [9–11]. However, evaluation of the role and requirement for satellite cell motility during muscle regeneration in vivo has been difficult, as even observing satellite cell motility in vivo directly is technically challenging [12].

Thus, the majority of work on satellite cell motility has necessarily been done in vitro, and multiple studies examining the cellular and environmental factors influencing total [13–16] and/or directional motility [17–20] of satellite cells can be found in the literature. Many of them use timelapse microscopy to record the total and directional displacement of single satellite cells, followed by quantitative analysis. This is a powerful technique, however due to differences in culture conditions, imaging parameters, and video analysis it can be difficult to compare results between different datasets. It also has the limitation that it is a two-dimensional representation of what is in fact a three-dimensional process, on an artificial substrate that is neither biochemically nor biophysically equivalent to the host myofiber.

Another factor to be considered when developing a model of satellite cell motility in vivo is the presumed difficulty of moving through the dense interstitial matrix (reviewed in [21]). Cell types that are notable for their ability to negotiate such environments, such as tumor cells and endothelial cells, do so by modifying the matrix via metalloproteinases [22, 23]. While comparatively little is known regarding the matrix invasion capacity of satellite cells, modifying satellite cell metalloproteinase activity in vitro increased invasive capacity [24, 25] and changes in metalloproteinase activity are associated with neuromuscular disease [26, 27]; augmenting metalloproteinase activity has also been shown to enhance the spread of exogenous myoblasts in vivo [28–31].

We describe here three independent techniques for evaluating satellite cell motility in vitro (1) in two dimensions; (2) in three dimensions on a physiologically appropriate substrate (the myofiber); and (3) in three dimensions through an artificial extracellular matrix. All three techniques are amenable to modification of cell type/genotype, exogenous factors, and (if appropriate reagents are available) cell-specific factors contributing to total, directional, and invasive motility. All three are designed for timelapse imaging, although the invasion assay can be evaluated quantitatively without timelapse microscopy.

2 Materials

2.1 Timelapse Video Analysis of Satellite Cells In Vitro on Purified Substrates

Ca^{++}, Mg^{++}-free Dulbecco's Modified Phosphate-buffered saline (D-PBS).

Collagenase Type I (Worthington Biochemical).

Ham's F-12 medium (Gibco).

Horse serum (Equitech).

Gelatin (USB, 0.66 %) or precoated plates.

Laminin (Sigma).

Optically transparent (i.e., Corning Costar) multiwell tissue culture plates.

Microscopy platform appropriate for timelapse capture (inverted microscope with motorized stage control and enclosure or stagetop incubator) and timelapse-capable image acquisition program [i.e., MetaMorph (Molecular Devices)].

2.2 Timelapse Video Analysis of Satellite Cells on Single Myofibers

Ham's F-12 medium (Gibco).

Horse serum (Equitech).

Penicillin/streptomycin (Gibco).

Recombinant human FGF-2 (R&D Systems).

Acid-extracted rat tail type I collagen (Sigma).

M-199 medium (Gibco).

5 N NaOH.

Growth factors, antibodies, pharmacological agents, etc. as desired.

Optically transparent (i.e., Corning Costar) 48-well tissue culture dishes.

Microscopy platform appropriate for timelapse capture (inverted microscope with motorized stage control and enclosure or stagetop incubator) and timelapse-capable image acquisition program [i.e., MetaMorph (Molecular Devices)].

2.3 Method for Assessing 3D Human Satellite Invasion In Vitro

2.3.1 Cell Culture

M-199 medium (Gibco)DMEM (Gibco).

Fetal calf serum (Gibco).

Penicillin/streptomycin (Gibco).

Recombinant human HGF (Miltenyi).

Ca^{++}, Mg^{++}-free Dulbecco's Modified Phosphate-buffered saline (D-PBS).

Trypsin-EDTA (Gibco).

Acid-extracted rat tail type I collagen (Sigma).

5 N NaOH.

Growth factors, antibodies, pharmacological agents, etc. as desired.

Optically transparent (i.e., Corning Costar) 96-well tissue culture dishes.

2.3.2 Staining

4 % Paraformaldehyde (PFA).

30 % and 50 % sucrose in PBS (filter sterilize).

1.5 ml Eppendorf tubes.

Biopsy cryomolds (Tissue-Tel).

Optimal cutting temperature O.C.T. (Tissue-Tek®).

(Superfrost™ Plus Gold Slides (Fisher)).

Humid chamber for staining.

Hydrophobic marker (Vector).

0.5 % TritonX-100 solution: 0.5 % Triton X-100, 0.2 % sodium azide in 1× TBS (filter sterilize).

Tris–glycine buffer: 1.5 g tris-base, 7.2 g glycine, 1 g sodium azide (0.2 %), in 500 ml 1× TBS (filter sterilize).

Tween-20 solution: 500 ml ddH_2O, 1.2 ml Tris (1 M), 4.5 g NaCl, 500 μl Tween-20 10 % goat serum (or other as appropriate for secondary antibodies used) with 1 % NP-40.

2.3.3 Timelapse Imaging

If timelapse analysis is to be done, you will also need a microscopy platform appropriate for timelapse capture (inverted microscope with motorized stage control and enclosure or stagetop incubator) and timelapse-capable image acquisition program [i.e., MetaMorph (Molecular Devices)].

3 Methods

3.1 Timelapse Video Analysis of Satellite Cells In Vitro on Purified Substrates

3.1.1 Preparing Cells (This wash step helps to eliminate most debris)

1. ~ 6 h before cells are needed on day 4 of a mass satellite cell prep [as described in [32] or elsewhere in this volume] remove plates and aspirate supernatant.
2. Wash 4–6 plates of cells with 5 ml each D-PBS and add to a 50 ml conical. Repeat.
3. Add 1 ml of 117.5 U/ml collagenase to each 10 cm plate and place in an incubator for ~10 min or until cells detach.
4. Rinse each plate with 5 ml D-PBS and add to conical, then rinse with 5 ml F-12 and add to conical.

5. Centrifuge for 5′ @ 180 × *g*.
6. Aspirate supernatant and resuspend pellet in F-12 media w/ 15 % HS (10 ml/plate).
7. Transfer cell suspension onto (4) gelatin-coated 10 cm plates. The gelatin-coated plate is essential for proper cell adhesion as satellite cells do not adhere well to plastic.

3.1.2 Preparing Tracking Plates

1. ~2–3 h before you plan to start prepare the plates you will seed with satellite cells and track. Our lab prefers to use laminin as a substrate for motility assays; however, other substrates can be used, provided the user is mindful that the choice of substrate will have an effect on motility.
2. Dilute laminin 1:100 in D-PBS and pipet sufficient volume to completely coat the bottom of each well into the inner wells of your plate. Do not coat or seed the outer wells of the plate as some mechanical stages have trouble aligning on these wells to properly track. It is also frequently helpful to add sterile DI water to the outer wells to normalize humidity in the plate during imaging.
3. Place laminin-coated plates in a standard tissue culture incubator (humidified, 5 % CO_2 at 37 °C) and wait 2–3 h before use.

3.1.3 Seeding Cells

1. Aspirate media off 10 cm plates from part A.
2. Wash with 5 ml D-PBS (you can use the same D-PBS for each plate) and transfer to 15 ml conical.
3. Add 1 ml of 117.5 U/ml collagenase to each plate and incubate until cells have detached (~5–10′).
4. Wash plates with 5 ml D-PBS and transfer to 15 ml conical from **step 2.**
5. Centrifuge cells for 5′ at 180 × *g*.
6. Aspirate supernatant and add 1 ml of F-12 to pellet. Triturate gently with a blue tip pipette 10–15 times to get a single cell suspension.
7. Count an aliquot of cells.
8. Aspirate laminin from plates prepared in part B and plate cells @ 5000 cells/cm^2 (or less) with appropriate media volume for the well size.
9. Incubate plate for 30 min to allow cells to attach before moving on to image acquisition.

3.1.4 Image Acquisition

1. Turn on the microscope, check that CO_2 and humidity in the incubation chamber match those of the tissue culture incubator, mount the plate, and close the chamber.
2. Set the image acquisition conditions; both the interval and duration of the capture are defined by the goal of the experiment.

Interval: in our hands, an interval of 5 min is sufficient for the stage to image all inner wells of a 48-well plate at multiple positions. Additionally, this interval is short enough that the cells movement is generally trackable from one frame to the next provided you have plated at the proper density. Shorter intervals run the risk of not completing an image cycle before the program begins a new cycle. This may need to be optimized for each mechanical stage. Duration: the viability of the cells is finite, and may vary depending on culture conditions and according to individual incubator/imaging setups. We recommend running trial experiments in which motility and cell viability is evaluated for 24 h. In our hands, a duration of 12 h gives sufficient data points to track motility while maintaining a high degree of cell viability.

3. Select multiple regions of interest to track in each well and focus the camera on a per position basis. Ensure that you have a naming convention for each position to keep track of the images.
4. For brightfield images an exposure time of ~30 ms produces clear images although optimization will be necessary depending on the individual setup. For fluorescent imaging, determine the optimum exposure time empirically.
5. Turn off all light in the room and acquire images. Watch the first acquisition cycle to ensure that the capture proceeds as expected and all images are in focus.
6. When the experiment is over, wash plates in PBS and fix in 4 % paraformaldehyde for reference.

3.1.5 Image Tracking

1. Ideally, use an automated tracking program (i.e., [33, 34]) with parameters appropriate to your cells/experiment. When tracking manually, we have found that even with the same initial set of "rules" for cell selection and tracking there is significant person-to-person variability. Automated tracking is faster, more reproducible, and generates larger datasets than manual tracking.
2. If tracking manually, the experimental conditions should be blinded to prevent unconscious selection bias. Either get someone to track that does not know the naming conventions, or have someone else record and change the naming convention of the original data file.
3. Before tracking, define a set of criteria that all cells must meet to be included in the dataset. For example:
   (a) All cells must be chosen at random initially and checked to see if they meet the remaining criteria. If the cell meets all the criteria, it is tracked.
   (b) All cells must stay within the field of view for the entire duration of the experiment.
   (c) All cells must remain viable for the entire duration of the experiment.

(d) All cells must move freely (i.e., do not get stuck on bits of debris on the plate).

(e) If a cell divides, choose the cell to the right of the last point to continue the track.

4. Open tracking program and load a suite of images that corresponds to one position in one well.
5. Look at the last frame taken for a given suite of images and pick a cell at random (criterion i). If the cell is present in this frame, it satisfies criterion iii that all cells remain viable for the entire timelapse.
6. Follow this cell backward through the suite of images until you reach the very first image. If the cell never left the screen then criterion ii has been satisfied.
7. Did the cell pass the other selection criteria you set forth? If so you have found a cell that should be tracked.
8. Starting at the beginning of the suite of images track the cell as it moves through each frame. Try to standardize where you click on the cell to track, as clicking sometimes on the leading edge and sometimes the trailing edge of a cell will result in a different distance traveled and therefore a different velocity. To overcome this issue zoom in on the images until they are sufficiently large to easily click on the nucleus in each frame.
9. Repeat **steps 4–7** for additional cells in a given ROI. We usually track 5+ cells per ROI to include a representative sample.

3.2 Timelapse Video Analysis of Satellite Cells on Single Myofibers

3.2.1 Myofiber Isolation

Isolate viable myofibers from mouse muscle using your preferred protocol and, if not imaging immediately, culture for 12–24 h in Ham's F-12 supplemented with 15 % horse serum, P/S, and 0.5 nmol/L recombinant human FGF-2. This culture step will help to eliminate any damaged or otherwise nonviable myofibers and increase the success rate during embedding and imaging.

3.3 Embedding

3.3.1 Important: prechill all tubes, plates, pipettes, etc. and keep all solutions, plates, and tubes on ice at all times until **Step 5**

1. Make up 5 ml of collagen gel: to a 15 ml conical tube add in order

   2500 μl acid-extracted rat tail type I collagen (Sigma) 2 mg/ml.

   250 μl 10× M-199.

   15 μl5 N NaOH.

   2235 μl growth media.

   All reagents should be cold and mixed on ice. An equal volume of water in a separate 15 ml conical tube should be compared to the volume of collagen since collagen is very viscous. While mixing make sure to pipet slowly to avoid bubbles. Make sure to mix until the solution has a homogenous appearance. Use low adherence tips whenever handling collagen.

2. If you are adding growth factors, antibodies, or other conditions to the gel add them prior to adding the fibers. If using multiple conditions, aliquot the collagen before adding and be sure to reserve at least 1–2 ml of the collagen for pre-equilibrating the fibers.
3. For this step, it helps if there are two people (one with the plate and one with the fibers). Using an optically clear 48-well plate (Corning Costar) add 200 μl of collagen to each experimental well; do not use the wells at the perimeter of the plate. With a fire-polished glass Pasteur pipette that has been prerinsed in medium with serum to prevent the myofibers from sticking, using a dissection scope pick about twice the number of fibers needed for the entire experiment into a 3 cm culture dish (on ice). Remove as much of the media as possible without drying out the fibers and add 1–2 ml of cold collagen to the edges of the pool of fibers, then move the myofibers carefully into the gel with the fire-polished pipette and swirl gently.
4. Using the same glass pipette, aspirate 3–5 myofibers in the minimum amount of collagen possible and gently dispense into the center of a well of collagen. Continue until fibers have been added to all wells, check in the microscope to be sure all wells have visible fibers and they are well dispersed in the gel.
5. Put the plate in a tissue culture (humidified, 5 % CO_2 at 37 °C) incubator to polymerize for 30′.
6. After 30′, overlay the wells containing the fibers with 300 μl medium and add 500 μl of sterile DI water to the wells on the perimeter. This will help both with humidity and heat dispersion while imaging.

3.3.2 Timelapse Imaging

1. Turn on the microscope, check that temperature, CO_2, and humidity in the incubation chamber match. Minimize the time taken to transfer the culture dish to the chamber. Mount the plate and close the chamber.
2. Set the image acquisition conditions; we generally use 10′ intervals and image for 24 h.
3. Select multiple regions of each myofiber to image, trying to focus on the surface (top or bottom). Ensure that you have a naming convention for each position to keep track of the images.
4. For brightfield images an exposure time of ~30 ms produces clear images although optimization will be necessary depending on the individual setup. For fluorescent imaging, determine the optimum exposure time empirically.
5. Turn off all light in the room and acquire images. Watch the first acquisition cycle to ensure that the capture proceeds as expected and all images are in focus.

3.4 Method for Assessing 3D Human Satellite Invasion In Vitro

All the procedures should be carried out in a Class II biosafety cabinet using sterile technique (human cells of any kind are considered Biosafety Level 2). Also make sure to wipe the hood down with ddH_2O and 70 %EtOH before and after use.

3.4.1 Culture of Human Myogenic Cells

Human myoblasts immortalized by expression of telomerase and cdk4 [35, 36] can be requested from MYOBANK, affiliated with EUROBIOBANK. These cells can be cultured in a multitude of medias, but in our hands these three are most effective:

X-Media: (25 % M-199 in DMEM) supplemented with 2.5 ng/ml hepatocyte growth factor, 10^{-7} M dexamethasone, and 20 % fetal calf serum.

SkGM™-2 (LONZA).

Skeletal Muscle Cell Growth Medium (PromoCell).

3.4.2 Seeding Onto Collagen Gels

1. The day of the experiment, prepare collagen gel (ml = number of wells/10 + 10 %). Mix in a 15 ml conical tube in order:

   50 % Collagen Type I (2 mg/ml final concentration).

   5 % 10× M199.

   0.6 % NaOH (5 N).

   45.4 % growth media.

   All reagents should be cold and mixed on ice. While mixing make sure to pipet slowly to avoid bubbles. An equal volume of water should be compared to the volume of collagen since collagen is very viscous. Make sure to mix until the solution has a homogenous appearance.
2. If you are adding growth factors, antibodies, or other conditions to the gel, aliquot the collagen as appropriate before adding.
3. Once the collagen solution has been mixed, add 100 µl of the solution per well to a prechilled 96-well tissue culture plate. Place the plate in a 37 °C tissue culture incubator for at least 30′ to allow ample time for polymerization. It can be left in the incubator for 1–2 h.
4. Human myoblasts should be grown on a $10cm^2$ tissue culture plate to approximately 60–80 % confluency prior to lifting. When ready, aspirate all of the media off the plate and gently rinse with 1× D-PBS. Afterward, add 2 ml of 0.25 % Trypsin to the plate and place it in a 37 °C incubator or on a 37 °C hot plate for 5–10 min. Checking frequently to see if the cells have lifted from the bottom. Collect the 2 ml of trypsin with cells into a 15 ml conical tube and centrifuge for 5′ at 260×*g* at 4 °C. After pelleting, aspirate the supernatant and resuspend the cell pellet in 1 ml D-PBS. Repeat the centrifugation step.
5. Aspirate the supernatant and resuspend the cell pellet in 1 ml of growth medium. Count cells and make up to 10^6 cells/ml with growth medium. Pipette 100 µl (100,000 cells) gently on

top of each collagen well and return the plate to the incubator. Warm growth media can be changed every 2–4 days by gently pipetting off the top layer of media. Cells should be allowed to penetrate the 3D matrix for roughly 90 h.

6. If timelapse video analysis of invasion (*see* Supplemental Movie 1) is planned, instead of returning the plate to the incubator refer to protocol II.C above. Select multiple planes in the *Z*-axis (even though there no cells currently visible) at 10-micron intervals. If cells have been seeded underneath a 3D layer of collagen (*see* Subheading 4) focus on the bottom of the monolayer and select multiple *Z* planes at 10-micron intervals going up.

3.4.3 Embedding Matrices for Cryosectioning

1. For the fixed analysis, the plate should gently be removed from the incubator. The culture media should gently be pipetted off without puncturing the 3D collagen. Gently pipet 100 μl of 1× DPBS on top of each gel and remove. Place the 96-well plate on ice and add 100 μl of 4 % paraformaldehyde (PFA) to each well containing the 3D collagen. Allow 3D gels to fix for 20–30 min on ice. Remove the PFA and rinse gently three times with 1× PBS.
2. Gently use forceps to circle 3/4 of the perimeter of the gel and pluck it out into a 1.5 ml Eppendorf tube with 1 ml of 30 % sucrose solution. Place the gel with the sucrose solution at 4 °C with gentle agitation. Wait 30′ and remove the solution and replace with fresh, cold 30 % sucrose. Repeat this process two more times and then remove the 30 % sucrose, add 1 ml 50 % sucrose, and incubate the matrices at 4 °C overnight with gentle agitation. The matrices should sink to the bottom of the 50 % sucrose solution after the overnight incubation.
3. Next, gently use forceps to remove the 3D matrix and place it into a biopsy-size cryomold with OCT. Ensue that the orientation of the matrix is as it would rest in the 96-well plate (with the initial monolayer of cells on top).
4. Place the cryomold with the embedded matrix carefully on dry ice and make sure that the gel remains in proper orientation. Once the OCT turns completely opaque the gel can be stored at −80 °C. The gel is now ready to be cryosectioned.

3.4.4 Cryosectioning and Staining

1. Cryosection the 3D matrix at 40 μm onto high-adhesion glass slides. Immediately place into a humid chamber; all subsequent steps are done in the humid chamber. Draw a hydrophobic perimeter around the sections using a hydrophobic pen. Next, cross-link the sections to the slide using 4%PFA for 10–15′ on ice. Rinse 2× gently with 1× PBS and add 0.5 % Triton X-100 solution to the top of the slides and incubate at 4 °C for 1 h.
2. Remove the 0.5 % Triton X-100 solution and add the Tris-glycine solution to neutralize the permeabilization.

3. Add blocking serum (usually 10 % serum from the source of the secondary antibody) to slide and incubate at 4 °C for 3–4 h. Remove blocking solution and incubate in primary antibody at 4 °C overnight (this is necessary).
4. Remove the 1° antibody and excess serum. Add an excess of Tween-20 solution and aspirate. Follow by four consecutive washes with Tween-20 solution for 30′ each at 4 °C.
5. Remove final wash of Tween-20 solution and add 2° antibody solution. Incubate at 4 °C for 1–2 h.
6. Remove 2° solution and perform four consecutive Tween-20 washes for 30′ each at 4 °C.
7. Add mounting medium and seal the slide, it is now ready for imaging. We calculate the maximum and average penetration of invading cells using in-software micrometers.

4 Notes

I. Timelapse Analysis of Satellite Cells In Vitro on Purified Substrates

1. If tracking manually, be aware of the potential for deviation (not necessarily error) in results between different people, even for the same movies. You can avoid this by using automated tracking software, or mitigate it by having only one tracker for each set of movies that are to be compared. Alternatively, you can try to "train" each other to select that same cells for tracking by comparing results from the same movie.
2. If you want to do immunohistochemistry to get expression data from the cells after the movies are analyzed, you should be able to stain the fixed plate, put it in the microscope you used for the timelapse, reopen the series, and have the microscope automatically return to the same fields for fluorescent imaging.

II. Timelapse Video Analysis of Satellite Cells on Single Myofibers.

The most likely cause of failure in this assay is death of the myofibers after they are embedded. This can be caused by rough handling during transfer to the collagen gel, shear forces due to uneven polymerization of the collagen gel, or changes in temperature during transfer to the microscope. Unfortunately, death of one myofiber in a well frequently leads to death of the remaining fibers, due to the effects of proteases released from the fiber that depolymerize the gel. Particularly when testing multiple conditions, we recommend at least three wells per condition to help ensure that duplicate wells will be viable for all conditions.

III. Method for Assessing 3D Human Satellite Invasion In Vitro

1. If you want to see if the cells will "invade up," let them first adhere to the bottom of a gelatin-coated 96-well plate for several hours then add the cold collagen on top and let polymerize. Continue as listed, except bear in mind the reversed orientation during embedding and sectioning. Importantly, make sure to set the *Z* plane deep enough to capture the bottom most visible cells.

References

1. Campion DR (1984) The muscle satellite cell: a review. Int Rev Cytol 87:225–251
2. Hawke TJ, Garry DJ (2001) Myogenic satellite cells: physiology to molecular biology. J Appl Physiol 91:534–551
3. Shi X, Garry DJ (2006) Muscle stem cells in development, regeneration, and disease. Genes Dev 20:1692–1708
4. Peault B, Rudnicki M, Torrente Y, Cossu G, Tremblay JP, Partridge T, Gussoni E, Kunkel LM, Huard J (2007) Stem and progenitor cells in skeletal muscle development, maintenance, and therapy. Mol Ther 15:867–877
5. Schmalbruch H (1978) Satellite cells of rat muscles as studied by freeze-fracturing. Anat Rec 191:371–376
6. Morgan JE, Coulton GR, Partridge TA (1987) Muscle precursor cells invade and repopulate freeze-killed muscles. J Muscle Res Cell Motil 8:386–396
7. Hardy D, Besnard A, Latil M, Jouvion G, Briand D, Thepenier C, Pascal Q, Guguin A, Gayraud-Morel B, Cavaillon JM, Tajbakhsh S, Rocheteau P, Chretien F (2016) Comparative Study of Injury Models for Studying Muscle Regeneration in Mice. PLoS One 11:e0147198
8. Schultz E, Jaryszak DL, Gibson MC, Albright DJ (1986) Absence of exogenous satellite cell contribution to regeneration of frozen skeletal muscle. J Muscle Res Cell Motil 7:361–367
9. Phillips GD, Hoffman JR, Knighton DR (1990) Migration of myogenic cells in the rat extensor digitorum longus muscle studied with a split autograft model. Cell Tissue Res 262:81–88
10. Hughes SM, Blau HM (1990) Migration of myoblasts across basal lamina during skeletal muscle development. Nature 345:350–353
11. McCormick KM, Schultz E (1992) Mechanisms of nascent fiber formation during avian skeletal muscle hypertrophy. Dev Biol 150:319–334
12. Ishido M, Kasuga N (2011) In situ real-time imaging of the satellite cells in rat intact and injured soleus muscles using quantum dots. Histochem Cell Biol 135:21–26
13. Chazaud B, Christov C, Gherardi RK, Barlovatz-Meimon G (1998) In vitro evaluation of human muscle satellite cell migration prior to fusion into myotubes. J Muscle Res Cell Motil 19:931–936
14. Villena J, Brandan E (2004) Dermatan sulfate exerts an enhanced growth factor response on skeletal muscle satellite cell proliferation and migration. J Cell Physiol 198:169–178
15. Bondesen BA, Jones KA, Glasgow WC, Pavlath GK (2007) Inhibition of myoblast migration by prostacyclin is associated with enhanced cell fusion. FASEB J 21:3338–3345
16. Siegel AL, Atchison K, Fisher KE, Davis GE, Cornelison DDW (2009) 3D timelapse analysis of muscle satellite cell motility. Stem Cells 27:2527–2538
17. Bischoff R (1997) Chemotaxis of skeletal muscle satellite cells. Dev Dyn 208:505–515
18. Germani A, Di Carlo A, Mangoni A, Straino S, Giacinti C, Turrini P, Biglioli P, Capogrossi MC (2003) Vascular endothelial growth factor modulates skeletal myoblast function. Am J Pathol 163:1417–1428
19. Griffin CA, Kafadar KA, Pavlath GK (2009) MOR23 promotes muscle regeneration and regulates cell adhesion and migration. Dev Cell 17:649–661
20. Stark DA, Karvas RM, Siegel AL, Cornelison DDW (2011) Eph/ephrin interactions modulate muscle satellite cell motility and patterning. Development 138:5279–5289
21. Lund DK, Cornelison DD (2013) Enter the matrix: shape, signal and superhighway. FEBS J:4089–4099
22. Deryugina EI, Bourdon MA, Reisfeld RA, Strongin A (1998) Remodeling of collagen matrix by human tumor cells requires activation and cell surface association of matrix metalloproteinase-2. Cancer Res 58:3743–3750
23. Hotary K, Allen E, Punturieri A, Yana I, Weiss SJ (2000) Regulation of cell invasion and morphogenesis in a three-dimensional type I collagen matrix by membrane-type matrix

metalloproteinases 1, 2, and 3. J Cell Biol 149:1309–1323

24. Allen DL, Teitelbaum DH, Kurachi K (2003) Growth factor stimulation of matrix metalloproteinase expression and myoblast migration and invasion in vitro. Am J Physiol Cell Physiol 284:C805–C815
25. Lund DK, Mouly V, Cornelison DDW (2014) MMP-14 is necessary but not sufficient for invasion of three-dimensional collagen by human muscle satellite cells. Am J Physiol Cell Physiol 307:140–149
26. Dehmel T, Janke A, Hartung HP, Goebel HH, Wiendl H, Kieseier BC (2007) The cell-specific expression of metalloproteinase-disintegrins (ADAMs) in inflammatory myopathies. Neurobiol Dis 25:665–674
27. Fukushima K, Nakamura A, Ueda H, Yuasa K, Yoshida K, Takeda S, Ikeda S (2007) Activation and localization of matrix metalloproteinase-2 and -9 in the skeletal muscle of the muscular dystrophy dog (CXMDJ). BMC Musculoskelet Disord 8:54
28. Torrente Y, El Fahime E, Caron NJ, Bresolin N, Tremblay JP (2000) Intramuscular migration of myoblasts transplanted after muscle pretreatment with metalloproteinases. Cell Transplant 9:539–549
29. Brzoska E, Kowalewska M, Markowska-Zagrajek A, Kowalski K, Archacka K, Zimowska M, Grabowska I, Czerwinska AM, Czarnecka-Gora M, Streminska W, Janczyk-Ilach K, Ciemerych MA (2012) Sdf-1 (CXCL12) improves skeletal muscle regeneration via the mobilisation of Cxcr4 and CD34 expressing cells. Biol Cell 104:722–737
30. Zimowska M, Olszynski KH, Swierczynska M, Streminska W, Ciemerych MA (2012) Decrease of MMP-9 activity improves soleus muscle regeneration. Tissue Eng Part A 18:1183–1192
31. Hindi SM, Shin J, Ogura Y, Li H, Kumar A (2013) Matrix Metalloproteinase-9 Inhibition Improves Proliferation and Engraftment of Myogenic Cells in Dystrophic Muscle of mdx Mice. PLoS One 8:e72121
32. Capkovic KL, Stevenson S, Johnson MC, Thelen JJ, Cornelison DDW (2008) Neural cell adhesion molecule (NCAM) marks adult myogenic cells committed to differentiation. Exp Cell Res 314:1553–1565
33. Chowdhury AS, Paul A, Bunyak F, Cornelison DDW, Palaniappan K (2012) Semi-automated tracking of muscle satellite cells in brightfield microscopy video, Proceedings – International Conference on Image Processing, ICIP. pp 2825–2828.
34. Al-Shanti N, Faulkner SH, Saini A, Loram I, Stewart CE (2011) A semi-automated programme for tracking myoblast migration following mechanical damage: manipulation by chemical inhibitors. Cell Physiol Biochem 27:625–636
35. Zhu CH, Mouly V, Cooper RN, Mamchaoui K, Bigot A, Shay JW, Di Santo JP, Butler-Browne GS, Wright WE (2007) Cellular senescence in human myoblasts is overcome by human telomerase reverse transcriptase and cyclin-dependent kinase 4: consequences in aging muscle and therapeutic strategies for muscular dystrophies. Aging Cell 6:515–523
36. Mamchaoui K, Trollet C, Bigot A, Negroni E, Chaouch S, Wolff A, Kandalla PK, Marie S, Di Santo J, St Guily JL, Muntoni F, Kim J, Philippi S, Spuler S, Levy N, Blumen SC, Voit T, Wright WE, Aamiri A, Butler-Browne G, Mouly V (2011) Immortalized pathological human myoblasts: towards a universal tool for the study of neuromuscular disorders. Skelet Muscle 1:34

Chapter 17

Effects of Macrophage Conditioned-Medium on Murine and Human Muscle Cells: Analysis of Proliferation, Differentiation, and Fusion

Marielle Saclier*, Marine Theret*, Rémi Mounier, and Bénédicte Chazaud

Abstract

Skeletal muscle is a highly plastic tissue, which is able to regenerate after an injury. Effective and complete regeneration requires interactions between myogenic precursor cells and several cell types such as macrophages. Bone marrow derived macrophages in mouse and monocyte-derived macrophages in human are useful tools to obtain macrophage populations that may be specifically activated/polarized in vitro (e.g., pro-inflammatory, anti-inflammatory, and alternatively activated macrophages). In vitro, human or murine primary myogenic cells recapitulate the adult myogenesis program through proliferation, myogenic differentiation, and fusion. Macrophages being highly secreting cells, they act on various biological processes including adult myogenesis. Here, we present protocols to analyze in vitro the effect of macrophage-secreted factors on muscle cell proliferation or differentiation in both mouse and human.

Key words Macrophages, Muscle precursor cells, Proliferation, Differentiation, Myogenesis, Human, Mouse, Conditioned media

1 Introduction

Macrophages are immune cells essential to skeletal muscle regeneration [1]. During this process, they adopt distinct and sequential phenotypes [1–3]. Soon after injury, circulating blood monocytes infiltrate the damaged muscle and differentiate into inflammatory macrophages. These macrophages stimulate the proliferation of muscle precursor cells (MPCs) through the secretion of pro-inflammatory molecules such as TNFα, IL-6, and IL-1β. Upon phagocytosis of muscle debris, inflammatory macrophages skew their phenotype to an anti-inflammatory profile, which sustains the differentiation and fusion of MPCs and the growth of myofibers through the secretion of anti-inflammatory effectors (e.g., IL-10, TGFβ) [1–2]. Thus, depending on their

*These authors contributed equally to this work.

Eusebio Perdiguero and DDW Cornelison (eds.), *Muscle Stem Cells: Methods and Protocols*, Methods in Molecular Biology, vol. 1556, DOI 10.1007/978-1-4939-6771-1_17, © Springer Science+Business Media LLC 2017

phenotype, macrophages support and act on the sequential phases of muscle regeneration. This sequence of inflammatory then anti-inflammatory macrophages has been also evidenced in human regenerating muscle, although both inflammatory types of macrophages may be present in the same regenerating muscle areas at the same time [2]. In vitro, mouse Bone Marrow Derived Macrophages (BMDM) [4] and monocyte-derived macrophages in human [1] have been shown to be useful to characterize macrophage functions in well-defined inflammatory conditions. As macrophages are highly secreting cells, culture of MPCs with conditioned media (CM) of macrophages is an efficient way to analyze the effects of macrophage populations on the different steps of myogenesis (proliferation, differentiation, and fusion). Here are presented the conditions that have been already used to stimulate macrophages in distinct inflammatory phenotypes: pro-inflammatory (stimulation with IFNγ and /or LPS), anti-inflammatory (IL10 and/or glucocorticoids), and alternatively activated macrophages (IL-4). Of course, a variety of stimuli may be tested to trigger specific inflammatory state in macrophages [5].

2 Materials

2.1 Culture of Murine Macrophages and MPCs

1. 2-well permanox Lab-Tek® (Nunc #177429, *see* **Note 1**).
2. 12- and 24-well plates.
3. Matrigel® low growth factor (Corning #356231).
4. Horse serum, heat inactivated at 56 °C for 30 min.
5. Ultroser™ G serum substitute (#15950-017, Pall Corporation) (*see* **Note 2**).
6. Serum-free macrophage medium: DMEM Glutamax, 4.5 g/l glucose, 1× pyruvate (#31966-021) containing 100 U/ml Penicillin/streptomycin (P/S).
7. Macrophage medium: DMEM Glutamax, 4.5 g/l Glucose, 1× Pyruvate (#31966-021), containing10 % Heat-inactivated Fetal Bovine Serum (FBS) and 100 U/ml P/S.
8. MPC growth medium: DMEM/F-12 Glutamax (#31331-028), 20 % FBS, 100 U/ml P/S. Filter through Stericup-GP 0.22 μm and add 2 % Ultroser G (*see* **Note 2**).
9. Cytokines (R&D system) are dissolved in PBS 1X Ca^{2+} and Mg^{2+} free (PBS): IFNγ (#485-MI, stock solution at 50 μg/ml), IL-4 (#404-MI, stock solution at 5 μg/ml), IL-10 (#417-ML, stock solution at 5 μg/ml).
10. Murine MPCs are isolated from *gastrocnemius* and *tibialis anterior* and cultured as previously described in [6].
11. Murine macrophages are isolated from bone marrow and prepared as previously described in [7].

2.2 Immunolabeling of Murine MPCs

1. 4 % Paraformaldehyde (PFA) (Boster Biological Technology #AR1068).
2. Anti-Ki67 antibody made in rabbit (Abcam #ab15580, use at 5 μg/ml).
3. Anti-desmin antibody made in rabbit (Abcam #ab6322, use at 0.085 μg/ml).
4. Donkey anti-rabbit Cy3 secondary antibodies (Jackson Immuno Research Inc. #711-165-152) (*see* **Note 3**).
5. Hoescht 33342 (Sigma Aldrich #B2261, dilute in PBS at 2 mM for stock, use at 2 nM).
6. Mounting medium (Interchim#FP-483331).
7. Coverslips 18 × 18 mm.

2.3 Culture of Human Macrophages and MPCs

1. 12-well plates.
2. T75 culture flasks.
3. Glass coverslips 18 mm diameter (Marienfeld #0111580) that have been sterilized.
4. MPC growth medium: Ham-F12 Glutamax (#31331) containing 15 % FBS, 100 U/ml P/S (#15140).
5. MPC differentiation medium: Ham-F12Glutamax (#31331) containing 5 % FBS, 100 U/ml P/S.
6. Macrophage medium: Advanced RPMI 1640 (#12633) containing 1× Glutamine, 100 U/ml P/S, 10 mM Hepes, 100× MEM vitamins, 0.5 mM2-mercaptoethanol, 15 % FBS.
7. Low serum macrophage medium: Advanced RPMI 1640 containing 1× Glutamine, 100 U/ml P/S, 10 mM Hepes, 100× MEM vitamins, 0.5 mM 2-mercaptoethanol, 0.5 % FBS.
8. Lipopolysaccharide (LPS) (#L4130 Sigma Aldrich, stock solution at 1 mg/ml in culture medium)
9. Dexamethasone (Dex) (D4902 Sigma Aldrich, stock solution at 20 μg/ml, add 1 ml of 100 % ethanol *per* mg, and then add 49 ml of culture medium *per* ml of ethanol).
10. Cytokines (R&D system): IFNγ (#285-IF/CF, stock solution at 0.1 mg/ml in deionized water), IL-4 (#204-IL/CF, stock solution at 0.1 mg/ml in PBS, IL-10 (#217-IL/CF, stock solution at 50 μg/ml are dissolved in PBS).
11. Cell Proliferation ELISA, BrdU (colorimetric) (Roche Diagnostics GmbH #11647229001).
12. Human MPCs are isolated from human muscle biopsies and cultured as previously described [8].
13. Human macrophages are isolated from human blood and cultured as previously described [8].

2.4 Immunolabeling of Human MPCs

1. 4 % PFA (Boster Biological Technology #AR1068).
2. Anti-Myogenin antibody made in mouse (BD Pharmingen #556358, use at 10 μg/ml).
3. Anti-desmin antibody made in rabbit (Abcam #32362, use at 60 μg/ml).
4. Donkey anti-rabbit Cy3 1/200 secondary antibodies (Jackson Immunoresearch Inc. #711-165-152), biotinylated anti-mouse antibodies (Vector Laboratories #BA-2000), and Streptavidin DTAF (Beckman Coulter #PN IM0307) (*see* **Note 3**).
5. Hoescht 33342 (Sigma Aldrich #B2261, dilute in PBS at 2 mM for stock, use at 2 nM).
6. Mounting medium (Interchim#FP-483331).
7. Microscopic slides.

3 Methods

3.1 Effect of Macrophage-CM on MPC Proliferation in Mouse (Fig. 1)

1. Day 0, macrophages are seeded at 71,000 cells/cm² in three wells for each condition (24-well plate) in macrophage medium (500 μl/well).
2. 6 h after plating, macrophages (that should have adhered) are activated with cytokines: IFNγ (50 ng/ml), IL-10 (10 ng/ml), IL-4 (10 ng/ml) in macrophage medium to obtain pro-inflammatory, anti-inflammatory, and alternatively activated macrophages, respectively.
3. Day 3, cells are washed three times with PBS 1× (500 μl/well) and once with 500 μl serum-free DMEM medium (*see* **Note 4**).
4. 250 μl of serum-free macrophage medium is added *per* well (three empty wells are used for the control) and the cells are incubated for 24 h for the preparation of macrophage-CM.

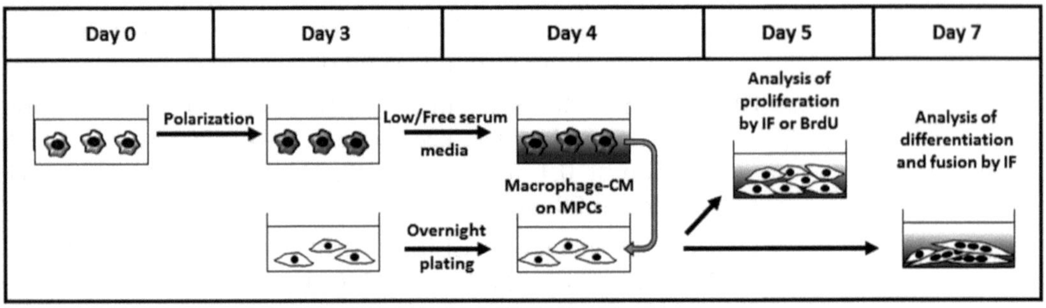

Fig. 1 Method to analyze the effects of macrophage-CM on MPCs. Day 0: macrophages are polarized with specific activators. Day 3: conditioned medium from macrophages is prepared through incubation of the cells for 24 h in serum-free or low-serum medium. MPCs are seeded. Day 4: macrophage-CM is recovered and added to MPC cultures. Day 5: Proliferation is analyzed by immunofluorescence or BrdU incorporation. Day 7: Differentiation/fusion is analyzed by immunofluorescence

5. Day 3 (*see* **Note 5**), MPCs are seeded at 10,000 cells/cm^2 on matrix-coated Lab-Tek®(Nunc #177429, *see* **Note 6**) in MPC growth medium.
6. Day 4, macrophage-CM is recovered. Triplicate-wells are pooled in 15 ml conical centrifuge tubes, FBS is added at 2.5 % (*see* **Note 7**) and macrophage-CM is centrifuged for 10 min at 800 × *g*, to eliminate any cell debris.
7. MPC cultures are carefully washed (*see* **Note 8**) three times with PBS 1× and 700 μl of macrophage-CM *per* well is added.
8. Day 5 and 24 h after addition of macrophage-CM, perform the Ki67 labeling (*see* **Note 9** and Subheading 3.3 for the protocol).

3.2 Effect of Macrophage-CM on MPC Differentiation in Mouse (Fig. 1)

1. Day 0, macrophages are seeded at 71,000 cells/cm^2 in two wells for each condition (12-well plate) in macrophage medium (1 ml/well).
2. After 6 h of plating, macrophages are activated with cytokines: IFNγ (50 ng/ml), IL-10 (10 ng/ml), IL-4 (10 ng/ml) in macrophage medium, to obtain pro-inflammatory, anti-inflammatory, and alternatively activated macrophages, respectively.
3. Day 3, cells are washed three times with PBS 1× (1 ml/well) and once with 1 ml serum-free macrophage medium (*see* **Note 4**).
4. 500 μl of serum-free macrophage medium is added *per* well (two empty wells are used for the control) and the cells are incubated for 24 h for the preparation of macrophage-CM.
5. Day 3 (*see* **Note 5**), MPCs are seeded at 30,000 cells/cm^2 on matrix-coated Lab-Tek®(Nunc #177429, *see* **Note 6**) in MPC growth medium.
6. Day 4, macrophage-CM is recovered. Duplicate-wells are pooled in 15 ml conical centrifuge tubes, 2 % horse serum is added (*see* **Note10**), and macrophage-CM is centrifuged for 10 min at 800 × *g*, to eliminate any debris.
7. MPC cultures are carefully washed (*see* **Note 8**) three times with PBS 1× and 700 μl of macrophage-CM per well is added.
8. Day 7, check that in the untreated condition MPCs are differentiated (*see* **Note 11**), then perform desmin labeling (*see* **Note 9** and Subheading 3.3 for the protocol).

3.3 Immunolabelings in Mouse

All steps are performed at room temperature (RT), unless otherwise indicated (*see* **Note 12**).

1. The above part of the Lab-Tek® is removed (walls and silicone joint).
2. Cells are washed three times for 5 min with PBS 1× in a Coplin jar.

3. Cells are fixed for 15 min with PFA 4 % in a Coplin jar.
4. Cells are washed three times for 5 min with PBS 1× in a Coplin jar.
5. Cells are permeabilized for 10 min with 0.5 % Tritonin PBS 1× in a Coplin jar.
6. Cells are washed three times for 5 min with PBS 1× in a Coplin jar.
7. 40 μl of primary antibodies diluted in PBS 1× (anti-Ki67 antibody for the proliferation assay or anti-desmin antibody for the differentiation assay) is dropped on each well and is covered with a piece of parafilm (*see* **Note 13**).
8. Cells are incubated with primary antibodies for 2 h at 37 °C in a humid chamber (*see* **Note 14**).
9. The parafilm is carefully removed and cells are washed three times for 5 min with PBS 1× in a Coplin jar.
10. 40 μl of secondary antibodies diluted in PBS 1× (anti-rabbit antibody and cover the wells with a piece of parafilm) is dropped on each well and is covered with a piece of parafilm (*see* **Note 13**).
11. Cells are incubated with secondary antibodies for 1 h at 37 °C in a humid chamber (*see* **Note 14**).
12. The parafilm is carefully removed and cells are washed three times for 5 min with PBS 1× in a Coplin jar.
13. Cells are quickly (10 s) incubated in a Coplin jar containing 2 nM Hoechst in PBS 1×.
14. Cells are washed once for 5 min with PBS 1× in a Coplin jar.
15. Cells are mounted in mounting medium with a coverslip and stored at +4 °C.
16. About 10 pictures of each well are recorded at 20× magnification with an epifluorescence microscope (*see* **Note 15** and Fig. 2).

3.4 Effect of Macrophage-CM on MPC Proliferation in Human (Fig. 1)

1. Day 0, macrophages are seeded at 36,000 cells/cm^2 in three wells for each condition (24-well plate) in macrophage medium (500 μl/well).
2. After 6 h (minimum) of plating, macrophages are activated with cytokines and hormone for polarization: LPS (1 μg/ml) and IFNγ (10 ng/ml), IL-10 (10 ng/ml) and Dex (80 ng/ml), and IL-4 (10 ng/ml), to obtain pro-, anti- and alternatively activated macrophages, respectively, in macrophage medium (500 μl/well).
3. Day 3, macrophages are washed three times with 500 μl PBS 1× (*see* **Note 4**).
4. 250 μl of low serum macrophage medium is added *per* well (three empty wells are used for the control) and cells are incubated for 24 h.

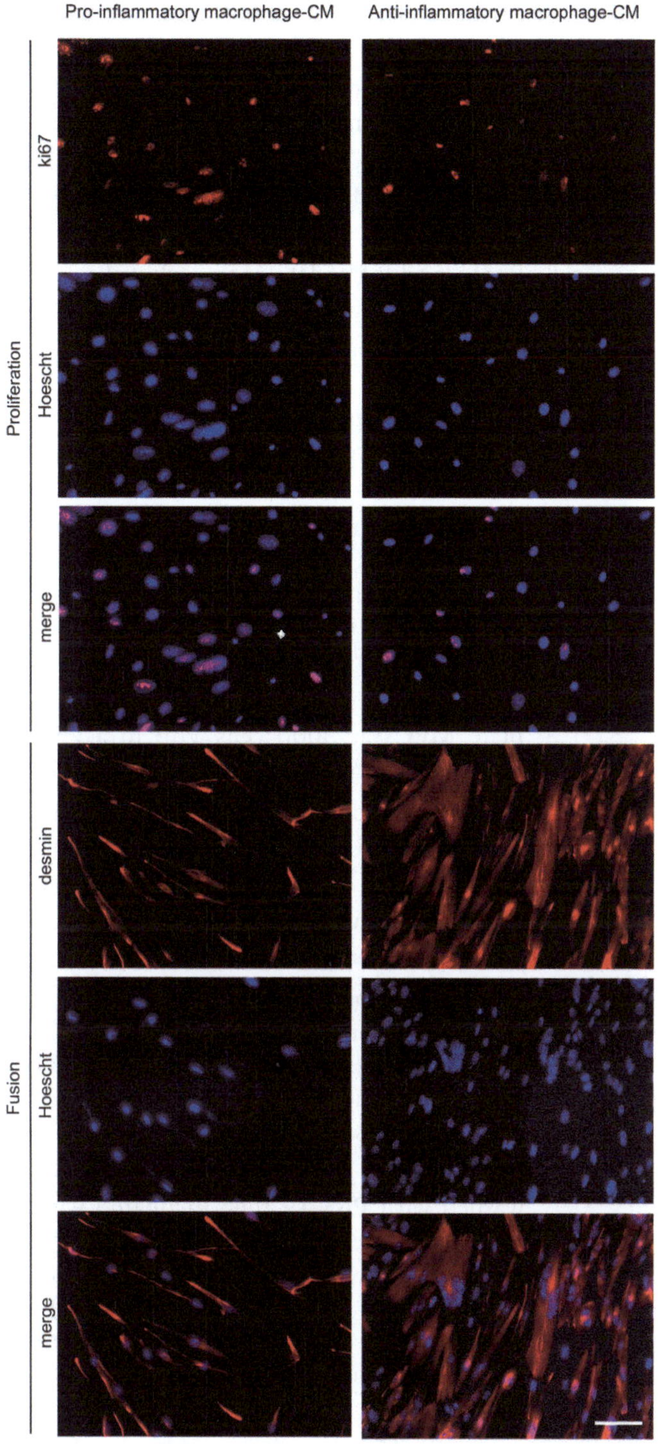

Fig. 2 Immunolabeling of murine MPCs cultured with inflammatory or anti-inflammatory macrophage-CM in proliferation or fusion conditions. Proliferation (*upper panel*): Ki67 (*red*) and nuclei (*blue*) labeling of MPCs in proliferation condition with inflammatory macrophage-CM (*left*) or anti-inflammatory macrophage-CM (*right*). Fusion (*lower panel*): Desmin (*red*) and nuclei (*blue*) labeling of MPCs in differentiation condition with inflammatory macrophage-CM (*left*) or with anti-inflammatory macrophage-CM (*right*). Inflammatory macrophage-CM activates proliferation of MPCs while anti-inflammatory macrophage-CM activates fusion of MPCs. Bar = 50 μm

5. MPCs are seeded at 12,000 cells/cm² in a 24-well plate in MPC growth medium (three wells *per* condition) (*see* **Note 16**).
6. Day 4, macrophage-CM is recovered: triplicate-wells are pooled in 15 ml conical centrifuge tubes and centrifuged for 10 min at 800 × *g*.
7. MPCs are washed three times MPCs with PBS 1× and 250 µl of macrophage-CM is added *per* well.
8. BrdU is added at 10 µM in each well.
9. Day 5, MPCs are fixed and BrdU is measured as recommended by manual supplier (colorimetric assay).

3.5 Effect of Macrophage- CM on MPC Differentiation in Human (Fig. 1)

1. Day 0, macrophages are seeded at 36,000 cells/cm² in three wells for each condition (12-well plate) in macrophage medium (1 ml/well).
2. After 6 h (minimum) of plating, macrophages are activated with cytokines and hormone for polarization: LPS (1 µg/ml) and IFNγ (10 ng/ml), IL-10 (10 ng/ml) and Dex (80 ng/ml), and IL-4 (10 ng/ml), to obtain pro-, anti- and alternatively activated macrophages, respectively, in macrophage medium (500 µl/well).
3. Plate MPCs at 3000 cells/cm² in a 75 cm² culture flask in MPC differentiation medium (*see* **Note 17**).
4. Day 3, macrophages are washed three times with 1 ml of PBS 1× (*see* **Note 4**).
5. 500 µl of low serum macrophage medium is added *per* well (three empty wells are used for the control) and cells are incubated for 24 h.
6. Glass coverslips are deposited in the bottom of 12-well plates (three wells for each condition), and 500 µl of MPC differentiation media is added (the air under the coverslip is removed by pressing the coverslip with a tip or a pipette so that the coverslip sticks in the bottom of the well).
7. MPCs (which are now myocytes) are trypsinated: cells are washed with PBS 1× then incubated with 1.5 ml of Trypsin-EDTA. Cell detachment is monitored under a microscope and should occur within few minutes. Cells are recovered in MPC differentiation media, centrifuged for 10 min at 250 × *g*, and seeded at 500 cells/cm² in MPC differentiation media on the coverslips in the 12-well plates.
8. Day 4, macrophage-CM is recovered. Triplicate-wells are pooled in 15 ml conical centrifuge tubes and centrifuged for 10 min at 800 × *g* to remove cell debris.
9. MPCs are washed three times with PBS 1× and 500 µl of macrophage-CM is added *per* well.

10. Day 7, MPCs are washed three times with PBS 1×. Myogenin-desmin immunolabeling is directly performed or coverslips. Alternatively, coverslips can be dry at RT and stocked at −20°C for further use (*see* Subheading 3.6 for the protocol).

3.6 Myogenin-Desmin Immunolabeling on Human MPCs

Steps are performed at RT unless otherwise indicated (*see* **Note 12**). Washes of coverslips are done in staining blocs (Vaccine dishes #2020, Glaswarenfabrik Karl Hecht GmbH & Co KG). Coverslips are handled with thin forceps, on the edges of the coverslips (*see* **Note 18**).

1. If the coverslips have been frozen, they must be rehydrated for 5 min in PBS 1×.
2. Cells are fixed with PFA 4 % for 15 min.
3. Cells are washed three times with PBS 1× for 5 min.
4. Cells are permeabilized with 0.5 % triton in PBS 1× for 10 min.
5. Cells are washed three times with PBS 1× for 5 min.
6. A big piece of parafilm is secured in the bottom of an incubation box. 50 μl of primary antibodies diluted in PBS 1× (anti-myogenin and anti-desmin antibodies) is dropped on the parafilm. The coverslip is laid on the antibody drop (cells facing the antibody) and incubated for 2 h at 37 °C in a humid chamber (*see* **Note 14**).
7. Cells are washed three times with PBS 1× for 5 min.
8. Using the same procedure as for primary antibodies, cells are incubated with the secondary antibodies diluted in PBS 1× (anti-mouse biotinylated and anti-rabbit secondary antibodies) for 1 h at 37 °C in a humid chamber (*see* **Note 14**).
9. Cells are washed three times with PBS 1× for 5 min.
10. Using the same procedure as for the antibodies, the cells are incubated with Streptavidin-DTAF for 20 min at 37 °C in a humid chamber (*see* **Note 14**).
11. Cells are washed four times with PBS 1× for 5 min.
12. Each coverslip is quickly (10 s) incubated with 2 nM Hoechst in PBS 1× (either the coverslip is laid on the top of 50 μl of Hoechst on a piece of parafilm, or the coverslip handed with a forceps is soaked for 10 s in a jar containing Hoechst).
13. Cells are washed once with PBS 1× for 5 min.
14. Coverslips are mounted with 50 μl of mounting medium on glass slides and stored at +4 °C.
15. About 10 pictures of each condition are recorded at 20× magnification with an epifluorescence microscope (*see* **Note 19**).

4 Notes

1. Lab-Tek® are made in glass, permanox, or Poly D-lysine coated. Murine MPCs adhere only on permanox, although it induces some background in immunofluorescence. 2-well Lab-Tek® (each well is 4.3 cm^2) are used to minimize impacts of the edges, which favor MPC fusion. Thus, areas near the edges should be excluded of the analysis.
2. 20 ml of sterile H_2O are added to one bottle of Ultroser G. After 15–20 min at RT, shaking the bottle ensures complete dissolution. Solution is filtered through 0.22 μm (Millipore #SLGP033RS) and kept aliquoted at −20 °C.
3. Antibodies and streptavidin are diluted in 400 μl of H_2O and 400 μl of glycerol and stored at −20 °C. They are used at 1/200 in PBS 1×.
4. Washes are here important to remove any trace of the reagents used for macrophage activation.
5. The best time to plate MPCs is the evening. Cells adhere during the night and thus do not stay too long in growth medium (thus not too high number of cells resulting in proliferation).
6. For Lab-Tek® coating, Matrigel® is diluted in cold serum-free DMEM/F-12 at 1/10. 1 ml is added per well and incubated for 30 min at 37 °C. Excess of Matrigel® is discarded and the well is washed three times with PBS 1× to remove Matrigel® aggregates.
7. Mouse MPCs do not proliferate well under very low FBS concentrations in the medium. On the other way, doing the coculture experiment in 10 % FBS would hide the effect of factors contained in macrophage-CM. 2.5 % has been shown to be the minimum concentration of FBS for an effective MPC proliferation.
8. Cells detach easily, carefully aspirate PBS.
9. Freezing the Lab-Tek® prevents good immunolabeling.
10. Low concentration (2 %) of horse serum, coupled with high cell density, induces rapid murine MPC differentiation and fusion.
11. The untreated MPCs should show signs of differentiation, i.e., the presence of myotubes. If not, the experiment is not valid.
12. The preparation must never dry during the whole procedure.
13. The drop of diluted antibodies is deposited in a corner of the well and a piece of parafilm is slowly laid. Parafilm size should be over that of the well. Ideally, one parafilm is used to cover the entire Lab-Tek®, i.e., the same labeling is realized in the two wells. Use separate smaller pieces of parafilm and avoid

mixing of antibodies in neighboring wells if doing different labelings in each well.

14. To avoid drying of the preparation during the incubation, wet papers are placed in the incubating box.
15. For the analysis of proliferation, the number of $Ki67^{+}$ nuclei is calculated as a percentage of the total number of nuclei. For the differentiation assay, fusion is evaluated as the percentage of nuclei into myotubes among the total number of nuclei (Fig. 2).
16. If macrophage-CM is recovered in the morning of day 4, MPCs should be seeded in the evening of day 3. If macrophage-CM is recovered in the afternoon, MPCs should be seeded in the early morning of day 4 (at least 6 h of plating).
17. Human MPCs are seeded at low concentration of FBS to induce differentiation and at low density to avoid their fusion.
18. The user must keep in mind during the whole procedure on which side of the coverslip the cells are. She/he should establish an own rule, e.g., "the cells are on the up side for the washings."
19. Desmin staining discriminates myogenic cells from non-myogenic cells (that may have raised in the culture with time). Only desmin positive cells and desmin-myogenin positive cells should be taken into account. Differentiation is evaluated as the percentage of the number of double myogenin-$desmin^{+}$ cells among the total number of $desmin^{+}$ cells.

References

1. Arnold L, Henry A, Poron F et al (2007) Inflammatory monocytes recruited after skeletal muscle injury switch into anti-inflammatory macrophages to support myogenesis. J Exp Med 204:1057–1069
2. Saclier M, Yacoub-Youssef H, Mackey AL et al (2013) Differentially activated macrophages orchestrate myogenic precursor cell fate during human skeletal muscle regeneration. Stem Cells 31:384–396
3. Mounier R, Theret M, Arnold L et al (2013) AMPKalpha1 regulates macrophage skewing at the time of resolution of inflammation during skeletal muscle regeneration. Cell Metab 18:251–264
4. Martinez FO, Sica A, Mantovani A et al (2008) Macrophage activation and polarization. Front Biosci 13:453–461
5. Xue J, Schmidt S, Sander J et al (2014) Transcriptome-based network analysis reveals a spectrum model of human macrophage activation. Immunity 40:274–288
6. Montarras D, Lindon C, Pinset C et al (2000) Cultured myf5 null and myoD null muscle precursor cells display distinct growth defects. Biol Cell 92:565–572
7. Stanley ER (1997) Murine bone marrow-derived macrophages. Methods Mol Biol 75:301–304
8. Chazaud B, Sonnet C, Lafuste P et al (2003) Satellite cells attract monocytes and use macrophages as a support to escape apoptosis and enhance muscle growth. J Cell Biol 163: 1133–1143

Chapter 18

Optimization of Satellite Cell Culture Through Biomaterials

Sadegh Davoudi and Penney M. Gilbert

Abstract

Hydrogels, a type of biomaterial, are an invaluable part of biomedical research as they are highly hydrated and properties such as elasticity, porosity, and ligand density can be tuned to desired values. Recently, culture substrate stiffness was found to be an important regulator of muscle stem cell self-renewal. Polyethylene glycol (PEG), a synthetic polymer, can be fabricated into hydrogels that match the softness of skeletal muscle tissue, thereby providing a culture surface that is optimal for maintaining muscle stem cell self-renewal potential ex vivo. In this Chapter, we describe a method to produce flat PEG hydrogels across a range of stiffnesses, including a formulation that matches the bulk stiffness of healthy skeletal muscle (12 kPa), while maintaining a constant ligand density. Since PEG is inert to protein adsorption, the steps required to surface functionalize the hydrogel with an adhesive interface (e.g., laminin) are also described.

Key words Muscle stem cell, Hydrogel, Polyethylene glycol, Substrate stiffness, Culture substrate, Self-renewal

1 Introduction

The cell-material interface is an important regulator of stem cell behavior [1]. A plethora of studies have shown that substrate stiffness is one of a variety of material properties that can influence myogenic and stem cell fate [2–6]. Hydrogels including, but not limited to collagen, polyacrylamide, hyaluronic acid, alginate, and polyethylene glycol (PEG) gels can be produced across a wide range of physiological stiffnesses including 1 to 500 kPa [1]. The ability to tune the stiffness of these hydrogels has allowed researchers to mimic the elasticity of many tissues in the human body in 2D and 3D culture studies.

Skeletal muscle stem cells (MuSC), also known as satellite cells, can sense and respond to variations in substrate stiffness [4, 5, 7]. Studies suggest that ignoring substrate stiffness as a biophysical regulator of MuSC fate is a compounding factor resulting in the rapid depletion of MuSC regenerative potential when cultured

Eusebio Perdiguero and DDW Cornelison (eds.), *Muscle Stem Cells: Methods and Protocols*, Methods in Molecular Biology, vol. 1556, DOI 10.1007/978-1-4939-6771-1_18, © Springer Science+Business Media LLC 2017

in vitro. Compared to muscle stem cells cultured on stiff tissue culture plastic (10^6 kPa), MuSCs grown on soft (12 kPa) PEG hydrogels mimicking the Young's Modulus of healthy skeletal muscle, self-renew in vitro and maintain their regenerative capacity as revealed by transplantation assays [4, 5]. Notably, soft hydrogels also improve MuSC viability but do not seem to impact proliferation rate to a great extent [5].

In this Chapter, we describe in detail a method to fabricate flat PEG hydrogels [8] with tunable stiffness varying from ~3.5 to 128 kPa. The major steps described are the preparation of the hydrogel components (PEG-VS, PEG-SH), calculating the ratio of each component to achieve the desired stiffness, preparation of laminin as the functionalizing protein, and producing the hydrophobic slides required for casting the gels.

2 Materials

2.1 Hydrogel Preparation

1. 0.2 M, Triethanolamine (TEAO) Buffer Solution, store at 4 °C (Sigma Aldrich, St. Louis, MO, USA).
2. Distilled (DI) and MilliQ water.
3. 8-arm PEG $((CH_2)_2\text{-}VS)_8$, MW 10,000 polymer powder (Sunbright HGEO-100VS; NOF Corporation, Tokyo, Japan), store at −20 °C.
4. 4-arm PEG, $((CH_2)_2\text{-}SH)_4$, MW 10,000 polymer powder (Sunbright PTE-100SH; NOF Corporation, Tokyo, Japan), store at −20 °C.
5. 5 ml syringe.
6. 0.2 μm filter membrane syringe attachment.
7. Plain glass slides (i.e. no coating, no charged surface).
8. Sigmacote® (Sigma Aldrich, St. Louis, MO, USA).
9. Lab soap (e.g., Sparkleen, DriClean Labware Detergent, etc.).
10. Paper towels.
11. Fume hood.
12. 1.5 ml Eppendorf tubes.
13. Binder clips.
14. Teflon spacers (1 mm).
15. Flat spatula.
16. Biological safety cabinet.
17. Phosphate buffered saline (PBS) containing 1 % 100× Antibiotic-Antimycotic (Anti-Anti).
18. 24-well tissue culture plate.

2.2 Preparation of Protein to Hydrogel Surface Tethering

1. 0.5 mg/ml Laminin from EHS-sarcoma (mouse) filtered through 0.2 μm pores (Roche Diagnostics, Mannheim, Germany), store at −20 °C.
2. Sterile PBS.
3. Dialysis cassette (7000 MWCO; Pierce).
4. Syringe.
5. 18 G × 1½ syringe needle.
6. 1 L glass beaker.
7. Magnetic stir bar and stir plate.

3 Methods

All procedures should be carried out at room temperature unless otherwise specified.

3.1 Preparation of PEG-VS Solution

1. Remove the PEG-VS powder component from the −20 °C and allow to reach room temperature.
2. Prepare a 10 % weight to volume solution in 0.2 M TEAO: add 920 μl of 0.2 M TEAO per 100 mg of PEG-VS. Vortex until fully dissolved. Sonicate sample in a water bath sonicator at setting 5 for 10 s. Set tube on ice for 1 min. Sonicate a second time at setting 5 for 10 s.
3. Sterilize the solution by filtering it using a syringe and a 0.2 μm syringe filter.
4. Aliquot and store at −20 °C.

3.2 Preparation of PEG-SH Solution

1. Remove the PEG-SH powder component from −20 °C and allow to reach room temperature.
2. Prepare a 10 % weight to volume solution in ddH_2O: add 920 μl of ddH_2O water per 100 mg of PEG-SH. Vortex until fully dissolved. Sonicate sample in a water bath sonicator at setting 5 for 10 s. Set tube on ice for 1 min. Sonicate a second time at setting 5 for 10 s.
3. Keep solution on ice to prevent the PEG-SH from forming intermolecular disulfide bonds that prevent hydrogel formation.
4. Sterilize the solution by filtering it using the syringe and a 0.2 μm syringe filter.
5. Aliquot and store at −20 °C (*see* **Note 1**).

3.3 Sigmacoting Glass Slides

This step prepares hydrophobic glass slides for easy removal of the polymerized hydrogels (*see* **Note 2**).

1. Wash both sides of the glass slides with lab detergent and then rinse with water until detergent is fully removed.
2. Rinse each side of the glass slides with DI water at least six times.
3. Dry the slides on a paper towel inside the fume hood. Complete **steps 4–6** in a fume hood.
4. Cover one side of the slides with Sigmacote® until the whole slide is covered (*see* **Note 3**).
5. Allow the Sigmacote® to dry (~20 min) (*see* **Note 4**).
6. Flip the glass slides over and repeat **steps 4** and **5**.
7. Place slides in a beaker, cover with aluminum foil, and autoclave using the dry cycle.
8. At this point, the slides are sterile and ready to use. Only use the sterile slides in the biological safety cabinet to maintain sterility (*see* **Note 5**). Sigmacoted® slides can be reused until they lose their hydrophobic surface.

3.4 Laminin Preparation

In this methods chapter, we covalently tether laminin to the surface of partially polymerized PEG hydrogels to serve as an adhesive interface (*see* **Note 6**). Since the PEG-VS reacts with primary amines, it is important that proteins intended to be functionalized on the hydrogel surface (a) contain primary amines and (b) be stored in a buffer that lacks primary amines. Dialysis is a commonly used method to transfer a protein from a primary amine containing solution (e.g., TRIS buffer) into a buffer that lacks primary amines (PBS).

1. Remove laminin solution from −20° and allow to thaw on ice.
2. Transfer the laminin solution from the glass vial into the dialysis cassette using an 18 G 1½ syringe needle.
3. Place the dialysis cassette and a magnetic stir bar into 1 L of sterile PBS in a glass beaker. Place the beaker on a magnetic stir plate at 4 °C and stir for 1 h.
4. After 1 h, replace the PBS with 1 L of fresh sterile PBS and stir overnight at 4 °C.
5. The next day, replace the PBS with 1 L of fresh sterile PBS and stir at 4 °C for one additional hour.
6. In the biological safety cabinet, remove the laminin from the dialysis cassette using a new 18 G 1½ syringe and store at 4 °C (*see* **Note 7**).

3.5 Hydrogel Preparation

In this Subheading, the instructions for preparing *flat* hydrogels are explained (*see* **Note 8**). All of the steps in this Subheading, except for the first step, should be performed in a biological safety cabinet to minimize the possibility of contamination. The polymerization reaction is pH and temperature sensitive, so all solutions should be kept on ice to control the speed of polymerization.

The relationship between the hydrogel's stiffness and weight to volume percentage (w/v %) can be calculated using the formulas below (*see* **Note 9**):

$$\frac{\mathrm{w}}{\mathrm{v}}\% = -3\times 10^{-4}\times kPa^{2} + 0.0952\times kPa + 2.2086$$

$$kPa = 0.9247\times\left(\frac{\mathrm{w}}{\mathrm{v}}\%\right)^{2} + 5.0275\times\left(\frac{\mathrm{w}}{\mathrm{v}}\%\right) - 14.856$$

Once the w/v % of the hydrogels has been determined based on the desired hydrogel stiffness, use the formulas below to calculate the amount of each component to add. "V", "V_{VS}", "V_{SH}", "V_{TEAO}" represent the total hydrogel, PEG-VS solution, PEG-SH solution, and TEAO volumes, respectively (*see* **Notes 10** and **11**):

$$V_{VS} = V\times\left(\frac{\mathrm{w}}{\mathrm{v}}\%\right)\times 0.0336$$

$$V_{SH} = V\times\left(\frac{\mathrm{w}}{\mathrm{v}}\%\right)\times 0.0664$$

$$V_{\mathrm{TEAO}} = V - \left(V_{VS} + V_{SH}\right)$$

1. Remove PEG-VS and PEG-SH aliquots from the −20 °C and allow to thaw on ice (*see* **Note 12**).
2. Place 1 mm Teflon spacers at each end of a Sigmacoted glass slide (Fig. 1a) (*see* **Note 13**).
3. Combine the appropriate hydrogel components in an eppendorf tube by first adding TEAO, followed by PEG-SH, and finally PEG-VS (*see* **Note 14**).
4. Immediately vortex the components for 10 s (*see* **Note 15**).
5. Pipette 150 μl of the hydrogel onto the glass slide (Fig. 1a). There is enough space to cast two hydrogel onto each slide (*see* **Note 16**).
6. Place a second Sigmacoted slide on top of the spacers on the first slide and keep it in place using binder clips (Fig. 1b) (*see* **Note 17**).
7. Allow the gels to partially polymerize at 37 °C incubator for 7–10 min (*see* **Note 18**).
8. While the gels polymerize, dilute the dialyzed laminin to 50 μg/ml (10× dilution) in sterile PBS (*see* **Note 19**).
9. After 7–10 min, remove the hydrogel from the incubator and gently remove the top glass (Fig. 1d) (*see* **Note 20**).

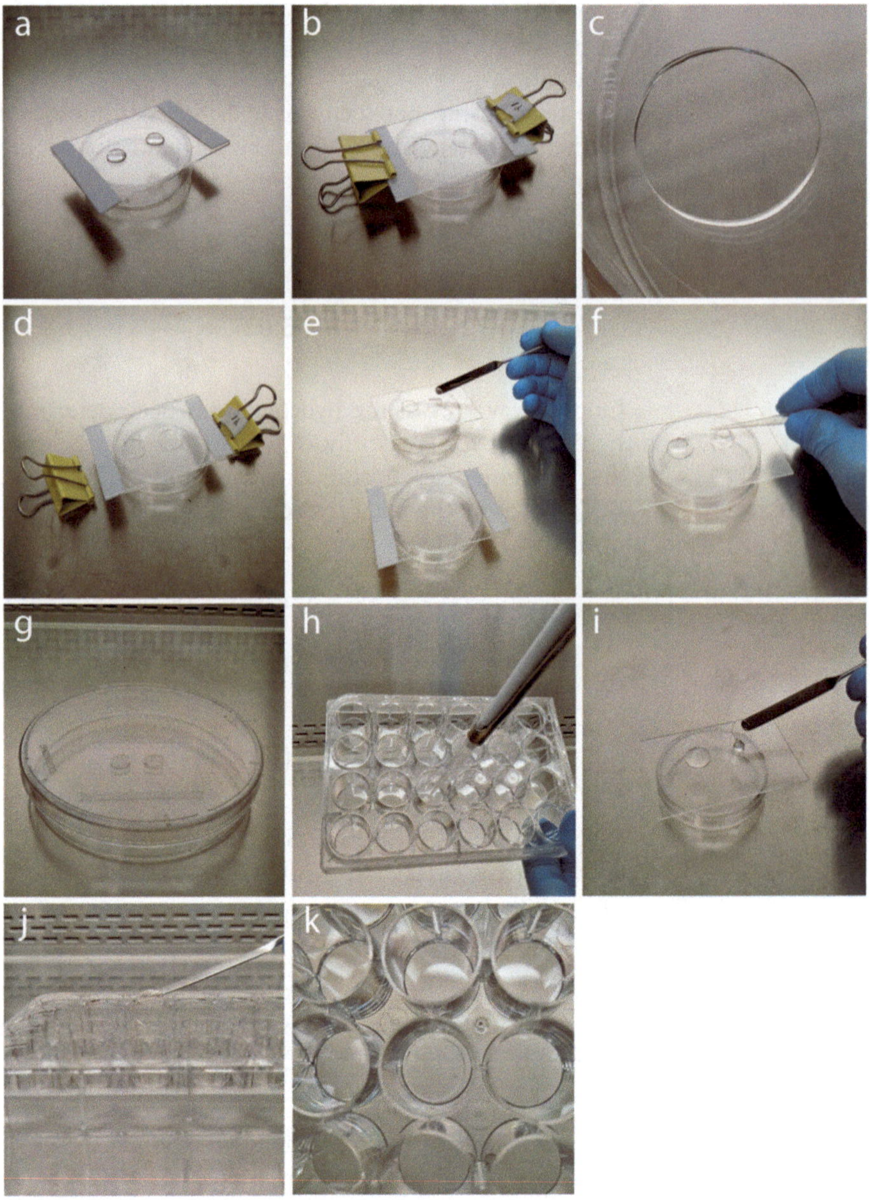

Fig. 1 Step by step images of flat hydrogel fabrication. (**a**) Teflon spacers are placed on the edges of Sigmacoted glass slide. Vortexed hydrogel mixture is pipetted onto glass slide. (**b**) Another Sigmacoted glass slide is placed onto the first slide and held in place using binder clips. (**c**) Examples of defects on the top side of the cast hydrogel. (**d**) Binder clips are removed after incubation at 37 °C. (**e**) Top glass is removed and hydrogels are transferred to the top glass with their top side facing down. (**f**) Hydrogel is surface tethered by adding a diluted laminin solution onto the hydrogels. (**g**) Hydrogels are incubated for 45 min at room temperature in a secondary container. (**h**) Hydrogel glue is added to the bottom of a well of a multiwell plate and spread around with the pipette tip. (**i**) The excess laminin on top of the hydrogels is removed by tilting the hydrogel. (**j**) Hydrogels are gently lowered into the well and placed on top of glue. (**k**) Image of glued hydrogel at the bottom of 24-well plate

10. Transfer the gels to the top glass with the side that was in contact with the bottom glass facing up (Fig. 1e). This might happen naturally if when removing the top glass, the hydrogel is stuck to it. If the hydrogel is still on the bottom glass, pick it up with a flat spatula and flip it over onto the top glass (*see* **Note 21**).
11. Pipette 70 μl of the diluted dialyzed laminin onto each hydrogel. Spread it out completely on the hydrogel using the pipette tip or a flat spatula (Fig. 1f). Avoid touching the hydrogel as it might acquire defects (*see* **Note 22**).
12. Allow the protein to incubate for 45 min at room temperature (*see* **Note 23**). During this time, the hydrogel will continue to polymerize (Fig. 1g).
13. After 45 min, prepare 15 μl of hydrogel glue by combining PEG-VS and PEG-SH at a 1:1 ratio (7.5 μl each). Vortex the mixture for 10 s.
14. Pipette 10 μl of the glue into a well of a 24-well plate. Spread it evenly into the bottom of the well with the pipette tip (Fig. 1h).
15. Gently pick up a hydrogel with a flat spatula. Tilt it to the side and remove excess laminin solution. Gently place it in the well with the laminin tethered side facing up (Fig. 1i–j). Gently remove any air bubbles that might form under the gel to ensure even contact with the glue.
16. Polymerize the glue by placing the plate into the 37 °C incubator for 10 min (Fig. 1k).
17. Add 1 ml PBS + 1 % 100×Anti-Anti to the well and store at 4 °C overnight (*see* **Note 24**).

3.6 Preparing Hydrogel for Cell Culture

1. Prior to culturing cells, equilibrate the hydrogels in culture media. Perform a minimum of four 30-min media exchanges at 37° prior to use.
2. Seed cells at appropriate density into the wells and treat the hydrogel culture substrates as you would a plastic cell culture plate (*see* **Note 25**).

4 Notes

1. Depending on the final hydrogel concentration, different amounts of the PEG-VS and PEG-SH components will be used. However, the amount of PEG-SH used will be roughly twice the amount of PEG-VS. Therefore, it might be useful to prepare aliquot volumes that contain twice as much PEG-SH as PEG-VS.

2. Be sure to use plain glass and not charged glass. The Sigmacoting will not prevent hydrogel adherence if applied to charged glass slides.
3. One method to do this task is to expel the Sigmacote® solution from the pipette tip in a zig-zag motion. Then move the pipette tip in vertical and horizontal directions to cover the entire glass slide.
4. In order to minimize dust or debris on the finished slides, move any particles that land on the slides to a corner while the Sigmacote® is still wet.
5. Avoid soap or ethanol coming in contact with the Sigmacoted slides, since they will remove the Sigmacote surface.
6. Other proteins, such as fibronectin, may also be used.
7. The dialyzed laminin can be stored for up to 1 month at 4 °C. For longer storage, flash freeze aliquots in liquid nitrogen and store at −20 °C.
8. The instructions in Subheading 3.5 are to prepare flat hydrogels. An advantage of the flat hydrogels is that since they are flat, they are more suitable for imaging cultured cells using a microscope. Additionally, although hydrogel adherence is robust enough to withstand media exchanges, flat hydrogels can be removed from the plate relatively easily and flipped upside down onto a glass slide, which makes high-resolution imaging much easier (*see* **Note 25**). However, if imaging the cells or having a flat surface is not required, an easier and faster method is to directly cast the hydrogels inside the well. This is especially important in the case of smaller wells such as 96-well plates in which making flat hydrogels is not possible. A caveat of this method is that a meniscus is formed, which is a challenge to imaging. The steps for directly casting the hydrogel in the well are as follows: **steps 1–3**: As described in **steps 1–3** in Subheading 3.5. **Step 4**: Pipette 200 μl of the hydrogel into the well. This volume is appropriate for a 24-well plate. Adjust as necessary for other well sizes. **Step 5**: Allow the gels to partly polymerize at 37 °C incubator for 7–10 min (*see* **Note 18**). **Step 6**: While the gels polymerize, dilute the dialyzed laminin to 50 μg/ml using sterile PBS (*see* **Note 19**). **Step 7**: Pipette 200 μl of the diluted dialyzed laminin onto each hydrogel (the plate can be placed onto an orbital shaker to reduce the amount of diluted dialyzed laminin required to cover the well). **Step 8**: Allow the protein to incubate for 1 h at room temperature. During this time, the hydrogel will continue to polymerize. **Step 9**: Remove dialyzed laminin and wash with PBS + 1% Anti-Anti to the well. Store at 4 °C overnight (*see* **Note 24**).
9. These formulas are valid for hydrogels composed of the PEG-VS and PEG-SH from NOF Corporation in Tokyo. To obtain the relationship between hydrogel stiffness and w/v %,

the stiffness of four different gel compositions was determined by compression testing as displayed in Fig. 2. The formulas were then calculated by drawing a binomial trendline between the data points. Due to variation in manufacturing as well as lot-to-lot variation, it is highly recommended to derive new formulas if the hydrogel components are purchased from a different vendor and upon obtaining a new lot of polymer.

10. The formulas are valid for hydrogels composed of PEG-VS and PEG-SH from NOF Corporation in Tokyo, which have been dissolved at a ten weight per volume percentage (w/v %) in TEAO or ddH_2O water.

11. The equations are only valid if the hydrogel is made using the components and from the vendors mentioned in the materials section. If not, the volumes of each component can be calculated from the material properties described below and through the following fashion:

$$PEG\text{-}VS \text{ properties}: \left(\frac{w}{v}\%\right)_{VS}, \ Molecular\,weight\left(MW_{VS}\right), \ Functional\,arms\left(F_{VS}\right), \ Subst\left(S_{VS}\right)$$

$$PEG\text{-}SH \text{ properties}: \left(\frac{\mathrm{w}}{\mathrm{v}}\%\right)_{SH}, \ Molecular\,weight\left(MW_{SH}\right), \ Functional\,arms\left(F_{SH}\right), \ Subst\left(S_{SH}\right)$$

$$\text{Surplus}\left(X\right) = \%\text{ of crosslinker / thiol group in excess}$$

$$A = \frac{S_{VS}}{S_{SH}} \times \frac{F_{VS}}{F_{SH}} \times \left(1 + \frac{X}{100}\right)$$

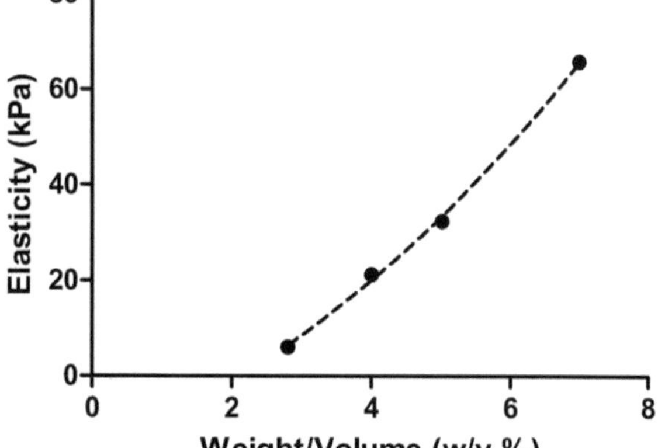

Fig. 2 Graph depicting stiffness measurements of four different hydrogel compositions and binomial trendline thereby establishing the relationship between hydrogel weight/volume percentage and stiffness

$$V_{VS} = \frac{V_{tot} \times \frac{MW_{VS}}{MW_{VS} + (A \times MW_{SH})} \times \left(\frac{\text{w}}{\text{v}}\%\right)}{\left(\frac{\text{w}}{\text{v}}\%\right)_{VS}}$$

$$V_{SH} = \frac{V_{tot} \times \frac{MW_{SH}}{MW_{SH} + \left(\frac{MW_{VS}}{A}\right)} \times \left(\frac{\text{w}}{\text{v}}\%\right)}{\left(\frac{\text{w}}{\text{v}}\%\right)_{SH}}$$

$$V_{\text{TEAO}} = V_{tot} - (V_{VS} + V_{SH})$$

The hydrogels used in this protocol have the following properties and the equations displayed in Subheading 3.5 are calculated below:

$$PEG\text{-}VS \text{ properties}: \left(\frac{\text{w}}{\text{v}}\%\right)_{VS} = 10\%,\ \textit{Molecular weight}(MW_{VS}) = 10^4\ Da,$$
$$\textit{Functional arms}\ (F_{VS}) = 8,\ \textit{Subst}(S_{VS}) = 0.9$$

$$PEG\text{-}SH \text{ properties}: \left(\frac{\text{w}}{\text{v}}\%\right)_{SH} = 10\%,\ \textit{Molecular weight}\ (MW_{SH}) = 10^4\ Da,$$
$$\textit{Functional arms}\ (F_{SH}) = 4,\ \textit{Subst}(S_{SH}) = 1$$

$$X = 10\%$$

$$A = \frac{0.9}{1} \times \frac{8}{4} \times \left(1 + \frac{10}{100}\right) = 1.98$$

$$V_{VS} = \frac{V_{tot} \times \frac{10^4}{10^4 + (1.98 \times 10^4)} \times \left(\frac{\text{w}}{\text{v}}\%\right)}{10} = V_{tot} \times \left(\frac{\text{w}}{\text{v}}\%\right) \times 0.0336$$

$$V_{SH} = \frac{V_{tot} \times \frac{10^4}{10^4 + \left(\frac{10^4}{1.98}\right)} \times \left(\frac{\text{w}}{\text{v}}\%\right)}{10} = V_{tot} \times \left(\frac{\text{w}}{\text{v}}\%\right) \times 0.0664$$

$$V_{\text{TEAO}} = V_{tot} - (V_{VS} + V_{SH})$$

As an example, to make V = 130 ml of 2.5w/v% PEG hydrogel, V_{VS} = 11.7 ml, V_{SH} = 20.93 ml, and V_{TEAO} = 97.37 ml are required.

12. Over time in the freezer, evaporation of the PEG-VS and PEG-SH components may occur. Be sure to pulse spin and vortex all aliquots prior to use as evaporation will concentrate the effective polymer concentration at the bottom of the tube and yield incorrect final hydrogel stiffness.
13. 1 mm hydrogels (constructed using 1 mm Teflon spacers) are difficult to image through due to the thickness of the substrate (unless removed and flipped upside down onto a glass slide as explained in **Note 25**). 0.75 mm hydrogels fabricated by using 0.75 mm Teflon spacers are thin enough to image through and are still relatively easy to handle.
14. The hydrogel polymerizes spontaneously via a Michael-type addition click-chemistry reaction. Hence, the hydrogel will begin polymerizing as soon as the PEG-VS and PEG-SH components are mixed together. The reaction is both pH and temperature sensitive. As a result, it is recommended that no more than four hydrogels (two if the user is new to the method) be prepared at any instance. All solutions should be kept on ice prior to mixing to slow the polymerization process. Also, it would be helpful to have a pipette set to the amount that is to be added onto the glass slides so as to prevent the hydrogel from polymerizing before it is placed between the two glass slides (especially in the case of high w/v % gels). To minimize the time between hydrogel preparations, the TEAO and PEG-SH components can be mixed together in several eppendorf tubes and kept on ice so that only the PEG-VS component remains to be added.
15. It is very important that the components be thoroughly mixed, which is the reason for the 10 s vortex. Improper mixing of the components will lead to un-polymerized hydrogels, which is a common mistake.
16. 150 μl is the volume required to fabricate 1 mm hydrogels for 24-well plates. For 48 well plates, add 45 and 50 μl to make 0.75 mm and 1 mm hydrogels, respectively.
17. Avoid moving the top glass side horizontally while in contact with the hydrogel and try to place it directly in its final position. Horizontal movements while in contact with the hydrogel will cause it to lose its circular shape and become smeared. To avoid bending the glass slides due to the binder clip force, which could result in the final hydrogel not being flat, clamp the glass on the Teflon spacers. Placing the two binder clips simultaneously will also help to obtain a more circular hydrogel.
18. Stiff hydrogels polymerize faster, so less time is required to partially polymerization a stiffer formulation. When making hydrogels of different stiffness simultaneously, it is suggested to cast the softer hydrogels first so the hydrogels become ready

at the same time. The excess gel remaining at the bottom of the eppendorf tube used to mix the components can be used as an indicator of the polymerization status of the hydrogel.

19. Lower amounts of surface-tethered laminin can be attained by increasing (i.e., 100×) the dilution of the stock dialyzed laminin solution.
20. To facilitate the release of hydrogels from the glass slides, remove the binder clips one at a time.
21. The hydrogel side contacting the bottom slide is generally smoother and contains fewer bubble defects compared to the top side (Fig. 1c). Bubbles are more common when using PEG aliquots that have been stored for a long time in the freezer. As a result, the hydrogels are placed so that their bottom side is facing up.
22. Add 30 μl of the diluted dialyzed laminin for hydrogels fitting a 48-well plate.
23. In order to prevent the laminin and the hydrogel from evaporating, place the glass slide inside a 15 cm tissue culture dish and close the lid.
24. Anti-Anti is used to minimize the risk of contamination that could occur due to the handling of the gel. The surface tethered hydrogel can be stored at 4 °C for up to 1 week.
25. Paraformaldehyde fixation, cell lysis for protein or RNA isolation, and routine cell culture procedures can be performed on hydrogels in a manner similar to methods used with standard tissue culture plates. Since the hydrogel is three-dimensional, it can be damaged and break upon physical contact with pipette tips; therefore, extra care must be taken so as not to disturb the hydrogel during various procedures. In addition, the hydrogel is porous, so it is important to work quickly with post-culture solutions (e.g., lysis buffer) as they will be soaked into the pores of the hydrogel. To enhance image quality after staining cells cultured on the hydrogel surface, flipping the hydrogel onto a glass slide is advised. The hydrogel can be easily removed from the bottom of the multiwell plate by gently inserting a flat spatula underneath the flat hydrogel. Once the gel has been picked up, flip it over upside down (cells contacting the glass) onto a glass microscope slide. During imaging, prevent the hydrogel from drying by adding drops of PBS or blocking solution onto the hydrogel.

Acknowledgment

The Natural Sciences and Engineering Research Council and the Canadian Institutes of Health Research supported this work.

References

1. Murphy W, McDevitt T, Engler A (2014) Materials as stem cell regulators. Nat Mater. doi: 10.1038/NMAT3937
2. Engler AJ, Griffin M, Sen S et al (2004) Myotubes differentiate optimally on substrates with tissue-like stiffness: pathological implications for soft or stiff microenvironments. J Cell Biol 166:877–887. doi:10.1083/jcb.200405004
3. Engler AJ, Sen S, Sweeney HL, Discher DE (2006) Matrix elasticity directs stem cell lineage specification. Cell 126:677–689. doi:10.1016/j.cell.2006.06.044
4. Cosgrove BD, Gilbert PM, Porpiglia E et al (2014) Rejuvenation of the muscle stem cell population restores strength to injured aged muscles. Nat Med 20:255–264. doi:10.1038/nm.3464
5. Gilbert PM, Havenstrite KL, Magnusson KEG et al (2010) Substrate elasticity regulates skeletal muscle stem cell self-renewal in culture. Science 329:1078–1081. doi:10.1126/science.1191035
6. Cosgrove BD, Sacco A, Gilbert PM, Blau HM (2009) A home away from home: challenges and opportunities in engineering in vitro muscle satellite cell niches. Differentiation 78:185–194. doi:10.1016/j.diff.2009.08.004
7. Morrissey JB, Cheng RY, Davoudi S, Gilbert PM (2015) Biomechanical origins of muscle stem cell signal transduction. J Mol Biol. doi:10.1016/j.jmb.2015.05.004
8. Lutolf MP, Hubbell JA (2003) Synthesis and physicochemical characterization of end-linked poly(ethylene glycol) co-peptide hydrogels formed by Michael-type addition. Biomacromolecules 4(3):713–722

Chapter 19

Systematic Identification of Genes Regulating Muscle Stem Cell Self-Renewal and Differentiation

Krishnamoorthy Sreenivasan, Thomas Braun, and Johnny Kim

Abstract

The hallmark of stem cells is their capability to either self-renew or to differentiate into a different cell type. Adult skeletal muscle contains a resident muscle stem cell population (MuSCs) known as satellite cells, which enables regeneration of damaged muscle tissue throughout most of adult life. During skeletal muscle regeneration, few MuSCs self-renew to maintain the muscle stem cell pool while others expand rapidly and subsequently undergo myogenic differentiation to form new myofibers. However, like for other stem cell types, the molecular networks that govern self-renewal and/or differentiation of MuSCs remain largely elusive. We recently reported a method to isolate sufficient amounts of purified MuSCs from skeletal muscle which enables us to study their cell autonomous properties. Here, we describe a lentiviral, image-based loss-of function screening pipeline on primary MuSCs that enables systematic identification of genes that regulate muscle stem cell function.

Key words Muscle stem cells (MuSCs), FAC sorting, RNAi screen, Lentiviral

1 Introduction

Adult muscle stem cells (MuSCs), also known as satellite cells, comprise a rare population of adult myogenic precursor cells that is required for the regeneration of skeletal muscle [1].

Under resting conditions MuSCs are dormant. Muscle injury can induce MuSC activation resulting in their exit out of quiescence and entry into the cell cycle. During regeneration most activated MuSCs proliferate and subsequently undergo myogenic differentiation to replenish the damaged muscle tissue while a subset of MuSCs self-renews and returns to quiescence to maintain the stem cell pool enabling future potential rounds of regeneration [2]. However, as for other stem cell types, the genetic pathways that regulate MuSC self-renewal and/or differentiation have remained in large parts enigmatic.

At the molecular level, undifferentiated MuSCs can be robustly identified by the expression of Pax7, a member of the paired box

Eusebio Perdiguero and DDW Cornelison (eds.), *Muscle Stem Cells: Methods and Protocols*, Methods in Molecular Biology, vol. 1556, DOI 10.1007/978-1-4939-6771-1_19, © Springer Science+Business Media LLC 2017

transcription factor family. Once committed to differentiation Pax7 expression is lost and coincident with the subsequent expression of later markers of myogenic differentiation, such as the basic helix-loop-helix transcription factors Myf5, MyoD, and MyoG. At this late stage of differentiation, myocytes fuse into mature syncytial multinucleated myotubes, negative for Pax7, both in vitro and in vivo [3–5].

We and others have generated methods to purify MuSCs from freshly dissociated skeletal muscle. Recently, we reported a method to isolate relatively large amounts of a pure population of MuSCs using transgenic Pax7 reporter mice and FAC sorting [6–9]. Significantly, the prospective purification of MuSCs together with the observation that Pax7 expression is lost during myogenic differentiation means that the relative generation rate of self-renewing $Pax7^{+}$ stem cells vs. $Pax7^{-}$ differentiated cells in a cultured population of pure MuSCs can be monitored by immunofluorescent staining for Pax7 itself [10]. Genome-wide loss of function screens via RNAi, and more recently via Crispr/Cas9 technology, have been established in both arrayed and pooled formats and have been harnessed to understand the molecular mechanisms governing various cellular events [11–16]. Lentiviral-based shrna screens have proven to be especially advantageous to stably manipulate gene expression in cells refractory to conventional transfection methods, including MuSCs. The possibility of isolating a large and pure population of MuSCs combined with the availability of genome-wide lentiviral RNAi and Crispr/Cas9 libraries suitable for efficient gene knockdown and knockout in these cells has offered opportunity for massive and parallel investigations on muscle stem cell function. Here, we describe a robust image-based lentiviral RNAi screening pipeline on purified MuSCs to simultaneously identify genes controlling muscle stem cell self-renewal and differentiation (*see* Fig. 1).

2 Materials

2.1 Cell Culture

Matrigel (BD Biosciences)

Reagents for Media

1. Tryptone.
2. Yeast extract.
3. NaCl.
4. DMEM-Glutamax-I, High Glucose, (Gibco-Life Technologies).
5. Fetal Bovine Serum (Sigma Aldrich).
6. Penicillin/Streptomycin (Gibco-Life Technologies).
7. Basic FGF (Pepprotech).
8. Polybrene (Sigma-Aldrich).

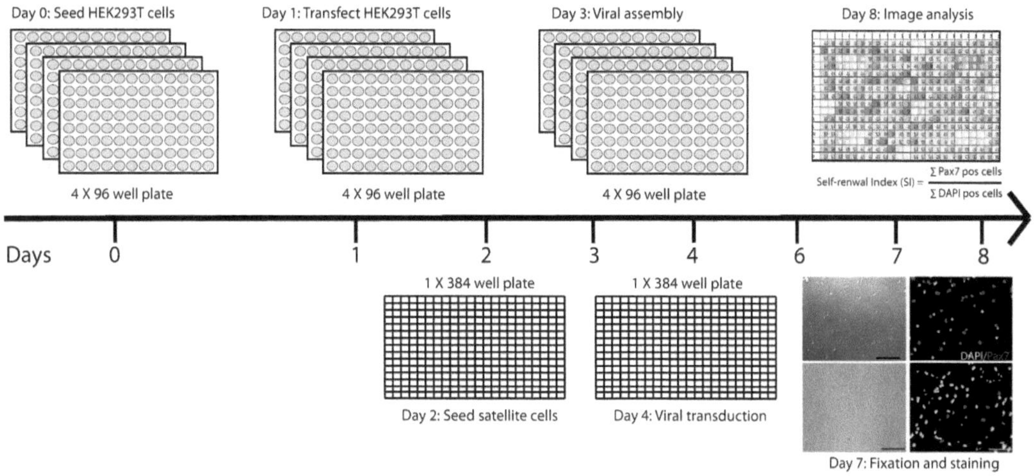

Fig. 1 Schematic illustrates the screening pipeline for one 384-well plate. On Day 0, HEK293T cells are seeded into four 96-well plates. On Day 1, HEK293T cells are transfected with the plasmid library for viral production. On Day 2, satellite cells are purified by FACS and seeded into the target 384-well plate. On Day 4 activated satellite cells are transduced with the supernatants containing lentivirus generated from the HEK293T cells. On Day 7 the infected satellite cells are fixed and stained with Pax7 antibody. Subsequently, the 384-well plate is subjected to high-throughput image acquisition and analysis. Refer to main text for media exchanges that are to be performed throughout the pipeline

9. Puromycin (Life Technologies).
10. Ampicillin.
11. LB media: Dissolve 10 g Tryptone, 5 g Yeast extract, and 10 g NaCl with 950 ml of distilled water. Adjust the pH to 7.00 with 0.5 N NaOH and bring the volume to 1l. Sterilize by autoclaving.
12. Satellite cell growth media (500 ml): 400 ml DMEM-Glutamax-I, supplemented with 100 ml of heat-inactivated fetal bovine serum (FBS), 1 % penicillin/streptomycin solution, and 5 ng/ml bFGF. Store at 4 °C.
13. HEK293T cell media (500 ml): 450 ml DMEM-Glutamax-I, supplemented with 50 ml of heat-inactivated fetal bovine serum (FBS) and 1 % penicillin/streptomycin solution. Store at 4 °C.
14. Virus production media: Satellite cell growth media containing 16 μg/ml polybrene.
15. Puromycin selection media: Satellite cell growth media containing 4 μg/ml puromycin.

2.2 Plasmid Isolation

1. 2.0 ml Deep Well Plates with Shared-Wall Technology (Thermo Scientific).
2. Breathable Sealing Film for Tissue Culture (Axygen).

3. GenElute™ HP 96 Well Plasmid Miniprep Kit (Sigma-Aldrich).
4. PvuII (NEB).

2.3 Virus Production and Transduction

1. Cell Culture Microplate, 96 Well, PS, F-Bottom, Clear (Greiner bio-one).
2. Cell Culture Microplate, 96 Well, PS, V-Bottom, Clear (Greiner bio-one).
3. Cell Culture Microplate, 384 Well, PS, F-Bottom, μClear (Greiner bio-one).

Buffers

1. HEPES buffered saline: Prepare 140 mM NaCl, 1.5 mM $Na_2HPO_4.2H_2O$, and 50 mM HEPES. For 10× solution, dissolve 4.0 g of NaCl, 0.108 g of $Na_2HPO_4.2H_2O$, and 4.8 g of HEPES in 90 ml of distilled water. Adjust the pH to 7.05 with 0.5 N NaOH, and adjust the volume to 100 ml with distilled H_2O. Sterilize the solution using a 0.22 μm filter and store at 4 °C.
2. 1.25 M $CaCl_2$.

Plasmids

1. psPAX2 (packaging plasmid containing gag, pol, and rev genes).
2. pMD2.G (envelope plasmid).
3. pLKO.1 (lentiviral shrna plasmid). Vector information is available from TRC website (http://www. broadinstitute.org/rnai/public/vector/summary).

2.4 Immunostaining

1. Fixation: 4 % Paraformaldehyde (Sigma-Aldrich).
2. Cell permeabilization buffer: 0.05 % PBST (1× PBS with 0.05 % Triton).
3. Blocking buffer: 5 % BSA in 0.05 % PBST.
4. Pax7 monoclonal antibody (R&D systems).
5. Goat anti-Mouse IgG1 Antibody, Alexa Fluor® 594 conjugated (Invitrogen).
6. DAPI (2 μg/ml in PBS).

2.5 Instruments

1. Nanodrop 2000c UV-Vis Spectrophotometer (Thermo Scientific).
2. AquaMax 4000 Microplate Washer (Molecular Devices).
3. ImageXpress Micro XLS Widefield High-Content Analysis System (Molecular Devices).

3 Methods

3.1 Library Generation for Screening

1. Prepare an arrayed library of shRNA vectors targeting a set of candidate genes. In the method described here, we arrayed bacterial glycerol stocks containing plasmids encoding for individual short hairpin RNA(shRNA) sequences from the TRC shRNA library in 96-well plates [15] (*see* **Notel**).
2. Prepare a 96-well deep-well plate containing 1.8 ml/well of LB medium supplemented with ampicillin (1 μg/ml). Transfer 5 μl of the bacteria containing the shRNA vectors to each well of the deep-well plate and culture at 37 °C overnight on a shaker (*see* **Note 2**).
3. Centrifuge the deep-well plates at 4000 × *g* for 10 min to pellet the bacteria containing the shRNA vector. Discard the supernatant. At this point, plates can be stored at −20 °C before plasmid isolation.
4. Isolate shRNA plasmids from the bacterial pellets using a 96-well plasmid isolation kit according to the manufacturer's protocol. We routinely use the Sigma GenElute™ HP 96 Well Plasmid Miniprep Kit kit from which it is possible to obtain 1–5 μg of high purity plasmid DNA sufficient for efficient lentiviral production (*see* **Note 3**).
5. Validate concentration, purity, and integrity of the plasmids for each plate from several randomly selected wells. We measure DNA concentration using a Nanodrop device. Integrity of plko.1 vectors can be diagnostically tested with a PvuII digest that results in the generation of 3 bands of 4506, 1993, and 1217 base pairs.

3.2 Lentiviral Assembly and Virus Production

1. Seed low passage HEK293T cells at an optimal cell density (80 % confluence) in HEK293T growth media on a flat-bottom 96-well plate one day before transfection (**Day 0**) (*see* **Note4**).
2. Using a multi-channel pipette transfect the HEK293T cells with the plko.1 shRNA plasmid, the psPax2 packaging plasmid and the pMD2. Genvelope plasmid in the ratio of (1):(9):(0.08), respectively. We routinely use the calcium phosphate transfection method [17] that is very cost-effective and results in high transfection rates. Refer to Table 1 for the detailed buffer and plasmid dilutions (**Day 1**) (*see* **Note 5**).
3. The transfected HEK293T cells start to produce lentivirus 18 h post-transfection. Using a multi-channel pipette replace HEK293T growth media with media in which the cell of interest will be transduced and cultured. In the case of satellite cells, HEK293T growth media should be exchanged with 50 μl of satellite cell growth media 15 h post-transfection. When removing the HEK293T growth media take care to discard the tips in an appropriate waste container (**Day 2**) (*see* **Note 6**).

Table 1
Volumes for transfection reactions. The master mix for transfection for four 96-well plate includes 10x HBS, envelope vector (pMD2.G), and packaging vector (psPax2). Prepare the master mix as described in Note 5

	Stock concentration [ng/μl]	Volume for 1 well [μl]	Volume for 1 × 96 well [μl]	Volume for 4 × 96 well [μl]
shrna vector	50	10	–	–
psPax2	250	1.85	178	889
pMD2.G	250	0.06	6	29
10X HBS		1.19	114	572

4. On the second day after transfection add an additional 50 μl of fresh satellite cell growth media to each well of the 96-well plate (**Day 3**).
5. Spin down the 96-well plates at 300 rpm for 5 min to pellet any HEK293T cells or debris. Carefully collect 50 μl of the supernatant containing virus that will be used for the transduction of satellite cells.

3.3 Preparation of Target Muscle Stem Cells for Lentiviral Transduction

1. Isolate a pure population of muscle stem cells by FACS purification from 8-week old $Pax7^{ZsGreen}$ reporter mice as previously described (**Day 2**) [6–9].
2. Prior to satellite cell isolation pre-coat a 384-well plate with 35 μl of Matrigel diluted 1:50 with ice-cold satellite cell growth media. Matrigel serves as an adhesion matrix for the satellite cells in culture (*see* **Note 7**).
3. Seed freshly isolated satellite cells in satellite cell growth media. For 384-well Greiner microclear plates seed satellite cells at a density of 300–500 cells/well in 50 μl of medium (The total volume in the well is then 85 μl). Centrifuge the 384-well plate at 100 × *g* for 10 min to facilitate their adhesion to the Matrigel matrix.
4. One day after seeding, aspirate 50 μl of media using a microplate washer and add 50 μl of fresh satellite cell growth media to each well (**Day 3**). Satellite cells activate and begin to proliferate in vitro 24–36 h post-isolation and are now ready for transduction (*see* **Note 8**).

3.4 Lentiviral Transduction and Selection

1. Add polybrene to the 384-well plate containing satellite cells shortly before the viral transduction. Aspirate 40 μl of media using a microplate washer and add 40 μl of fresh media containing polybrene at a concentration of 16 μg/ml. Incubate for 30 min at 37 °C and then aspirate 40 μl of media using the plate washer. Satellite cells are now ready for viral transduction.

2. Using a multichannel pipette carefully transfer 50 μl of viral supernatant from the 96-well plate (*see* Subheading 3.2, **step 5**) in which the virus was produced to corresponding wells of the 384-well plate containing the satellite cells (**Day 4**).
3. Replace the media containing virus with fresh satellite cell growth media 6–12 h post-transduction (**Day 4**).
4. Check for viral transduction efficiency in wells transduced with the control lentiviruses expressing GFP/RFP. 48 hours after transduction GFP/RFP expression should be apparent in most, if not all, cells. It is also highly recommended to have positive controls for the phenotype of interest (**Day 6**) (*see* **Notes 1** and **9**).
5. Select the satellite cells with satellite cell growth media containing puromycin (4 μg/ml) and exchange media every day (**Days 5–6**) to select for transduced cells.

3.5 Cell Fixation and Staining

1. Wash the plate three times with 1X PBS using a microplate washer and fix the cells using 4 % PFA for 10 min (**Day** 7).
2. Permeabilize the fixed cells using 0.05 % Triton in 1X PBS for 10 min and block with 1XPBS containing 5 % BSA for 30 min before immunostaining.
3. Freshly isolated and proliferating satellite cells express Pax7 protein but cannot be detected in differentiated myoblasts and myotubes. Therefore, Pax7 staining faithfully reflects the undifferentiated state of satellite cells [10]. Incubate wells with Pax7-antibody diluted 1:1000 in 1X PBS containing 5% BSA and 0.05 % Triton overnight at 4 °C. Using a plate washer, wash the cells three times with 1X PBS to remove excess unbound antibody. Incubate with fluorescently labeled secondary anti-mouse IgG1 antibody diluted 1:2000 in 5 % BSA, 0.05 % PBST for 1 h at room temperature. Wash three times with 1X PBS and stain nuclei with DAPI (2 μg/ml) for 5 min and wash again three times with 1X PBS.
4. Image the plates using a high-content imaging microscope (*see* **Note 10**).
5. After image acquisition quantify for Pax7 and DAPI positive cells to calculate the rate of self-renewing $Pax7^{+}$ satellite cells and differentiated $Pax7^{-}$ cells (*see* **Note 11**).

4 Notes

1. Each 96-well plate should contain several internal controls to monitor the efficiency of plasmid transfection and viral transduction (GFP and RFP vectors), as well as positive controls for the phenotype of interest.

2. The deep-well plates should be covered with a breathable foil for efficient bacterial growth.
3. High purity of plasmid DNA is essential for optimal transfection efficiency. Therefore, column-based isolation of plasmid shRNA is necessary. As a general rule, a low transfection efficiency will inevitably result in the production of a low viral titer and ultimately low transduction efficiency. We routinely use GenElute™ HP 96 Well Plasmid Miniprep Kit (Sigma-Aldrich, #NA9604) resulting in high transfection and transduction efficiencies as well as low well-to-well variability. It is *very* important that the columns are completely devoid of residual ethanol after the washing step, which will otherwise result in low plasmid yields.
4. It is important to use HEK293T cells that express attenuated Large T antigen. This promotes the enhanced replication of plasmids containing a SV40 origin of replication in dividing cells. Significantly, transfected HEK293T cells will thus yield higher viral titers. We cannot stress enough that it is mandatory to have Safety level 2 working conditions and permission when conducting lentiviral assembly and transduction.
5. Prepare a master mix of the 10X HBSS buffer, packaging, and envelope vectors as described in Table 1. Distribute 5 μl of the master mix across a fresh 96-well plate (V-bottom) using a multichannel pipette. Add 10 μl of plasmid DNA from the library plate. Mix well before placing a drop of $CaCl_2$ on the side of the well. When this is done, gently tap the plate to "drop" the $CaCl_2$. This results in uniform precipitation of the plasmid DNA across the plate. Mix the precipitate by gently pipetting up and down before adding it to the 96-well plate containing the HEK293T cells.
6. Remove the media from the edge of the plate from one side to minimize aspiration of cells. This step will also remove dead cells. Immediately add fresh media to the wells to avoid cell detachment.
7. One day before satellite cell isolation thaw the appropriate amount of Matrigel on ice overnight. Matrigel is liquid at 0–15 °C but will form a gel at higher temperatures. Dilute Matrigel 1:50 with ice-cold satellite cell growth media and precoat cell culture dishes of the desired format with a volume so that the surface is just covered. Incubate at room temperature for 1 h. It is not necessary to aspirate the residual solution after coating the cell culture dish. Subsequently, incubate plates at 37 °C in a humidified incubator until cells are ready to be plated.
8. For media exchange, we use an AquaMax Microplate Washer (Molecular Devices) which greatly diminishes variability and enhances speed of media exchange. Aspirate only 40 μl of

media (50 % of total media) as to not remove the satellite cells. We have successfully performed infection of satellite cells before and during activation of satellite cells, hence 24 and 48 h after seeding, respectively. Maintain the timing of steps consistent throughout the screening pipeline to minimize phenotypic variability across different plates.

9. As for any screen, it is important to have positive and negative controls. We monitor satellite cell transduction with lentiviruses encoding for GFP or RFP in separate wells. If a good transduction efficiency is achieved most satellite cells, if not all, will express GFP/RFP 48 h post-transduction. For phenotypic positive controls in the method described here we employ shRNAs against *pax7* and *nf1* genes that result in enhanced differentiation and proliferation and less or more Pax7-expressing cells, respectively, 72 h post-transduction (Fig. 2a) [5, 18]. These controls are also good indications that the transduction was successful and that the cells are ready for fixation. Notably, in the method described here, the satellite cells in the 384-well plate will be targeted by viruses generated from four 96-well plates. Hence, since each

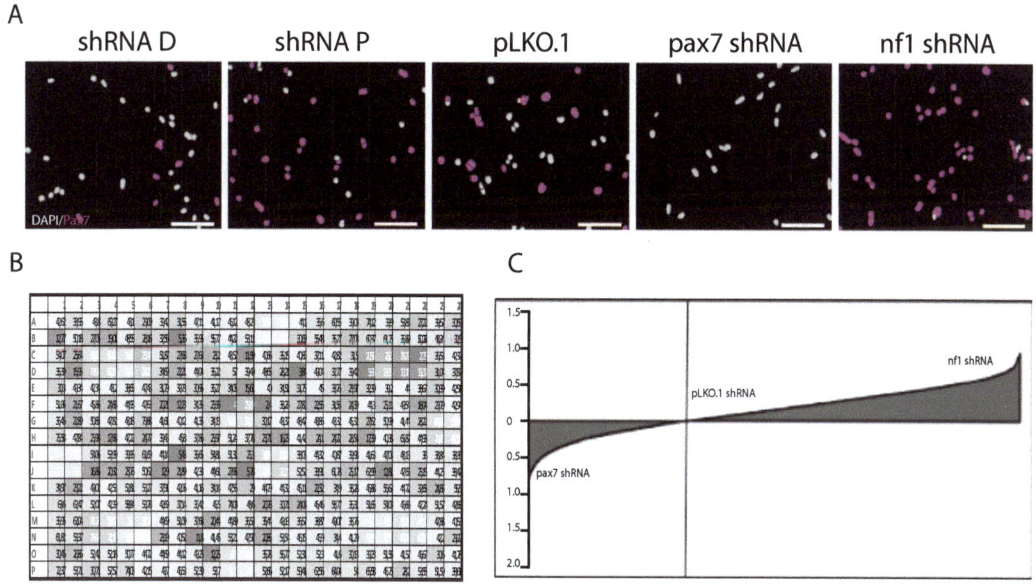

$$\text{Self-renwal Index (SI)} = \frac{\sum \text{Pax7 pos cells}}{\sum \text{DAPI pos cells}}$$

Fig. 2 (**a**) Representative segmented images of individual wells stained for Pax7 (*violet*) and total nuclei (*gray*). Controls include empty vector as negative control and knockdown of *Pax7* and *Nf1* as positive controls and representative hits from the screen for proliferation phenotype as shRNA P and differentiation phenotype as shRNA D. (**b** and **c**) Representative analyses of a 384-well plate scored for Pax7/DAPI ratios via heat mapping (in B) and Poisson distribution of Pax7/DAPI ratios for all targeting shRNAs across the plate (in C). Scale bars represent 100 μm

96-well plate contains the same controls one can assess variability across the plates producing lentivirus.

10. Imaging in high throughput can be achieved by HTS microscopes from different vendors. We employ the ImageExpress Micro high-throughput microscope setup that not only allows acquisition of fluorescent but also brightfield images. Thus, morphological effects upon gene knockdown can be additionally scored. This feature also helps in avoiding false positives during downstream analysis.
11. Here, we scored for DAPI positive cells that are $Pax7^{+}$or $Pax7^{-}$ (Fig. 2). A low Pax7/DAPI ratio indicates genes upon knockdown of which regulate maintenance of the undifferentiated state while a high Pax7/DAPI ratio indicates genes upon knockdown of which regulate differentiation of satellite cells.

$$Self-renewal\ Index = \frac{\sum Pax7\ positive\ cells}{\sum all\,cells} = \frac{\sum nuclei^{Pax7\ pos}}{\sum nuclei^{DAPI}}$$

Acknowledgments

This work was supported by the DFG (Excellence Cluster Cardio-Pulmonary System (ECCPS), Br1416, and SFB TR81), the German-Israeli Fund, and the LOEWE Centre for Cell and Gene Therapy, the German Centre for Cardiovascular Research, and the Universities of Giessen and Marburg Lung Centre (UGMLC).

References

1. Braun T, Gautel M (2011) Transcriptional mechanisms regulating skeletal muscle differentiation, growth and homeostasis. Nat Rev Mol Cell Biol 12:349–361
2. Kuang S, Kuroda K, Le Grand F, Rudnicki MA (2007) Asymmetric self-renewal and commitment of satellite stem cells in muscle. Cell 129:999–1010
3. Zammit PS et al (2006) Pax7 and myogenic progression in skeletal muscle satellite cells. J Cell Sci 119:1824–1832
4. Kuang S, Charge SB, Seale P, Huh M, Rudnicki MA (2006) Distinct roles for Pax7 and Pax3 in adult regenerative myogenesis. J Cell Biol 172:103–113
5. Gunther S et al (2013) Myf5-positive satellite cells contribute to Pax7-dependent long-term maintenance of adult muscle stem cells. Cell Stem Cell 13:590–601
6. Bosnakovski D et al (2008) Prospective isolation of skeletal muscle stem cells with a Pax7 reporter. Stem Cells 26:3194–3204
7. Kim J, Braun T (2014) Skeletal muscle stem cells for muscle regeneration. Methods Mol Biol 1213:245–253
8. Montarras D et al (2005) Direct isolation of satellite cells for skeletal muscle regeneration. Science 309:2064–2067
9. Sherwood RI et al (2004) Isolation of adult mouse myogenic progenitors: functional heterogeneity of cells within and engrafting skeletal muscle. Cell 119:543–554
10. Zhang T et al (2015) Prmt5 is a regulator of muscle stem cell expansion in adult mice. Nat Commun 6:7140
11. Boutros M et al (2004) Genome-wide RNAi analysis of growth and viability in Drosophila cells. Science 303:832–835

12. Berns K et al (2004) A large-scale RNAi screen in human cells identifies new components of the p53 pathway. Nature 428:431–437
13. Ngo VN et al (2006) A loss-of-function RNA interference screen for molecular targets in cancer. Nature 441:106–110
14. Wang T, Wei JJ, Sabatini DM, Lander ES (2014) Genetic screens in human cells using the CRISPR-Cas9 system. Science 343:80–84
15. Moffat J et al (2006) A lentiviral RNAi library for human and mouse genes applied to an arrayed viral high-content screen. Cell 124: 1283–1298
16. Moffat J, Sabatini DM (2006) Building mammalian signalling pathways with RNAi screens. Nat Rev Mol Cell Biol 7:177–187
17. Sambrook J,Russell DW(2006)Calcium-phosphate-mediated transfection of eukaryotic cells with plasmid DNAs. CSH protocols 1150–1155
18. Kossler N et al (2011) Neurofibromin (Nfl) is required for skeletal muscle development. Hum Mol Genet 20:2697–2709

Chapter 20

Bioinformatics for Novel Long Intergenic Noncoding RNA (lincRNA) Identification in Skeletal Muscle Cells

Xianlu Peng, Kun Sun, Jiajian Zhou, Hao Sun, and Huating Wang

Abstract

Long intergenic noncoding RNAs (lincRNAs) have emerged as critical participators in gene regulation in myriads of cell types. The development of the whole transcriptome sequencing technology, or RNA-seq, has enabled novel lincRNA detection, but the bioinformatics analysis toward distinguishing reliable ones remains a challenge. Here, we describe the bioinformatics workflow developed for identifying novel lincRNAs step by step, including read alignment, transcriptome assembly and transcript filtering.

Key words lincRNA, RNA-seq, Bioinformatics, Skeletal muscle cells

1 Introduction

Long intergenic noncoding RNAs (lincRNAs), which are larger than 200 nucleotides and transcribed from the intergenic regions of protein coding genes, have been shown by accumulating findings to be widely expressed and extensively functional in many cellular processes [1]. Nevertheless, given their cell/tissue-specificity, there is a need of identifying novel lincRNAs in a given system. To fulfill this purpose, we recently described the bioinformatics workflow for detecting novel lincRNAs using a RNA-seq dataset [2]. As shown in Fig. 1, aligning reads, reconstructing transcriptome, and filtering are three main steps. After preparing RNAs and performing RNA-seq, sequence information of single-end or paired-end reads are obtained. Reads are then mapped or aligned back to a reference genome to identify locations where the sequences originate from. Fulfilling this purpose, multiple programs have been designed to map reads across splice junctions in RNA-seq data, such as TopHat [3], GSNAP [4], and STAR [5]. For transcriptome reconstruction, Cufflinks [6] and Scripture [7] are commonly used for *ab initio* transcript assembly based on the aligned reads. In the final step, several filters need to be applied to discriminate real lincRNA transcripts from assembly artifacts, including filters of transcript length,

Eusebio Perdiguero and DDW Cornelison (eds.), *Muscle Stem Cells: Methods and Protocols*, Methods in Molecular Biology, vol. 1556, DOI 10.1007/978-1-4939-6771-1_20,

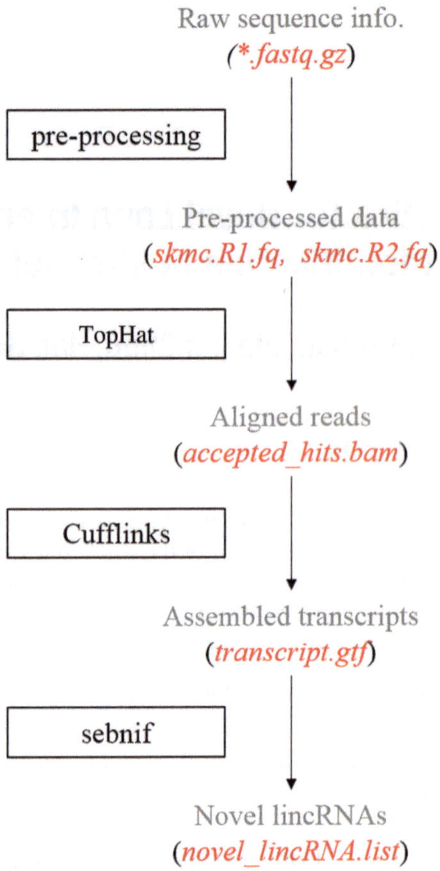

Fig. 1 Overview of the bioinformatics workflow for the identification of novel lincRNAs

expression level and coding potential, etc. To this end, an integrative bioinformatics pipeline, sebnif (self-estimation-based novel lincRNA filtering pipeline) [2] is implemented to identify *bona fide* novel lincRNAs with high quality. Furthermore, sebnif utilities enable the annotation of high-confidence lincRNAs using additional datasets such as ChIP-seq of histone modifications and CAGE (Cap analysis gene expression) tags when available.

To walk through the process, in this protocol, a RNA-seq dataset generated in human skeletal muscle cells (HSkMC) by ENCODE project [8] is used as input. After the above-described sequencing reads mapping, transcriptome reconstruction, and sebnif filtering, a catalog containing 917 novel lincRNAs is successfully built [2].

2 Materials

To complete software installation and lincRNA identification, a computing machine with 64-bit GNU/Linux operating system installed and at least 4GB memory as well as Internet access are required.

2.1 Datasets

1. Raw RNA-seq reads of human skeletal muscle cells (SkMCs) from ENCODE project. (GEO accession: GSM984615).
2. Aligned H3K4me3 ChIP-seq reads of human SkMCs from ENCODE project. (GEO accession: GSM945214).
3. Aligned CAGE tags of human SkMCs from ENCODE project. (GEO accession: GSM979653).

2.2 Programs (See Note 1)

1. TopHat (version 2.0.6) (http://ccb.jhu.edu/software/tophat/index.shtml).
2. Bowtie (version 0.12.9) (http://bowtie-bio.sourceforge.net/index.shtml).
3. Cufflinks (version 2.1.1) (http://cole-trapnell-lab.github.io/cufflinks/).
4. sebnif (version 1.2.2) (http://sunlab.lihs.cuhk.edu.hk/sebnif/).

3 Methods

The primary output of most sequencers consists of *.bcl* basecall files, which store the sequence information and can be converted to *.fastq.gz* files on the sequencer. Publicly available datasets are usually provided as *.fastq.gz* files or *.sra* files, which can be decompressed or converted to *.fastq* files that can then be used by alignment programs as input.

3.1 Data Preprocessing

1. Change to the directory where data are stored.
   $ cd /path-to-data/
2. Download raw RNA-seq data in HSkMC (two biological replicates). (*See* **Note 2**)
   $ wget

   http://hgdownload.cse.ucsc.edu/goldenPath/hg19/encodeDCC/wgEncodeCshlLongRnaSew/wgEncodeCshlLongRnaSeqSkmc9011302CellTotalFastqRd1Rep1.fastq.gz

   http://hgdownload.cse.ucsc.edu/goldenPath/hg19/encodeDCC/wgEncodeCshlLongRnaSew/wgEncodeCshlLongRnaSeqSkmc9011302CellTotalFastqRd2Rep1.fastq.gz

   http://hgdownload.cse.ucsc.edu/goldenPath/hg19/encodeDCC/wgEncodeCshlLongRnaSew/wgEncodeCshlLongRnaSeqSkmc812190217CellTotalFastqRd1Rep2.fastq.gz

   http://hgdownload.cse.ucsc.edu/goldenPath/hg19/encodeDCC/wgEncodeCshlLongRnaSeq/wgEncodeCshlLongRnaSeqSkmc812190217CellTotalFastqRd2Rep2.fastq.gz

3. Decompress *.fastq.gz* files in the current directory.

   $ gunzip -c wgEncodeCshlLongRnaSeqSkmc9011302CellTotalFastqRd1Rep1.fastq.gz > replicate1.read1.fastq

   $ gunzip -c wgEncodeCshlLongRnaSeqSkmc9011302CellTotalFastqRd2Rep1.fastq.gz > replicate1.read2.fastq

   $ gunzip -c wgEncodeCshlLongRnaSeqSkmc812190217CellTotalFastqRd1Rep2.fastq.gz > replicate2.read1.fastq

   $ gunzip -c wgEncodeCshlLongRnaSeqSkmc812190217CellTotalFastqRd2Rep2.fastq.gz > replicate2.read2.fastq

4. Merge two replicates.

   $ cat replicate1.read1.fastq replicate2.read1.fastq > merged.read1.fq

   $ cat replicate1.read2.fastq replicate2.read2.fastq > merged.read2.fq

5. Trim adaptors and remove duplication. (*See* **Note 3**)

   $ Ktrim merged.read1.fq merged.read2.fq trimmed_out

   $ Krmdup trimmed_out.R1.fq trimmed_out.R2.fq skmc

   The output files from the above step are *skmc.R1.fq* and *skmc.R2.fq*.

3.2 Read Alignment Using TopHat

1. Change to program installation directory.

   $ cd /path-to-programs/

2. Download the binaries for Bowtie and TopHat. Bowtie is an aligner adopted in TopHat.

   $ wget

   http://sourceforge.net/projects/bowtie-bio/files/bowtie/0.12.9/bowtie-0.12.9-linux-x86_64.zip

   $ wget

   http://ccb.jhu.edu/software/tophat/downloads/tophat-2.0.6.Linux_x86_64.tar.gz

3. Unpack the packages of TopHat and Bowtie separately. (*See* **Note 4**)

   $ unzip bowtie-0.12.9-linux-x86_64.zip

   $ tar -xzf tophat-2.0.6.Linux_x86_64.tar.gz

4. Change to the Bowtie index directory.

   $ cd /path-to-programs/bowtie-0.12.9/indexes

5. Download a pre-built Bowtie index of hg19 into the Bowtie index directory. (*See* **Note 5** and **Note 6**)

   $ wget

   ftp://ftp.ccb.jhu.edu/pub/data/bowtie_indexes/hg19.ebwt.zip

6. Unzip the index.

   $ unzip hg19.ebwt.zip

7. Change to the directory where genome or transcriptome references are stored.

   $ cd /path-to-references/

8. Download gene annotation of known transcripts [9]. (*See* **Note** 7)

   $ wget

   ftp://ftp.sanger.ac.uk/pub/gencode/Gencode_human/release_16/gencode.v16.annotation.gtf.gz

9. Unzip the annotation file.

   $ gunzip -d gencode.v16.annotation.genes.gtf.gz

10. Change to project directory.

    $ cd /path-to-projects/

11. Make a new directory for the lincRNA identification project.

    $ mkdir lincRNA_skmc

12. Change to the new directory.

    $ cd lincRNA_skmc

13. Run TopHat. (*See* **Notes 8** and **9**)

    $ /path-to-programs/tophat-2.0.6.Linux_x86_64/tophat --bowtie1 -o tophat_output --phred64-quals -G /path-to-references/gencode.v16.annotation.genes.gtf /path-to-programs/bowtie-0.12.9/indexes/hg19 /path-to-data/skmc.R1.fq,/path-to-data/skmc.R2.fq

accepted_hits.bam in the output folder (*tophat_output*) will be used by Cufflinks as input in the following step.

3.3 Transcriptome Assembly Using Cufflinks

1. Change to program directory.

   $ cd /path-to-programs/

2. Download the binaries for Cufflinks.

   $ wget

   http://cole-trapnell-lab.github.io/cufflinks/assets/downloads/cufflinks-2.1.1.Linux_x86_66.tar.gz

3. Unpack the package of Cufflinks.

   $ tar -xzf cufflinks-2.1.1.Linux_x86_66.tar.gz

4. Change to the lincRNA project directory.

   $ cd /path-to-projects/lincRNA_skmc

5. Run Cufflinks. (*See* **Note 10**)

   $ /path-to-programs/cufflinks-2.1.1.Linux_x86_64/cufflinks -o cufflinks_output tophat_output/accepted_hits.bam

   transcripts.gtf in the folder of *cufflinks_output* is the output file that will be used by sebnif in the following step.

3.4 Novel lincRNA Identification Using sebnif

1. Change to program directory.

   $ cd /path-to-programs/

2. Download the binaries for sebnif.

   $ wgethttp://sunlab.lihs.cuhk.edu.hk/sebnif/sebnif-1.2.2.tar.gz

3. Unpack the package of sebnif.

   $ tar -xzf sebnif-1.2.2.tar.gz

4. Configure iSeeRNA used by sebnif, which filters out transcripts with coding potential [10].

   $ cd /path-to-programs/sebnif-1.2.2/iSeeRNA

   $ sh auto_download_data.sh hg19

5. Change to the lincRNA project directory.

   $ cd /path-to-projects/lincRNA_skmc

6. Run sebnif. (*See* **Notes 11 and 12**)

   $ /path-to-programs/sebnif-1.2.2/sebnif -g Homo -o sebnif_output cufflinks_output/transcripts.gtf

   After the above steps, 9,812 novel transcripts are obtained as a primary catalog of novel lincRNAs, which can be further annotated by H3K4me3 ChIP-seq and CAGE data to increase the reliability using sebnif utilities.

7. Download H3K4me3 ChIP-seq and CAGE data in HSkMC.

   $ cd /path-to-data/

   $ wget

   http://hgdownload.cse.ucsc.edu/goldenPath/hg19/encodeDCC/wgEncodeUwHistone/wgEwgEwgEncodeUwHiston3k04me3StdAlnRep1.bam

   $ wget

   http://hgdownload.cse.ucsc.edu/goldenPath/hg19/encodeDCC/wgEncodeRikenCage/wgEwgEncodeRikenCageSkmcCellPapAln1.bam

8. Change to the lincRNA project directory.

   $ cd /path-to-projects/lincRNA_skmc

9. Run sebnif utilities.

   $/path-to-programs/sebnif-1.2.2/util/transcript_annotater/transcript_annotater -tss /path-to-data/wgEwgEwgEncodeUwHiston3k04me3StdAlnRep1.bam,/path-to-data/wgEwgEncodeRikenCageSkmcCellPapAln1.bam -o annotation_output.info -t sebnif_output/novel.final.gtf

10. Only lincRNAs with at least 10 H3K4me3 reads and 5 CAGE tags in their promoter regions (i.e., 2 kbp upstream to 1 kbp downstream of the transcript start site (TSS)) are selected. (kbp: kilobase pair)

```
$ sed '1d' annotation_output.info | perl -lane 'my @h=split(/;/,$F[3]); @h=sort {$b<=>$a} @h; my @c=split(/;/,$F[4]); @c=sort {$b<=>$a} @c; if($h[0]>=10 && $c[0]>=5) {print $F[0],"\t",$F[1];}' > novel_lincRNA.list
```

The final output file is *novel_lincRNA.list* containing the information of 917 identified novel lincRNAs, including transcript IDs (assigned in the Cufflinks assembly), genomic coordinates, and strands.

4 Notes

1. Several extra program dependencies should be noted. In order to use the binary package of TopHat, the Boost library (http://www.boost.org/) needs to be installed. Moreover, Samtools (http://samtools.sourceforge.net/) is also a prerequisite. And the execution of sebnif requires Perl (version 5.8 or higher) (https://www.perl.org/) and R (version 2.10 or higher with MASS library) (http://www.r-project.org/) in the path. Installation of these programs should follow the standard procedure thus no further details will be provided here.
2. Sufficient depth of the sequencing data can ensure high-quality assembly. As specified by ENCODE consortium, a minimal depth of 100–200 million 2 × 76 bp or longer reads is currently recommended to detect reliable novel transcribed elements in a mammalian sample [11]. In our case, the dataset contains 425 million paired-end reads in total.
3. Ktrim and Krmdup are in-house programs. Ktrim is designed to perform quality trimming and adapter trimming for paired-end reads generated by Illumina sequencers and Krmdup is for the removal of duplicated reads. Both programs are available upon request.
4. Export newly installed Bowtie path using the following command, or else TopHat will search for and employ Bowtie in the default path.

   ```
   $ export PATH=$PATH:/path-to-programs/bowtie-0.12.9
   ```
5. Note that Bowtie and Bowtie2 use different indexes.
6. Users can build Bowtie or Bowtie2 index when the pre-built index is not available for a given species.
7. Transcript annotation can be downloaded from other databases, e.g., Ensembl, UCSC, and NCBI. The annotation file must be converted to GTF/GFF format.
8. Because Bowtie, instead of Bowtie2, was used, specify the parameter of *--bowtie1* in the command. Note that if a sequencing platform like Illumina is used, Bowtie and Bowtie2 are both suitable; but for some other platforms like SOLiD, only Bowtie is applicable.

9. Before alignment, base quality scoring system must be taken into consideration. PHRED scores with an ASCII offset of 33 or 64 can be set.
10. Neither *-G* nor *-g* should be set for Cufflinks to ensure ab initio assembly.
11. Multiple parameters of sebnif can be adjusted to achieve optimal results. For example, when users intend to retain more lincRNAs, less stringent parameters (e.g., smaller value for "-F" and broader range for "-E") can be set; it may however cause decreased reliability thus is not recommended.
12. Sebnif also provides a user-friendly webserver where filtering for novel lincRNAs can be done without installing a local program. (http://sunlab.lihs.cuhk.edu.hk/sebnif/webserver/)

Acknowledgment

The work is substantially supported by seven General Research Funds (GRF) to H.W. and H.S. from the Research Grants Council (RGC) of the Hong Kong Special Administrative Region, China (Project Code: 14133016, 14100415, 14102315, 14116014, 14 113514, 476113 and 473713).

References

1. Mercer TR, Dinger ME, Mattick JS (2009) Long non-coding RNAs: insights into functions. Nat Rev Genet 10(3):155–159
2. Sun K, Zhao Y, Wang H et al (2014) Sebnif: an integrated bioinformatics pipeline for the identification of novel large intergenic noncoding RNAs (lincRNAs)--application in human skeletal muscle cells. PLoS One 9(1):e84500.3
3. Trapnell C, Pachter L, Salzberg SL (2009) TopHat: discovering splice junctions with RNA-Seq. Bioinformatics 25(9):1105–1111
4. Wu TD, Nacu S (2010) Fast and SNP-tolerant detection of complex variants and splicing in short reads. Bioinformatics 26(7):873–881
5. Dobin A, Davis CA, Schlesinger F et al (2013) STAR: ultrafast universal RNA-seq aligner. Bioinformatics 29(1):15–21
6. Trapnell C, Williams BA, Pertea G et al (2010) Transcript assembly and quantification by RNA-Seq reveals unannotated transcripts and isoform switching during cell differentiation. Nat Biotechnol 28(5):511–515
7. Guttman M, Garber M, Levin JZ et al (2010) Ab initio reconstruction of cell type-specific transcriptomes in mouse reveals the conserved multi-exonic structure of lincRNAs. Nat Biotechnol 28(5):503–510
8. ENCODE Project Consortium (2012) An integrated encyclopedia of DNA elements in the human genome. Nature 489(7414): 57–74
9. Harrow J, Frankish A, Gonzalez JM, Tapanari E, Diekhans M et al (2012) GENCODE: the reference human genome annotation for The ENCODE Project. Genome Res 22:1760–1774 10. Sun K, Chen X, Jiang P et al (2013) iSeeRNA: identification of long intergenic non-coding RNA transcripts from transcriptome sequencing data. BMC Genomics 14 Suppl 2:S7
10. Sun K, Chen X, Jiang P et al (2013) iSeeRNA: identification of long intergenic non-coding RNA transcripts from transcriptome sequencing data. BMC Genomics 14(Suppl 2):S7
11. The ENCODE Standards, guidelines and best practices for RNA-Seq. http://genome.ucsc.edu/ENCODE/protocols/dataStandards/ENCODE_RNAseq_Standards_V1.0.pdf

Correction to: Muscle Stem Cells

Eusebio Perdiguero and DDW Cornelison

Correction to:
Eusebio Perdiguero and DDW Cornelison (eds.), *Muscle Stem Cells: Methods and Protocols*, Methods in Molecular Biology, vol. 1556, https://doi.org/10.1007/978-1-4939-6771-1

1) A typographical error in the title for chapter 4 had appeared on SpringerLink and Springer.com websites, and has been corrected to: "Isolation, Culture, and Immunostaining of Skeletal Muscle Myofibers from Wildtype and Nestin-GFP Mice as a Means to Analyze Satellite Cells" on these websites.

2) A wrong video was posted for Chapter 16 "Methods for Observing and Quantifying Muscle Satellite Cell Motility and Invasion In Vitro", on Springerlink and has been corrected now.

The updated online version of these chapters can be found at
https://doi.org/10.1007/978-1-4939-6771-1_4
https://doi.org/10.1007/978-1-4939-6771-1_16
https://doi.org/10.1007/978-1-4939-6771-1

Eusebio Perdiguero and DDW Cornelison (eds.), *Muscle Stem Cells: Methods and Protocols*, Methods in Molecular Biology, vol. 1556, https://doi.org/10.1007/978-1-4939-6771-1_21, © Springer Science+Business Media LLC 2019

INDEX

Eusebio Perdiguero and DDW Cornelison (eds.), *Muscle Stem Cells: Methods and Protocols*, Methods in Molecular Biology, vol. 1556, DOI 10.1007/978-1-4939-6771-1, © Springer Science+Business Media LLC 2017